E. Seibold, W.H. Berger • The Sea Floor

Springer

Berlin
Heidelberg
New York
Barcelona
Budapest
Hong Kong
London
Milan
Paris
Santa Clara
Singapore
Tokyo

E. Seibold   W.H. Berger

The Sea Floor

An Introduction to Marine Geology

Third Edition

With 209 Figures and 9 Tables

 Springer

Prof. Dr. Dr. h.c. mult. EUGEN SEIBOLD

Geologisches Institut
Universität Freiburg
79104 Freiburg
Germany

Prof. WOLFGANG H. BERGER, Ph.D.

Scripps Institution of Oceanography
Geological Research Division
La Jolla, California 92093, USA

Cover photo:
A new kind of deep sea benthos. Pogonophore-like tubeworms, several feet long, and a
host of other creatures crowd around the hot vents of the Galápagos, spreading center.
"Rose Garden" vent site, Alvin Dive 984. (Photo courtesy R. Hessler, S.I.O.)

ISBN 3-540-60191-0 3rd ed Springer-Verlag Berlin Heidelberg NewYork

ISBN 3-540-56884-0 2nd ed. Springer-Verlag Berlin Heidelberg New

Library of Congress Cataloging-in-Publication Data. Eugen Seibold, Wolfgang H. Berger,
The Sea Floor – An Introduction to Marine Geology. p. cm. Includes Bibliographical
references and index. ISBN 3-540-60191-0 (alk. paper) I. Geology. II. Marine Sciences.
GRQ 359.G.233 1996 618.7'25073-dc21 DNLM/DDCL for Library of Congress 95-9738 CIP

© Springer-Verlag Berlin Heidelberg 1993, 1996
Printed in Germany

Cover design: Springer-Verlag, Design & Production

SPIN: 10503107   32/3136 – Printed on acid free paper

Preface

Man's understanding of how this planet is put together and how it evolved has changed radically during the last 30 years. This great revolution in geology – now usually subsumed under the concept of *Plate Tectonics* – brought the realization that convection within the Earth is responsible for the origin of today's ocean basins and continents, and that the grand features of the Earth's surface are the product of ongoing large-scale horizontal motions. Some of these notions were put forward earlier in this century (by A. Wegener, in 1912, and by A. Holmes, in 1929), but most of the new ideas were an outgrowth of the study of the ocean floor after World War II. In its impact on the earth sciences, the plate tectonics revolution is comparable to the upheaval wrought by the ideas of Charles Darwin (1809–1882), which started the intense discussion on the evolution of the biosphere that has recently heated up again. Darwin drew his inspiration from observations on island life made during the voyage of the *Beagle* (1831–1836), and his work gave strong impetus to the first global oceanographic expedition, the voyage of *HMS Challenger* (1872–1876). Ever since, oceanographic research has been intimately associated with fundamental advances in the knowledge of Earth. This should come as no surprise. After all, our planet's surface is mostly ocean.

This book is the result of our conviction that to study introductory geology and oceanography and environmental sciences, one needs a summary of the tectonics and morphology of the sea floor, of the geologic processes active in the deep sea and in shelf seas, and of the climatic record in deep-sea sediments.

Our aim is to give a brief survey of these topics. We have endeavored to write for all who might be interested in the subject, including those with but little background in the natural sciences. The decade of the 1980s was characterized by an increasing awareness of man's dependency on natural resources, including the ocean as a weather machine, a waste bin, and a source of energy and minerals. This trend, we believe, will persist as resources become ever more scarce and as the impact of human activities on natural cycles escalates in the coming decades. An important part of this awareness will be an appreciation for the elementary facts and concepts of marine geology, especially as they apply to processes within hydrosphere and atmosphere.

In what follows, we shall first give a brief overview of the effects of endogenic forces on the morphology of the sea floor. Several excellent summaries for the general reader are available for this topic, which is closely linked to the theory of continental drift, and has been a focus of geologic discussion for the last three decades. For the rest, we shall emphasize the exogenic processes, which determine the physical, chemical, and biological environment on the sea floor, and which are especially relevant to the intelligent use of the ocean and to an understanding of its role in the evolution of climate and life.

The results and ideas we report on are the product of the arduous labors of many dedicated marine geologists. We introduce some particularly distinguished scientists by portrait (Fig. 0.1). Of course, there are many more, and most of them are alive today. We have occasionally mentioned the authors of important contributions. However, we did not find it possible in a book like this to give credit systematically where it is due. We sincerely apologize to our colleagues for this unscholarly attitude, citing necessity in defense. For those who wish to pursue the subjects discussed in greater depth, we append suggested readings at the end of each chapter, as well as a list of key references.

For this second edition, we have extensively rewritten those parts of the first edition where substantial and fundamental progress has occurred in the fields of interest. Also, we have incorporated many of the suggestions for improvements that were communicated to us by several colleagues and reviewers. There are, however, limitations to the scope of subjects that can be treated in a short introduction such as this: we attempted neither a balanced nor an encyclopedic survey of all of marine geology with its many ramifications. We tried to keep highly technical information to a minimum, relegating certain necessary details to the Appendix.

Both authors wish to express their profound gratitude to collaborators and students who, over the years, have shared the excitement of discovery and the toil of research on numerous expeditions and in the laboratory. We also owe special thanks to the colleagues who helped us put this book together, by sending reprints and figures, or by offering advice.

Freiburg and La Jolla, Spring 1993   *E. Seibold*
                                                      W. H. Berger

Contents

Introduction

Pioneers of Marine Geology: Identifying the Basic Questions

Marine geology is a young offshoot of geology, a branch of science which begins with James Hutton (1726–1797) and his *Theory of the Earth* (Edinburgh, 1795). Among other things, Hutton studied marine rocks on land. Changes of sea level ("encroachment of the ocean", and "the placing of materials accumulated at the bottom of the sea in the atmosphere above the surface of the sea") was a central tenet of his "Theory". Thus, the question of what happens at the bottom of the sea was raised at the beginning of systematic geologic investigation. This question had to be attacked if the marine deposits on land were to be understood. Hutton was not alone in these concerns. A few years before the *Theory of the Earth* appeared, the great chemist Antoine Laurent Lavoisier (1743–1794) distinguished two kinds of marine sedimentary layers, that is, those formed in the open sea at great depth, which he called *pelagic* beds, and those formed along the coast, which he termed *littoral* beds. "Great depth" for Lavoisier was everything beyond wave base. Lavoisier supposed that sediment particles would settle quietly in deep water and reworking would be much less in evidence here than near the shore.

Marine geology starts in earnest with geologists going to sea and looking for the processes which helped produce the marine rocks with which they were familiar on land. Such work, obviously, began in the intertidal and in easily accessible shallow waters, and it was undertaken by a great many investigators. The German geologist Johannes Walther (Fig. 0.1.2) was very familiar with the observations resulting from this work and was himself a pioneer in these types of studies. Firmly rooted in classical geology, he set an outstanding example of applying *uniformitarianism,* that is, Hutton's doctrine that observable processes are sufficient to explain the geologic record (*Lithogenesis of the Present,* Jena, 1894, in German). In his book *Bionomie des Meeres* (Jena, 1893), he gave an account of marine environments and the ecology of their inhabitants, with emphasis on shell-forming organisms and the deposits they produce. Research on marine sedimentation in the following four decades is summarized in P. D. Trask (ed.) *Recent Marine Sediments* (AAPG, Tulsa, 1939). This symposium spans the range of sedimentary environments, from beach to deep sea, and many of the articles were written by pioneers in marine geology.

As marine geologic studies progressed and moved further out to sea, there was a gradual change of emphasis in the set of problems to be solved. The sea floor itself became the focus of attention, not for the sake of the clues it would yield for the

Fig. 0.1. Founders of marine geology. *Upper row, left* to *right* John Murray (1841–1914),
Johannes Walther (1860–1937), N. I. Andrusov (1861–1924).
Lower row Alfred Wegener (1880–1930), Jaques Bourcart (1891–1965),
Francis P. Shepard (1897–1985).

purposes of land geology, but for the clues it contained for its own evolution and its
role in the history of Earth. The new emphasis is first evident in the works of the
Scotsman, John Murray (Fig. 0.1.1), naturalist on the HMS *Challenger* Expedition
(1872–1876). This expedition, led by the biologist Charles Wyville Thomson (1830–
1882), marks the beginning of modern oceanography. It established the general morphology of the deep-sea floor and the types of sediments covering it. John Murray's
chief opus, *Deep Sea Deposits* (written with A. F. Renard and published in 1891), laid
the foundation for the sedimentology of the deep ocean floor (Chap. 8). Murray's
studies established a fundamental dichotomy of shallow water and shelf sediments on
the one hand, and of deep-sea deposits on the other, and it became a commonplace

Upper row, left to *right* N. M. Strachow (1900–1978), Philip H. Kuenen (1902–1976),
Maurice Ewing (1903–1976). Lower row Harry H. Hess (1906–1969), Sir E. C. Bullard (1907–1980),
Bruce C. Heezen (1924–1977).

textbook truism that no true deep-sea deposits are found on land anywhere. This
doctrine was challenged when Ph. H. Kuenen (Fig. 0.1.8) demonstrated by experiment that clouds of sediment could be transported downslope on the sea floor at great
speed and to great depth, due to the fact that muddy water is heavier than the clear
water surrounding it. Sediment transported in suspension in this fashion settles out at
its site of deposition, heavy and large grains first, fine grains last. The resulting layer
is *graded,* and such layers are indeed common in the geologic record (e. g., Alpine
flysch deposits). Strong circumstantial evidence that Kuenen's concept is important in
deep-sea sedimentation was first presented by B. C. Heezen and M. Ewing, in 1952
(see Chap. 2.10). Kuenen made many other important contributions to marine geo-

logy, attacking a broad range of subjects in his book, *Marine Geology* (New York, 1950), and in numerous publications.

Studies in marine sedimentation eventually led into ocean history when geologists started to take cores. The pioneering expeditions were those of the German vessel *Meteor* (1925–1927), which first established deep-sea sedimentation rates, and of the Swedish vessel *Albatross* (1947–1948), led by Hans Petterson. The *Albatross* results established the presence, in all oceans, of cyclic sedimentation due to the climate fluctuations during the last million years, which included several ice ages (Chap. 9). The last (and biggest) effort in this line is the Deep Sea Drilling Project (now Ocean Drilling Program), using the vessel *Glomar Challenger* (1968–1983; Fig. 0.6), which made Tertiary and Cretaceous sediment sequences available for systematic study.

The investigation of the sea floor proceeded parallel to that of marine sedimentation. Coastal landforms were the most accessible, and considerable information had been accumulated early in this century (D. W. Johnson, *Shore Processes and Shoreline Development,* Wiley, New York, 1919). Much additional work on these topics was done by F. P. Shepard (Fig. 0.1.6) and also by J. Bourcart (Fig. 0.1.5), who were able to test earlier concepts against field data collected in shallower water. These two pioneers of marine geology especially attacked the problems of the origin of continental margins, and of submarine canyons, by studying their morphology and associated sedimentary processes.

F. P. Shepard's wide-flung field areas include the shelf off the US East Coast, shelf and slope off the US West Coast, and the floor of the Gulf of Mexico. His textbook *Submarine Geology* (New York, 1948 and subsequent editions) summarizes the results of this work and presents global statistics on sea-floor morphology. In the same year, M. B. Klenova's textbook *Geology of the Sea* was printed. Amongst other Russian pioneers, N. I. Andrusov (Black Sea) and N. M. Strachov (lithogenesis) should be mentioned (Figs. 0.1.3 and 0.1.7). Other representative works of Shepard are *Recent Sediments, Northwest Gulf of Mexico* (Tulsa, 1960, with F. B. Phleger and Tj. H. van Andel) and *Submarine Canyons and Other Sea Valleys* (Chicago, 1966, with R. F. Dill). J. Bourcart carried out similar geomorphologic and sedimentologic studies off the shores of France, especially in the Mediterranean. His concept of continental margin "flexure", with uplift landward of a *hingeline* and downwarp seaward of it, proved useful in explaining the migrations of sea level across the shelf, and in studying the nature of sediment accumulation on the continental slope (see Chap. 2).

The marine geomorphologist par excellence was B. C. Heezen (Fig. 0.1.12), whose *physiographic diagrams* of the sea floor (drawn with his collaborator, Marie Tharp) show great insight into the tectonic and sedimentologic processes of the sea floor. His graphs now appear in virtually all textbooks of geology and geography (see Figs. 1.3 and 2.2). Of B. C. Heezen it has been said (somewhat facetiously, by E. C. Bullard) that he perfected the art of drawing maps of regions where no data are available. Much of his work is summarized in the beautifully illustrated book *The Face of the Deep* (New York, 1971, with C. D. Hollister).

A satisfying explanation of the overall morphology of ocean basins and of the various types of continental margins could only come from geophysics, which deals with the motions and forces deep within the Earth. It is no coincidence that a geophysicist first formulated a global hypothesis for ocean-margin morphology which proved to be viable: the meteorologist Alfred Wegener (Fig. 0.1.4).

A. Wegener became intrigued with the parallelism of the coast lines bordering the Atlantic Ocean (a phenomenon noted already in 1801 by the famous naturalist-explorer, Alexander von Humboldt). It seemed to Wegener that the continents looked like puzzle pieces which belong together. He then learned quite by accident that paleontologists had invoked former land bridges between the shores facing each other across the Atlantic to explain striking similarities between the fossil records of both sides of the ocean. After an extensive literature search, he became convinced that the continents had once been joined, and had broken up after the Paleozoic. Replacing the concept of land bridges with his hypothesis of *continental drift*, he started the "debate of the century" in geology with an article in 1912 (Geol. Rdsch. 23: 276), and especially with his book *The Origin of Continents and Oceans* (Braunschweig, 1915, in German). He envisioned granitic continents floating in basaltic mantle magma like icebergs in water (Fig. 0.2a), and drifting about on the surface of Earth in response to unknown forces derived from the rotation of the planet.

Wegener's hypothesis, in modified form, is now an integral part of *sea-floor spreading and plate tectonics* (Chap. 1), the ruling theories explaining the geomorphology and geophysics of the ocean floor and of the crust of the Earth in general. It was the work of sea-going geophysicists which eventually led to the acceptance of *plate tectonics*. The pioneering efforts of E. C. Bullard (Fig. 0.1.11) on earth magnetism, heat flow, and seismic surveying, and of M. Ewing (Fig. 0.1.9) and his associates in all aspects of marine geophysics, were of central importance in this development (although M. Ewing did not himself advocate sea-floor spreading). These scientists also were instrumental in elucidating the structure of continental margins (M. Ewing et al., 1937, Geophysical Investigations in the Emerged and Submerged Atlantic Coastal Plain, Bull Geol Soc Am, 51, p 909; E. C. Bullard and T. F. Gaskell, 1941, Submarine Seismic Investigations, Proc Royal Soc, Ser A 177, p 476).

The turning point in the scientific revolution which shook the earth sciences, and which culminated in *plate tectonics* is generally taken to be the seminal paper of H. H. Hess, *History of Ocean Basins,* published in 1962. Hess (Fig. 0.1.10) started his distinguished career working with the Dutch geophysicist, F. A. Vening-Meinesz, on the gravity anomalies of deep-sea trenches. These investigations resulted in the hypothesis that trenches may be surface expressions of the downgoing limbs of mantle convection cells. As a naval officer, Hess discovered and mapped a great number of flat-topped seamounts, whose morphology suggested widespread sinking of the sea floor. Stimulated by subsequent discoveries on the Mid-Ocean Ridge (rift morphology, heat flow, and others) he proposed his idea of the sea floor being generated at the center of the Mid-Ocean Ridge (Fig. 0.3 d), moving away and downward as it ages, finally to disappear into trenches. The term "sea-floor spreading" was introduced by R. Dietz (in 1961) for this postulated phenomenon. *Sea-floor spreading,* and its offspring *plate tectonics* have since become the basic framework within which the data of marine geology are interpreted.

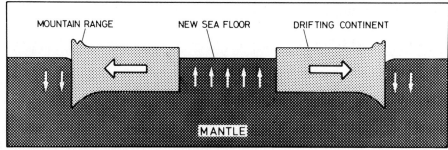

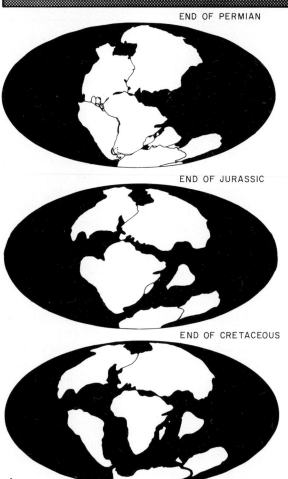

END OF PERMIAN

END OF JURASSIC

END OF CRETACEOUS

Fig. 0.2 a, b. Diagrammatic representation (ours, not his) of the Hypothesis of Continental Drift of Alfred Wegener. **a** Continental blocks made of light "sial" float iceberg-like in the heavier mantle "sima". As they drift apart, mountain ranges form at the bow, while new sea floor forms in between. **b** The breakup of Pangaea, first envisaged by A. Wegener, in a modern reconstruction by R. S. Dietz and J. C. Holden (1970, J Geophys Res 75:4939, simplified)

Seafloor Spreading and Plate Tectonics: the New Paradigm

Only some 35 years ago, it was still possible for geologists to think that the sediments on the deep-sea floor might contain the entire Phanerozoic record and lead us back even into the Precambrian (for stratigraphic terms see Appendix A3). Today, there would be few indeed who would harbor such fond hopes. The most ancient sediments recovered from the sea floor are about 150 million years old, which is less than 5 % of the age accorded to fossil-bearing sedimentary deposits on land. Where, then, is the debris which must have washed into the deep sea, for the several billion years that continents have existed?

According to the hypothesis of sea-floor spreading, all sediments accumulating on the ocean bottom are swept towards the trenches, as on a conveyor belt (Fig. 1.20). Here, some of the sediments are *subducted* into the mantle, others are scraped off against the inner wall of the trench. Thus, the ocean floor is cleaned of sediments by constant renewal and destruction.

This remarkable idea did not exactly find a warm welcome when first proposed by H. H. Hess (1960) and by R. S. Dietz (1961). Somewhat similar ideas had been advanced earlier (Fig. 0.3 a, b) and had likewise been discounted as premature speculations. Gradually however, as more facts became available to test the hypothesis, opposition weakened, and by 1970 there were only a very few defenders of tradition who challenged the concept of a moving sea floor.

The opposition to large-scale horizontal movement on Earth's surface had a long history. It all began with *Continental Drift,* an idea that arose early in the present century, and challenged the prevailing assumption of *"fixism"* with the new arguments of *"mobilism"*. The most serious challenge, as already mentioned, came from the German geophysicist A. Wegener, who proposed drastic changes in the distribution of continents and ocean basins within the last 200 million years. He also proposed that the continents actively plowed through the magma which carries them, an idea that proved to be wrong. In addition, he had a time table for the drifting of certain land masses which was quite unrealistic. Skeptical geophysicists recognized these weaknesses in Wegener's proposal and argued forcibly for the rejection of the theory of continental drift. Thus, despite the support from geologists familiar with the incredible similarities of ancient rocks and fossils in South America and South Africa, Wegener's hypothesis did not find general acceptance before the 1960s.

In essence, evidence for continental drift was dismissed because the proposed mechanism was wrong. Only in the late 1950s, through the work on *geomagnetism* and *polar wandering* by E. Irving and S. K. Runcorn, did it become possible again to talk about continental drift without being thought ignorant of physical principles. The most compelling evidence, however, came from the magnetism of the sea floor itself, and final proof came from deep-sea drilling, as we shall see in Chapter 1.

Toward the end of the 1960s, the new hypothesis of sea-floor spreading had metamorphosed into the theory of *plate tectonics.* The theory is based on the mapping of magnetic anomalies and earthquakes, by which it is possible to define large regions of the Earth's surface that move as units ("*plates*"), with earthquakes occurring at the boundaries of these units. The mathematical tool allowing efficient description of the

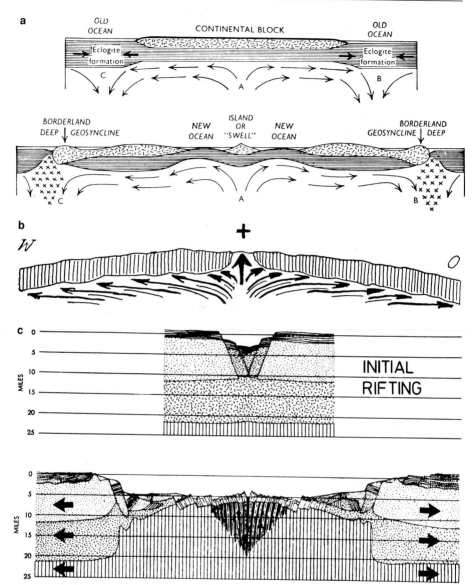

Fig. 0.3 a–c. Models of sea-floor spreading proposed to explain the origin of the Mid-Atlantic Ridge. **a** Hypothesis of A. Holmes (1929, Trans Geol Soc Glasgow 18:559). Note the left-over continental piece in the center (which does not in fact exist). **b** Hypothesis of the alpine geologist O. Ampferer, who first discussed magmatic "undercurrents" in 1906. In 1941, he supposed that rising magma currents break apart continents and eventually produce a mid oceanic ridge through symmetrical drifting apart of both continental margins. Also, he explained the Atlantic island arcs of the Caribbean and their trenches using the concept of subduction. (1941, Sitzungsber. Akad. Wiss. Wien, 150:20–35). **c** Hypothesis of B. C. Heezen (1960, Sci Am 203:98). Note the exaggerated mobility of continental margins, and the postulated diffuse injection of mantle material in a broad ridge area. (In fact, injection occurs in the narrow zone at the center only.)

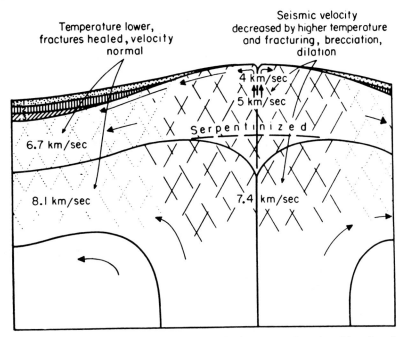

Temperature lower,
fractures healed, velocity
normal

Seismic velocity
decreased by higher temperature
and fracturing, brecciation,
dilation

4 km/sec

5 km/sec

Serpentinized

6.7 km/sec

8.1 km/sec

7.4 km/sec

Fig. 0.3 d. Hypothesis of H. H. Hess (1962). Note narrow injection area, and stratigraphic onlap of deep-sea sediments. This is the favored model, although serpentinization (a type of chemical alteration of basalt) was found to be less important than visualized by Hess.

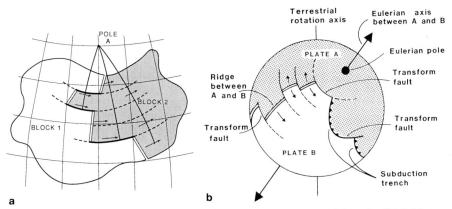

POLE
A

BLOCK 2

BLOCK 1

Terrestrial
rotation axis

Eulerian axis
between A and B

PLATE A

Eulerian pole

Ridge
between
A and B

Transform
fault

Transform
fault

Transform
fault

PLATE B

Subduction
trench

a   b

Fig. 0.4. a. The principle of plate tectonics: rotation of blocks about an Euler pole. [W. J. Morgan, 1968, J. Geophys. Res. 73: 1959] **b** The classical scheme (**a**) is illustrated in more detail in (**b**). Compare with the block diagram of Fig. 1.12. Blocks (plates) rotate around the "Eulerian" axis (*arrows*). This axis is independent of the terrestrial rotation axis. At the "divergent" plate boundary (the ridge between *A* and *B*) spreading occurs symmetrically. At the *right*, at "convergent" plate boundaries, plate *B* is subducted into the Earth's mantle beneath trenches. Both ridge and trench are segmented and offset by multiple "transform" faults. All these transform faults follow small circles about the Eulerian pole of relative movement between plates *A* and *B*. [C. Allègre 1988, Fig. 38, p. 96]

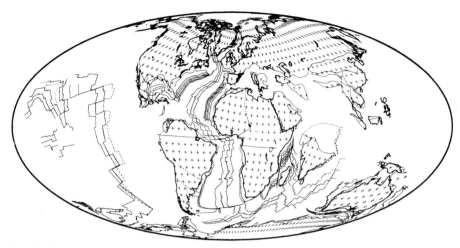

Fig. 0.5. Reconstruction of Paleocene geography, based on plate tectonics principles. Note the nearly circum-equatorial seaway connecting Atlantic and Pacific, the *Tethys,* with India far in the south. The opening of a seaway from the Atlantic to the Arctic is just beginning. [J.-Y. Royer et al., 1992. Univ. Texas Inst. Geophys. Tech. Rept. 117: 1–38. By courtesy of R. D. Müller and J. G. Sclater]

movement of the plates is a theorem of the Swiss mathematician Leonhard Euler (1707–1783), which states that uniform motion on a sphere is uniquely described by defining rotation about a *"pole"*. The path of migration of any point on a plate appears as a portion of a circle about that pole (Fig. 0.4). E. C. Bullard and associates introduced Euler's theorem to global tectonics, in 1965, to produce a new "fit" of the continents bordering the Atlantic (Phil Trans Roy Soc London 258: 41) (see Fig. 1.19). In the same year, J. T. Wilson explained the nature of what he called *"transform" faults* as boundaries between *rigid plates*, where there is lateral motion which ends at a spreading center or a trench (Nature 207: 343).

For the present, and at any time in the geologic past, plate motions on the surface of Earth are defined by giving the outlines of the dozen or so plates, the geographic location of the poles of rotation, and the associated angular velocities. The principles of this approach were presented by W. J. Morgan early in 1967 (and published in 1968, J. Geophys. Res. 73:1959), and by D. P. McKenzie and R. L. Parker, also in 1967 (Nature 216:1276). Using this methodology, X. Le Pichon, in 1968, made a global map of the relative motions of the major plates (J. Geophys. Res. 73:3661). The same concepts and methods allowed the large-scale reconstruction of the past geographic configurations of continents and ocean basins, back into the Cretaceous period. An example for the Paleocene, some 60 million years ago, is given in Fig. 0.5.

Deep-Sea Drilling: Discovering New Worlds

The emergence of the new paradigm greatly stimulated the beginning of a major venture in marine geology: the systematic exploration of the deep-sea floor by drilling. Up to 1968, before the drilling vessel *GLOMAR Challenger* (Fig. 0.6) set out on her first cruise (from Galveston, Texas), knowledge of sediments older than one million years was entirely based on cores taken in regions of greatly reduced sedimentation and erosion, where younger sediments had been removed. Such a core would typically represent a short stretch of time somewhere within the long history of the ocean. Detailed comparison with other cores, from other regions, was hardly ever possible, because of the difficulty of exact age assignment. The task of reconstructing a global ocean history for the pre-Quaternary could only be seriously attempted when more or less continuous sequences of samples became available, through drilling, from many different regions.

The materials recovered by drilling resulted in a quantum jump in biostratigraphic resolution – soon it became possible to date samples within sequences to within less than a million years relative to some standard. The first major result was full confir-

Fig. 0.6. The 120-m-long deep-sea drilling vessel *GLOMAR Challenger*, which was working from 1968 to 1983, sampling the sea floor of the world's ocean. During this period 1092 holes were drilled on 624 sites, many in water depths of more than 5 km. The length of core material recovered measures over 90 km. The ship was managed by the Scripps Institution of Oceanography within the Deep Sea Drilling Project (DSDP), which was the most important earth science project of the 1970s, funded by the USA. Since the mid-1970s the U.S. National Science Foundation, the USSR, Germany, Japan, the United Kingdom, and France have participated in the Project and helped in providing support. [Photo courtesy DSDP, S.I.O.]

mation of the theory of sea-floor spreading: sediments overlying the "basement" basalt showed the exact age predicted by the geophysicists from counting magnetic anomalies (Fig. 1.18). Occasionally, sediments were somewhat younger, which means that the basaltic rock was bare for a while, before it collected sediments. Other major results were more subtle, but have had far-reaching implications for our understanding of the co-evolution of life and climate, and for the role of the ocean in climatic change over long periods of time.

In the early years of deep-sea drilling, present sedimentation patterns were assumed to be static, so that changes down-core would be interpreted mainly in terms of motions of the sea floor. It soon became clear, however, that the ocean's productivity changed markedly through geologic time, which produced large changes in sedimentation patterns. Also, it became evident that changes in ocean conditions (as reflected in sediments) were quite sudden at certain times, and that such "steps" in climatic change were associated with reorganizations of the marine biosphere, including extinctions and subsequent radiations. The most intriguing period in the ocean's history, in this regard, proved to be the transition between the Cretaceous and the Tertiary, which witnessed large-scale extinction of tropical planktonic organisms.

The importance of the deep-sea record for the reconstruction of Earth's history for the last one hundred million years or so became obvious very quickly. The record on land is patchy and incomplete by nature – land is eroded and delivers sediment to the sea. This renders suspect all arguments about the pace of *evolution* that depend on a continuous record. (Darwin pointed this out a long time ago). Only the record in the deep ocean can promise complete sequences, and even here gaps prove to be quite common in many settings. The gaps are not distributed randomly, but occur preferentially just where conditions change. This illustrates a certain obstinacy of the ocean in yielding up the secrets of its history, something that many marine geologists have come to appreciate.

One of the most notable aspects of deep-ocean drilling – now with the larger drilling vessel *JOIDES Resolution* (Fig. 0.7) – is the creation of an international community of marine geologists – hundreds of scientists from around the world who were shipmates at one time or another, eagerly listening for the driller's call "core-on-deck", and sharing the thrill of exploring where nobody had gone before.

Further Reading

General Background
(least specialized literature first)

Press F, Siever R (1982) Earth, 3rd edn. Freeman, San Francisco
Emiliani C (1992) Planet Earth: cosmology, geology, and the evolution of life and environment. Cambridge University Press
Glen W (1975) Continental drift and plate tectonics. Merrill, Columbus
Open University Course Team (1989) The ocean basins: their structure and evolution. Pergamon Press, Oxford

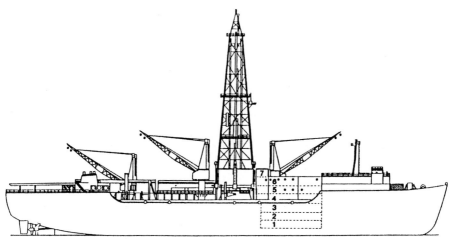

Fig. 0.7. Layout of the deep-sea drilling vessel *JOIDES Resolution.* It is the successor of the *GLOMAR Challenger,* but is larger, 143 m long. Therefore it can work in rougher seas and in higher latitudes. The capacity of the many modern laboratories on board allows 50 scientists and technicians to participate. From Level 7 downhole measurements are carried out. At Level 6 drilled cores are received and distributed to the sedimentological, petrological, paleontological, physical, and chemical laboratories at Levels 3–6. One half of each core goes to the archives. In 1983 the Ocean Drilling Program (ODP) began as continuation of DSDP with the participation of 10 US institutions, as well as Germany, France, the UK, Canada & Australia, Japan, and a consortium of the European Science Foundation. *JOIDES Resolution* has been active since 1985. Texas A&M University is managing the program. [Joides Journal 1985]

Allègre C (1988) The behavior of the earth – continental and seafloor mobility. Harvard University Press, Cambridge, Mass

Emiliani C (ed) (1981) The sea, vol 7. The oceanic lithosphere. Wiley Interscience, New York

Berger WH, Crowell JC (eds) (1982) Climate in Earth history. Studies in geophysics. Natl Acad Sci, Washington DC

Imbrie J, Imbrie KP (1979) Ice ages – solving the mystery. Enslow Short Hills NJ

Turekian KK (1976) Oceans, 2nd edn. Prentice-Hall, Englewood Cliffs NJ

Broecker WS (1974) Chemical oceanography. Harcourt Brace Jovanovich, New York

LePichon X, Convenor (1988) Report of the Second Conference on Scientific Ocean Drilling [Cosod II]. European Science Foundation and JOIDES, Strasbourg

Historical Perspectives

Wegener A (1929) The origin of continents and oceans. Translation by J. Biram, 1966. Dover, New York

Holmes A (1945) Principles of physical geology. Nelson, London

Kuenen PhH (1950) Marine geology. Wiley, New York

Heezen BC, Tharp M, Ewing M (1959) The floors of the oceans. I. The North Atlantic. Geol Soc Am Spec Pap 65

Dietz RS (1961) Continent and ocean basin evolution by spreading of the sea floor. Nature 190: 854–857

Hess HH (1962) "History of Ocean Basins", in Petrologic Studies: A Volume in Honor of AF Buddington, pp 599–620, ed. AEJ Engel et al. Boulder, Colorado: Geol Soc Am (ms circulated in 1960)

Runcorn SK (1962) Continental drift. Academic Press, New York

Menard HW (1964) Marine geology of the Pacific. McGraw-Hill, New York

Phinney RA (ed) (1968) The history of the Earth's crust. Princeton University Press, New Jersey

Takeuchi H, Uyeda S, Kanamori H (1970) Debate about the Earth, revised ed. Freeman Cooper, San Francisco

Wyllie PJ (1971) The dynamic earth. Wiley, New York

Vacquier V (1972) Geomagnetism in marine geology. Elsevier, Amsterdam

Tarling DH, Runcorn SK (eds) (1973) Implications of continental drift to the Earth sciences. Academic Press, New York

Hallam A (1973) A revolution in the Earth sciences. Clarendon Press, Oxford

Kahle CF (ed) (1974) Plate tectonics – assessments and reassessments. AAPG Mem 23, Am Assoc Petrol Geol, Tulsa Okla

1 Origin and Morphology of Ocean Basins

1.1 The Depth of the Sea

The obvious question to ask about the sea floor is how deep it is and why. The overall depth distribution first became known through the voyage of HMS *Challenger* (Fig. 1.1). We see that there are two most common depths: a shallow one near sea level (shelf seas) and a deep one between 1 and 5 km (normal deep ocean). The sea floor connecting shelves and deep ocean is of intermediate depths and makes up the continental slopes and rises.There is a portion of sea floor which is twice as deep as normal: such depths occur only in narrow trenches, mainly in a ring around the Pacific Ocean (Table 2.1).

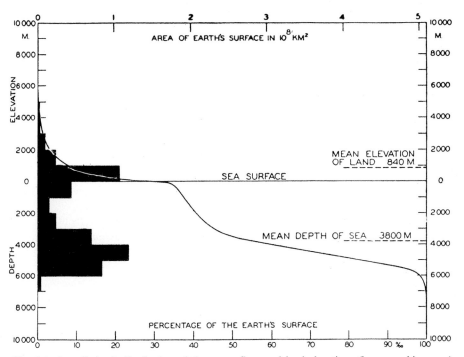

Fig. 1.1. Overall depth distribution of the ocean floor and land elevations (hypsographic curve). *Left:* frequency distribution of elevations. [H. U. Sverdrup et al. 1942, The oceans: 18]

On HMS *Challenger,* depth soundings were done by laboriously sending a weight to the ocean floor and measuring the length of the wire paid out. When the scattered soundings were connected in drawing depth contours, the ocean floor looked smooth. Only when echo sounding was used routinely did it become obvious that large parts of the ocean floor consist of immense mountain ranges whose cragginess rivals that of the Alps and the Sierra Nevada. Perhaps the most impressive of these ranges is the Mid-Atlantic Ridge, first discovered by the famous *Meteor* Expedition (1925–1927) (Fig. 1.2 b).

More recently, it has been shown, largely through the work of M. Ewing and co-workers, at Lamont Geological Observatory, that the seemingly endless Mid-Atlantic Ridge (Fig. 1.3) is itself only a portion of a world-encircling Mid-Ocean Ridge.

This was, of course, a discovery of immense importance. It identified the one unifying morphological feature of the planet, the central template to which the various scattered puzzle pieces of knowledge about the sea floor had to be fitted. The only other feature of the ocean floor of comparable magnitude is the line of trenches ringing the Pacific (Figs. 1.4 and 1.13). The complementary significance of these two features – the Ridge System and the Trench System – became obvious in the 1960s, from the study of magnetic properties of the sea floor, from earthquakes, and from heat flow distribution. In the late 1960s, the hypothesis that new sea floor forms at the center of the Mid-Ocean Ridge and that it travels toward the trenches where it sinks, gained general acceptance, as outlined in the introduction.

The hypothesis, called *sea-floor spreading,* explains in an elegant fashion the major features of the depth distribution of the sea floor.

Before we discuss this striking concept of sea-floor spreading in some detail, however, let us go back to consider the basic processes which shape Earth's surface, including the floor of the ocean.

1.2 Endogenic Processes

As is true of all of the face of the Earth, the sea floor is shaped by two kinds of processes, those deriving their energy from inside the Earth, called *endogenic,* and those driven by the Sun, called *exogenic.*

The forces inside the Earth produce volcanism and earthquakes; we meet them in the eruptions on Hawaii, in the geysers of Yellowstone Park, in the quakes in California. Working over long periods of time, the endogenic forces, fueled by heat sources within the Earth, build mountain ranges such as the Sierra Nevada and the Himalayas, or create gigantic rifts, such as Death Valley and the Rhine Graben. It is reasonable to suppose that the undersea mountains represent uplift, and that the great trenches result from downwarping of the sea floor by endogenic forces. Such motion, of course, requires flow of material within the Earth. Thus, matter has to rise to make the undersea mountain ranges, and must sink to make the trenches. The mental jump in formulating the hypothesis of sea-floor spreading was to see these necessary motions as part of a convection system (Fig. 1.5).

The exact nature of the forces and motions deep within the Earth is not accessible to direct observation by sampling. Even the deepest drill holes (down to 12 km, on land) only scratch the surface of the Earth. They do not penetrate the *crust*, which is defined as the uppermost layer of the solid Earth. Continental crust is about 20 to 50 km thick, while oceanic crust is much thinner, 5–10 km. The *mantle* below the crust is the source of the endogenic forces working on the crust. It is 2850 km thick

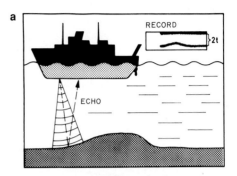

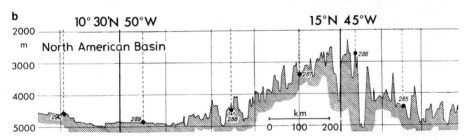

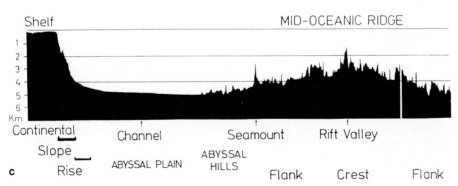

Fig. 1.2 a–c. Topography of the Mid-Atlantic Ridge and adjacent sea floor. **a** Principle of continuous echo profiling. The depth is given by $s = v \times t$, where v is the sound velocity and $2t$ is the time the sound takes to travel to and fro. *Record* output of the echo sounding device, on a moving strip of paper. **b** Results of echo sounding as obtained by the German *Meteor* Expedition (1925–1927). Soundings were taken every 4.5 km. *Numbers* refer to sampling stations. Note central rift. **c** Modern topographic profile with labels of physiographic features. [Modified after B. C. Heezen et al., 1959, Geol. Soc. Amer. Spec. Paper 65.]

Fig. 1.3. Physiographic diagram of the Atlantic ocean floor, from a painting by H. C. Berann (National Geographic Society), based on bathymetric studies of B. C. Heezen and M. Tharp

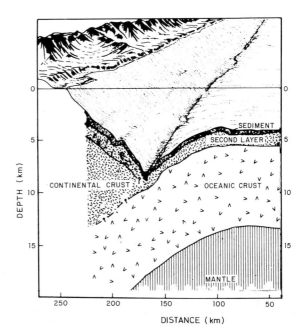

Fig. 1.4. Crustal structure of ocean margin off northern Chile, based on seismic refraction studies. [R. L. Fisher, R. W. Raitt, 1962, Deep-Sea Res 9: 423]

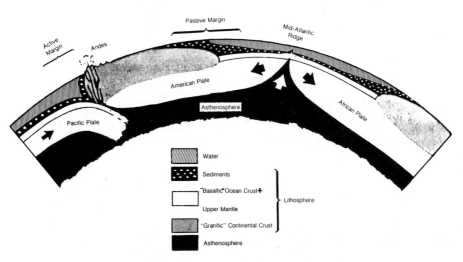

Fig. 1.5. Hypothetical convection currents in the upper mantle, producing sea-floor spreading and continental drift. This is one of a number of convection models which have been proposed to explain the origin of mid-ocean ridges and trenches.

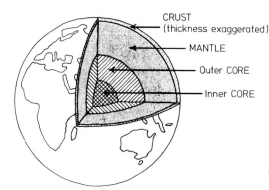

CRUST
(thickness exaggerated)

MANTLE

Outer CORE

Inner CORE

Fig. 1.6. Onion structure of the Earth: crust, mantle, and core. Typically, figures showing crust and mantle are not to scale. On a globe of diameter 1.75 m (the height of a person) the average ocean depth would shrink to 1 mm, just the thickness of human skin, and the lithosphere to a chalk line. This illustrates the great importance of the mantle and of mantle processes.

and makes up about two thirds of the mass of the Earth (Fig. 1.6). The remaining one third is mainly *core* (radius = 3470 km). Only 0.4 % of the mass of Earth is in the crust.

Thus, the mountain ranges and great trenches are but small wrinkles on the planet. The motions of sea floor and continents are perturbations on the very surface of a vast globe of hot rock. What kind of rock? And why hot? We do not know for sure. The kinds of rock dredged from deep clefts in the Mid-Ocean Ridge, and those recovered by deep drilling into the Ridge basalt, presumably most closely resemble the upper mantle material (see Appendix A 6). As mantle material pushes up at the Mid-Ocean Ridge, it changes its character through chemical fractionation, pressure release, degassing, and especially through reaction with seawater.

The heat most probably comes from the decay of radioactive elements. Another possible source of heat is the gravitational segregation of heavy and light material, which produced the onion structure of Earth in the first place. To receive the "messages from the mantle", that is, information about its structure and the processes in the interior, geophysicists make seismic measurements, and study the characteristics of magnetic and gravity fields. Mineralogists and petrologists investigate the behavior of rocks under high pressures and temperatures, and geochemists collect indirect evidence on the interior, derived from elemental abundances in the solar system in combination with the density distribution inside the Earth.

1.3 Exogenic Processes

Although the large-scale features of the ocean floor are shaped by endogenous forces, the sea floor also reflects the workings of the exogenic forces, that is, the processes of erosion and sedimentation. The classic example is the type of sea floor called abyssal plain – incredibly flat areas hundreds of miles in diameter (Fig. 1.2c). On land, the playas surrounding the Great Salt Lake in Utah can convey a feeling for the nature of these features.

The abyssal plains are vast undersea playas collecting debris from the continents, which is produced by the ever-present agents of weathering: rain, wind, ice. The chippings made by these sculptors, which carve canyons and wear down mountains, are carried to the ocean by rivers and winds. Here they build up continental margins, with the left-overs accumulating in abyssal plains.

Much of the sediment comes to the sea through quasi-catastrophic events: floods, storms, earthquakes or – on a longer time scale – ice advances. Other types of sediment arrive at the sea floor as a more or less continuous rain of particles: shells of plankton organisms, wind-borne dust, cosmic spherules. Through geologic time these gradually accumulating pelagic sediments built up a layer of a few hundred meters thickness, which forms a veneer on the oceanic crust. This veneer contains a detailed history of the evolution of ocean circulation and of pelagic organisms, for the last 100 to 150 million years.

Although exogenic processes tend to *level* the Earth, by erosion and deposition, they also can build mountains. The outstanding example is the Great Barrier Reef off eastern Australia, whose mountain tops rise thousands of meters above the floor of the Coral Sea. The mesa-like reef mountains are made from the calcium carbonate secreted by coralline algae, stony corals, mollusks, and small unicellular organisms called foraminifera. The algae, of course, depend on sunlight. The corals and foraminifera contain unicellular algae within their bodies, in symbiosis: they too depend on sunlight for growth.

With this briefest of all introductions to the opposing effects of endogenic processes (which wrinkle Earth's surface) and exogenic processes (which mainly smooth it), let us now return to the nature of the grand morphology of the sea floor – and to sea-floor spreading.

How does this concept explain the major features of the sea floor?

1.4 Morphology of the Mid-Ocean Ridge

Ultimately, it took geophysical evidence based on crustal magnetism to compel acceptance of the mobility of the ocean floor, and to turn the hypothesis of sea-floor spreading into the ruling theory. The most obvious achievement of the new theory is the explanation of the origin of the Mid-Ocean Ridge, the central morphologic feature of the sea floor. The Mid-Ocean Mountain Range is more than 60000 km long and takes up one third of the ocean floor, that is, about one fourth of the Earth's surface. In the Atlantic and along certain other portions, the crest is marked by a central rift, a 30- to 50-km-wide steep-walled valley 1 km deep, or more (see Figs. 1.2 and 1.3). The crestal morphology is usually very rugged and complicated, while the flanks tend to be smoothed by sediment (Fig. 1.7). The following is a brief account of how the theory of sea-floor spreading explains the character of the Mid-Ocean Ridge.

The crest is characterized by shallow earthquakes (centers at less than about 60 km depth), by active volcanism, and by high heat flow values. The upwelling and spreading of mantle material pulls apart the crust, producing the central rift, and generating the earthquakes. It also brings up heat from the Earth's interior. The *spreading rate*, that is, the rate at which sea floor on one side moves away from that on the other, is on the order of 1 to 10 cm per year. The hot mantle material filling the gap is less dense than old oceanic curst, because of thermal expansion. Away from the Central

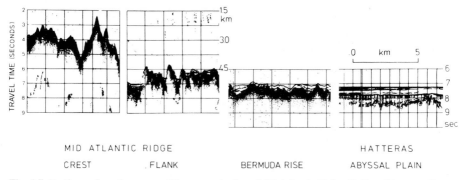

Fig. 1.7. Portions of continuous profiling records, from Mid-Atlantic Ridge (*left*) to Hatteras Abyssal Plain. Note the smoothing of the sea floor by a cover of sediment, with increasing age away from the central ridge area. [T. L. Holcombe, 1977, Geo Journal 1 6: 31] (For seismic profiling see Fig. 2.11.)

Ridge, it becomes denser as it cools, but on the whole, the new lithosphere and its sea floor float up higher on top of the mantle and thus protrude, forming the Ridge.

Generally, the Ridge crest has an elevation (i. e., depth) of – 2500 to – 3000 m, all around the world. From this similarity of elevations, we may conclude that the upwelling material and its temperature are rather uniform.

However, there is much variety on several scales. The ridge is much shallower than average in the North Atlantic, where it has, not so coincidentally, one of the most active "*hot spots*" associated with it, namely Iceland. Along its entire length, the Ridge is segmented between major *fracture zones* (Sect. 1.6) and each of the segments (300–500 km long) has its own history and morphology. More generally, everywhere along the Ridge axis there are subtle changes, comprising a few hundred meters of elevation, which are in large part due to the way the magma chambers along the Ridge axis are supplied with melt, from 30 to 60 km depth. This supply is discontinuous in time and space, giving rise to segmentation in scales of 50–100 and up to 300 km, with magmatically sated and starved sections. Detailed investigations of axial morphology, using side scan mapping, submersibles, high resolution seismics, and deep-sea drilling, have greatly advanced our knowledge of the significance of this segmentation in terms of Ridge dynamics. Mushroom-shaped *magma chambers* (with roofs only 1.5 to 2.5 km below the sea floor) cause local uplift and the formation of narrow axial graben structures (Fig. 1.8). These valleys can then be filled, episodically, by lava flows associated with dike instrusions from the underlying chamber. On fast-spreading Ridge areas, as on the East Pacific Rise (Fig. 1.13), the supply of lava is such that no big rift valley develops. Instead, there is an axial summit with or without a narrow graben. However, big deep rift valleys are generated along the axis of slow-spreading Rigde areas, as seen at the Mid-Atlantic Ridge.

The upwelling material forms pillow basalts and lava sheets after contact with the cold seawater. From comparing the seismic sequence of basaltic layers at the presently forming mid-ocean ridges with equivalent (but ancient) slabs of oceanic crust on land (such as the "ophiolites" of the Troodos Massif, Cyprus, or in Oman), we

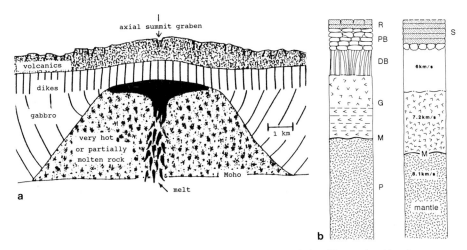

Fig. 1.8. a Schematic cross-section of the East Pacific Rise. A halo of hot rock with a few percent partial melt surrounds a much smaller mushroom-shaped axial magma chamber (*black*) with more than 50 % melt. [K. C. MacDonald et al., 1989, Nature, 339: 178]. **b** Comparison of field observations (*left*) and seismological investigations (*right*) regarding the structure of oceanic crust. Ophiolite sequence: *R* Radiolarites; *PB* pillow basalt; *DB* basalt in dike sheets; *G* gabbro (note layering in lower portion); *P* peridotite. *M* denotes position of the Mohorovicic discontinuity ("Moho", where sound velocity changes suddenly). The Moho is typically at 6–10 km. Corresponding sections in the seismic stratigraphy assigned by inference. *S* sediments. [C. Allègre, 1988, modified]

know that these pillow lavas are underlain by basaltic dikes. Seismically, they are known as "layers 2 A and 2 B". Magmatic gabbros ("layer 3") and peridotites ("layer 4") are separated by the *Moho-discontinuity*, normally at about 6 to 10 km depth, defining the thickness of the oceanic crust (Fig. 1.8b).

With few exceptions, the volcanic rocks forming the oceanic crust on the ridges are *olivine tholeiites*. They are dense, heavy silicate rocks rich in iron and magnesium, and belong to the basalts. Their minerals are essentially plagioclase feldspar, pyroxene, and olivine (Appendix A4). Compared with the more familiar basalt on land, they have low contents of potassium, titanium, and phosphorus (Appendix A6).

Also, they are depleted in those trace elements (rubidium, cesium, barium, lanthanum) which tend to be preferentially concentrated in the liquid phase during melting or during fractional crystallization. It is difficult, however, to deduce directly the composition of the mantle material from that of the tholeiite basalts. Many processes affect the magmas on their way up toward the sea floor. These processes include *differentiation* through partial melting; mixing of, and reactions between various types of melts; reactions with solutions including seawater, and escape of gases. The reactions of seawater with hot basalt, and the associated hot springs with their deposits are discussed in Chapter 10.4.4. Newly discovered life forms are of special interest in this context (Chap. 6.9; see also book cover).

As the sea floor spreads away from the crest, the lithosphere cools and sinks, about 1000 m during the first 10 million years. The next 1000 m of sinking takes about

26 million years (Fig. 1.9). It can be argued from physical principles that the depth of the sea floor on the spreading flanks of the Ridge should be a simple function of age:

$$\text{depth below crest} = k \cdot \sqrt{\text{age}}. \tag{1.1}$$

From the relationship given above, we can evaluate k as follows:

$$k = \frac{1000}{\sqrt{10}} \text{ and } k = \frac{2000}{\sqrt{10+26}},$$

which comes out as $k \approx 320$, when depth is in meters and age in million years. If this is correct, we can calculate the average age of the deep-sea floor from its average depth (after correcting for sediment cover). For an average (corrected) basement depth of 5000 m (= 2400 m below Ridge crest), we obtain an age of 60 million years, which is indeed close to the average age of the sea floor.

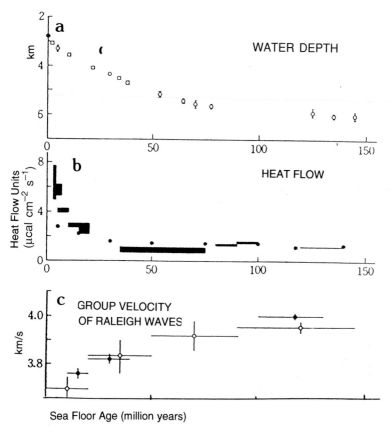

Fig. 1.9. a–c. Observations illustrating the cooling of the lithosphere at mid-oceanic ridges. **a** After J. G. Sclater et al., 1971, "reliable mean values" **b** After Sclater et al., 1976 (*bars*); and Sclater and Francheteau 1971 (*dots*). **c** Seismic waves velocities increase with cooling, after Yoshii (1975) and Forsyth (1977) *circles* and *dots*. [E. Seibold et al., 1986, The sea floor, Japanese edition]

Equation (1.1) occupies a special place in marine geophysics, geology, and pale-oceanography. It is one of the very few statements about Earth history that is both quantitative and simple, and at the same time also valid and useful.

During sinking of the sea floor, the rough topography produced by volcanism and faulting moves down the ridge flanks and is then gradually smoothed by the sediment cover (see Fig. 1.7). However, abyssal hills with a relief in the 50 to 1000 m range and with slopes of 1° to 15° remain as expressions of the underlying basement morphology over large regions. Abyssal-hill morphology is the most common type of landscape on the face of Earth: in the Pacific Ocean about 80 % of the sea floor belongs to this category.

1.5 Morphology of the Trenches

In general, trenches are found near the margins of ocean basins, notably the Pacific Basin. It is not obvious why there are not more mid-ocean trenches. To answer this question, we need to learn more about processes within the mantle. First, some observations: trenches are roughly 100 km wide (in their shallower part) and from hundreds to thousands of kilometers long. For example, the Aleutian Trench is 2900 km long. The cross-section is usually V-shaped (Fig. 1.10.a), and the deepest part may be flat due to ponded sediment. Such sediments generally show undisturbed horizontal layering – an observation which was sometimes used as an argument against subduction when the concept of sea-floor spreading was new. The trench walls usually have slopes between 8° and 15°. However, steep sides (up to 45°) as well as steps have also been mapped. In cases, outcrops of basalt have been observed by dredging and deep-sea photography.

The greatest depths are in the western Pacific, in sediment-starved trenches off island arcs: Mariana Trench maximally 10 915 m; Tonga Trench, 10 800 m; Philippine Trench, 10 055 m; Japan Trench, 9700 m; Kermadec Trench, 10 050 m. The values are not exact; they were determined from echo soundings which include corrections for effects of regional temperature and salinity distribution on the velocity of sound in the water. Any errors in the determinations, however, cannot mask the similarity in these depths. As in the similarity of ridge-crest elevations, this coherence points to the action of similar processes in each of the trenches in the western Pacific. Elsewhere, trenches are shallower: Puerto Rico Trench about 8600 m; South Sandwich, 8260 m, Sunda 7135 m. The eastern Pacific is characterized by trenches directly adjacent to continents, without intervening island arcs. These trenches are filled with continental debris, and this, presumably, is the reason why they are distinctly shallower than their western counterparts.

The ring of trenches girdling the Pacific is the site of most of the earthquakes on Earth: more than 80 % of the shallow earthquakes (< 60 km deep), 90 % of the intermediate ones (60 to 300 km deep), and almost all of the deep quake centers (300 to 700 km) are concentrated here (Fig. 1.11).

The rest of the deep and intermediate quakes also occur largely in trenches; some occur in the Mediterranean, in Iran, and in central Asia at the northern boundary of

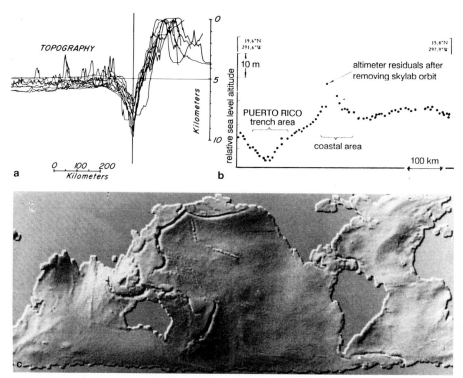

Fig. 1.10. a Topographic profiles across trenches in various regions of the ocean. Land or island arcs to the right. [M. Talwani, 1970, in The sea 4 [1]: 282]. **b** Deformation of the sea surface measured by satellite altimeter. Sea level is influenced by regional gravity. Trenches are well-known gravity minimum areas. Oceanic high areas not in isostatic equilibrium show positive gravity anomalies, therefore "attracting" sea water and thus raising sea level. [Skylab Data Catalog NASA 1974, p. 133]. **c** Map of the mean ocean surface reflecting ocean bottom features such as seamounts, trenches, ridges, and fracture zones. Land masses have been masked out. The map was developed by the US. National Aeronautics and Space Administration (NASA), 1982. [Courtesy Jet. Prop. Lab, Pasadena]

the Himalayas. When the depths of the quake centers are plotted below the epicenters, they are seen to occur on planes which intersect the surface near the trenches and dip underneath the arcs or the continents at an angle of between 15° and 75°, to a depth of about 700 km. It is reasonable to conclude that the earthquakes are caused by friction on the upper surface of the downgoing slab of sea floor, and that this friction ceases when the temperature becomes high enough to permit flow (Fig. 1.12).

Fig. 1.12. Relationship of plate motions to plate boundaries and earthquakes. *Divergent plate boundary* Mid-Ocean Ridge, shallow earthquakes; *convergent plate boundary, trench,* shallow, intermediate, and deep earthquakes. Lateral boundary: *fracture zone,* shallow earthquakes on active part only. Horizontal motions are of the order of 1 to 10 cm/yr. [Based on a diagram by B. Isacks, J. Oliver, and L. R. Sykes, 1968, J Geophys Res 73: 5855]

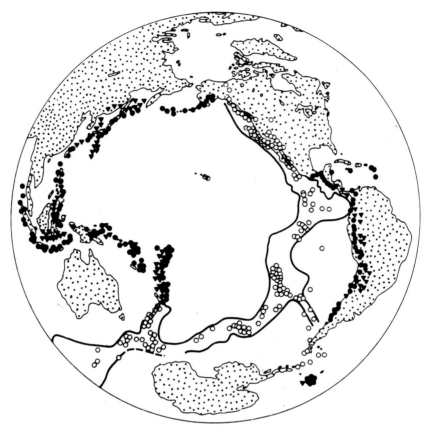

Fig. 1.11. Earthquake belts in the Pacific. Shallow earthquakes (open circles) characterize spreading centers (South Pacific) and transform faults (California). Deep and intermediate earthquakes (filled triangles, filled circles) are restricted to trench regions. [R. W. Girdler, 1964, Astron Soc Geophys J 8: 537]

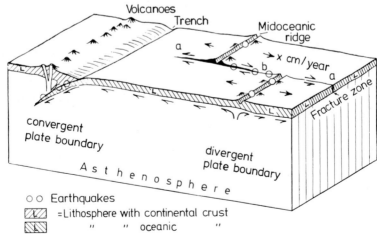

The *"ring of fire"* around the Pacific is closely associated with the trenches: volcanoes sit on top of the dipping earthquake planes, that is, the downgoing lithosphere. Of 800 active volcanoes, 75 % are in the "ring". Where the lithosphere descends before reaching the continents, the vulcanoes form island arcs; where it descends under a continent (South America), mountain ranges are formed. The rising magmas produced by partial melting of the downgoing slab mix with overlying materials on their way up and form characteristic volcanic rock types, the *andesites* – named after the Andes Mountains (Appendix, Fig. A 6.1).

According to the theory of sea-floor spreading, the trenches are produced by the subduction of the sea floor. Much of the descending lithosphere, some 100 km thick, sinks through the *asthenosphere,* that is, the "soft" part of the upper mantle. The descending lithosphere offers up materials to the continent, by the scraping off of sediment, and by partial melting. These materials contribute to continental accretion, that is, to the growth of continents.

Much of such growth, apparently, also depends on the accretion of *"terranes"*, slabs of oceanic or continental-type crust which arrive at the trench, but refuse to go down. Instead, they become part of the adjacent continent. The trench, now clogged, must move seaward when this happens. Much of the west coast of the USA is thought to consist of "terranes" moved in from elsewhere. The pieces of foreign real estate are mapped as having dimensions on the order of 100 to 1000 km on a side.

Continental accretion is one way in which the endogenic forces oppose the wearing down of the continents by exogenic agents. Thus, the continued existence of continents which rise high above the sea floor is intimately tied to the processes of sea-floor spreading.

1.6 Fracture Zones and Plate Tectonics

We have earlier mentioned that the Ridge Crest is not continuous but segmented. It occurs in more or less straight portions which are offset from each other. The consequence of such offset is that a lateral fault must form at the two ends of each crestal portion (Fig. 1.12). Since there is motion along this fault during active spreading, there are earthquakes on it. These earthquakes are shallow and define the *active* part of the *fracture zone,* that is, the *ridge-ridge transform fault.* Beyond this active part, the fracture zone is the frozen trace of the fault; the scarps subside as the sea floor ages on both sides of the zone. These extensive linear zones have an unusually irregular topography with large seamounts, steep-sided or asymmetrical ridges, troughs, or escarpments (Fig. 1.3).

Some fracture zones connect the end of a ridge crest portion to a trench. These zones are seismically active and constitute the third type of boundary defining a lithospheric slab or *plate.* The other two, of course, are *spreading center* and *trench.* The fact that these boundaries form "plates" was first pointed out by J. T. Wilson in 1965. On the basis of earthquake distributions and first motion studies (that is, observing which way the ground moves upon initiation of a quake), it is possible to outline a number of large lithospheric slabs dividing the surface of the globe. Each of the plates has its own particular motion, which can be read from the magnetism of the sea floor, as we shall see. The quantitative development of these concepts was initiated in the late 1960s by W. J. Morgan, D. P. McKenzie, and R. L. Parker; X. LePichon; and by B. Isacks, I. Oliver, and L. R. Sykes. (See A. Cox, 1973)

The motions are generally uniform and do not result in deformation of the plates; hence, they can be described as rotations on a sphere, according to a theorem of the famous mathematician, Leonhard Euler (1707–1783). The fracture zones provide traces for the latitudinal circles around the pole of rotation (which need not coincide with that of the rotation of the Earth, see Fig. 0.4). Thus, the pole of rotation can be determined for each plate. Geometry requires that spreading rates must increase away from the pole of rotation for separating plates, and this is indeed observed. It will be noted in Fig. 1.13 that a plate can contain both oceanic and continental lithosphere. In fact, the continents share the motions of the mobile ocean floor. Thus, the continents do drift, as Wegener had supposed, but not by plowing through the mantle magma.

Since Morgan proposed his scheme of plates making up the globe, new plates have been discovered, and several modifications of plate geometry and kinematics made. One recent model that summarizes progress in this field of research (C. de Mets et al., 1990, Geophys. J. 101) distinguishes 12 major plates including Philippine (PH), Cocos (North of Nazca), and Caribbean Plates, and also separates a North American Plate from a South American one and an Indian Plate from an Australian one by transition zones with diffuse deformation. The position of the boundary of the North American Plate in the Arctic realm likewise is not well defined. More than a hundred mantle plumes, spread around the oceans and on continents as well have now been defined by different authors. These hot spots (Fig. 1.13 b) are of interest in the context of plate movements, as we shall see later.

1.7 Seamounts, Island Chains, and Hot Spots

With few exceptions, oceanic islands are made of volcanic rock, with or without a crown of reef carbonate. A crown of reef carbonate, of course, can only be precipitated in shallow water because it depends on algal growth. Thus, if a seamount is found with a top of reef carbonate, and deeply submerged below the present sea level, it must have sunk. Such seamounts are common in the western Pacific.

It has been said that the discovery of flat-topped seamounts held the key to the new understanding of the origin of ocean basins. Flat-topped seamounts were first described in the 1940s by H. H. Hess (Fig. 0.1) in the central Pacific. Hess proposed that these table mounts, the *guyots,* as he named them, had formed as volcanic islands, were truncated by wave erosion, and then sank to their present depths. He also initially thought they might be of Precambrian age, with lots of time available for subsidence. However, no rocks older than Cretaceous were ever dredged from the guyots.

In essence, Hess' hypothesis of guyot formation was an extrapolation of Charles Darwin's hypothesis of atoll formation (see Chap. 7.4.3). The idea of seamount subsidence was easily reconciled with Hess' later concept of sea-floor spreading

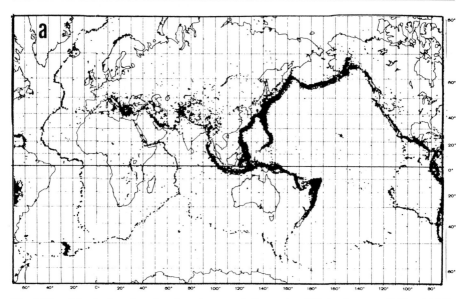

Fig. 1.13. a. Distribution of earthquake epicenters shallower than 700 km between 1961 and 1967. Note the close relationship to the plate boundaries in *b*. However, mid-plate earthquakes are also quite common. [M. Barazangi and J. Dorman, 1969, ESSA, Coast and Geodetic Survey, and Seismol. Soc. Amer. Bull. 59]

(Fig. 0.3d). Thus, he discovered both a major problem – the origin of guyots – and its solution: sea-floor spreading. Seamounts in general have elevations of more than 1000 m and typical slopes of 5° to 15°. The Pacific has about 10 000 of such seamounts.

There are a number of striking instances where seamounts – either flat-topped or not – occur in linear chains. The Hawaiian chain is a prime example. How are such chains generated? One possible explanation which has been proposed is to assume that the volcanoes making the chain lie on a long zone of weakness in the crust, a deep fracture, along which magma can rise to form volcanic islands. Yet, at least in the case of the Hawaiian Islands, there is clearly a progression from high, large islands with active volcanoes at the tip of the line, to sunken islands with extinct volcanoes at the end (Fig. 1.14). It certainly looks as though the big islands are young and the sunken ones old. Dating of the rocks, by radioactivity, confirms this impression. Thus, in the fracture hypothesis, we have to postulate a propagating crack, which opens at one end and closes at the other.

A more satisfactory explanation of the origin of island chains was given by J. T. Wilson (in 1965: A new class of faults and their bearing on continental drift. Nature 207: 343–347) and by W. J. Morgan (in 1968: Rises, trenches, great faults and crustal blocks. J. Geophys. Res. 73: 1959–1982). They proposed a stationary source of hot magma, deep in the mantle, over which the lithosphere rides. Volcanoes build up on top of the crust over the *"hot spot"*, a site of a *plume* of ascending magma. As the plate moves, a trail of extinct volcanoes forms behind the active tip of the line (Fig. 1.14).

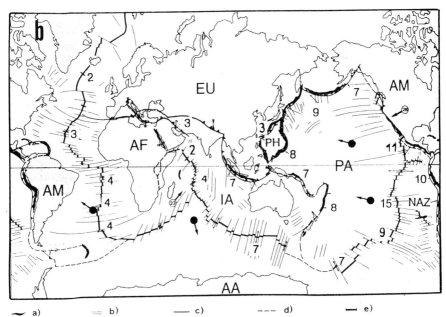

— a)   === b)   —— c)   --- d)   — e)

Fig. 1.13. b. Major lithosphere plates of the Earth, as proposed by W. J. Morgan (1968, J. Geophys. Res. 73: 1959). *EU* Eurasian plate; *AM* American; *PA* Pacific; *IA* Indo-Australian; *AF* African; *AA* Antarctic; *PH* Philippine; *NAZ* Nazsca Plate. Convergent plate boundaries (*a*) occur mostly around the Pacific. Convergent plate boundaries, with both sides having continental crust, occur in the Himalayas (*c*). Divergent plate boundaries, mostly as midoceanic ridges (*e*), are dissected by transform faults (*b*). Uncertain boundaries (*d*) occur in the Southern Ocean and between the Eurasian and American plates. Some of the many hot spots are indicated by *black dots. Arrows* show direction of plate movement relative to hot spots. Spreading rates in cm/year are generalized. Newer models have modified this picture somewhat, as mentioned in the text. [R. Trümpy, 1985, Z. Nat. forsch. Ges. Zürich, 5:13, modified]

This trail, then, indicates the direction of movement of the lithospheric plate, with respect to the (more or less stationary) mantle source. Changes in direction of the trail, as between Hawaiian islands and Emperor seamounts, would indicate changes in the direction of plate motion.

Hot spot plumes (with diameters of a few hundred kilometers) are thought to originate from the lower mantle. They constitute an important component of the mantle's convection. Basalts from this source are enriched in so-called *incompatible elements* (potassium, rubidium, cesium, strontium, uranium, thorium, and rare earth elements), compared with Ocean Ridge basalts. Presumably, the upper mantle was stripped of these elements in its long history of making continents, and this process continues.

The global distribution of the more than one hundred hot spots and a few larger plumes with diameters of 2000 km, so-called superplumes, is unexplained as yet. Is the mantle in control of these patterns? A relationship seems to exist between the general distribution of hot spots and mantle-wide upward convection as indicated by low

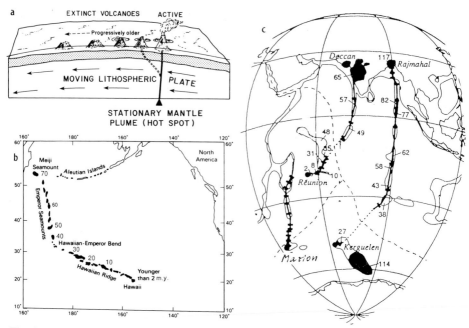

Fig. 1.14. a–c Origin of Haiwaiian islands, submerged coral banks, and Emperor Seamounts, according to the "hot spot" hypothesis as envisaged by J. T. Wilson (1963, Can J Phys 41: 863) and W. J. Morgan (1971, Nature, London 230: 42). **a** Sketch of hypothesis. **b** Ages in million years along Hawaiian Ridge [K-Ar determinations, summarized by S. Uyeda 1978 and biostratigraphic ages of basal sediment in *Glomar Challenger* Sites, 430–433 Leg 55, 1977]. **c** Selected hot spot tracks from the Indian Ocean basin. *Numbers* are radiometric ages in million years determined on basaltic rocks from islands, seamounts, continental locations, and underneath sediments penetrated by deep-sea drilling. The linear chains of volcanoes grow older in the direction of plate motion. In 1987–1989, four legs of ODP were devoted to the investigation of hot spots in the Indian Ocean. The results indicate that plumes may be stationary over periods as long as 100 million years as seen on the "Ninety-east Ridge", and that the Kerguelen and Réunion hot spots appear to have begun with massive outpouring of flood basalts (Deccan and Rajmahal). [R. A. Duncan, 1991, GSA Today (1, 10: 213–219), simplified]

seismic velocities in the lower mantle in these regions. Or are weaknesses in the lithosphere responsible? One suspects that both aspects will prove to be important.

1.8 Evidence for Sea-Floor Spreading: the Magnetic Stripes

We use the concept of sea-floor spreading for explaining many or most of the important morphological features of ocean basins. What proof do we have that the theory is sound? This question was raised well into the 1970s.

What is "proof" in the geologic sciences? Can we prove that a fossil was once part of a living organism? Can we prove that the Earth has an age of 4.6 billion years? Can we prove that large continental glaciers once covered vast areas of North America and northern Europe?

It all depends on what one is willing to accept as proof. Certainly, all the above questions were once vehemently answered with NO by experts, and would be answered with disbelief today – disbelief that anyone would ask such a silly question.

What, then, about proof for sea-floor spreading? Since 1968, the year in which several major articles appeared on the subject, "sea-floor spreading" has joined the various other propositions about the Earth which are accepted as fact.

Why can we be so confident that the sea-floor spreading story is correct?

Our confidence derives from the global pattern of magnetic anomalies on the sea floor (Fig. 1.15). For each of the observations on morphology, heat flow, seismic activity, etc., which are so nicely explained by sea-floor spreading, it is possible to conceive of some other way to produce the phenomenon. However, for the magnetic anomalies no reasonable alternative to sea-floor spreading has ever been proposed – and not for lack of trying.

The patterns, the *"magnetic stripes"*, were first discovered by geophysicists at Scripps Institution of Oceanography (R. G. Mason, A. D. Raff, V. Vacquier). However, their origin remained a complete mystery for several years. One of the problems was that the area for which they had been mapped is tectonically complicated, and the symmetry of the patterns about the Ridge, which holds the key to the explanation, is not obvious there.

The first successful attempt to account for the "stripes" was by F. J. Vine (then a graduate student at Cambridge University) and D. H. Matthews (his advisor), in 1963. Their suggestion was strikingly simple: put together the ideas on sea-floor spreading of H. H. Hess and of R. S. Dietz, and combine them with the evidence for periodic reversals in the Earth's magnetic field, as presented by A. Cox, Doell, R. R., and Dalrymple, G. B. in 1963 (Geomagnetic polarity epochs and Pleistocene geochronometry. Nature 198: 1049–1051.) The newly upwelled, hot material at the Ridge Crest (or within the central rift) is magnetized upon cooling below 525° (the Curie Point), in accordance with the prevailing magnetic field. If this field reverses periodically, and, they said, "if spreading of the ocean floor occurs, blocks of alternately normal and reversely magnetized material would drift away from the center of the ridge and parallel to the crest of it." Here, in a nutshell, was the key to proving the reality of sea-floor spreading (Fig. 1.16). The sea floor could be seen to act as a tape recorder of the earth's magnetic field.

Eventually, the magnetic anomaly sequence on each side of the ridge, in every major ocean basin, was found to be exactly the same as in the lava flows studied (and dated) on land.

The tape recorder is perhaps not of the finest quality, but it works remarkably well (Fig. 1.16a). With a time scale at hand (Fig. 1.16c), we can now read off the spreading rate, simply by matching the sea-floor anomalies to the magnetic reversal scale. Thus, we can make an age map of the sea floor (Fig. 1.17).

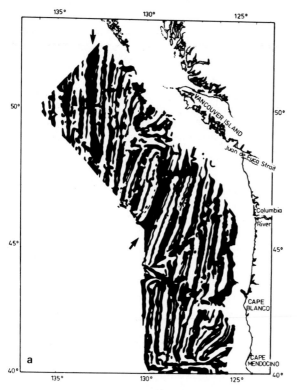

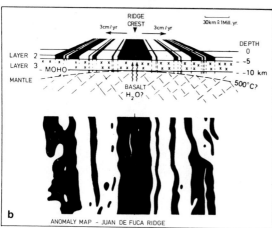

Fig. 1.15 a, b Magnetic lineations on the sea floor. **a** Anomaly patterns of A. D. Raff and
R. G. Mason (1961, Geol Soc
Am Bull 72: 1267). These anomalies remained unexplained for
several years. They are now
recognized as being generated at
the "spreading centers" (*arrows*).
b Mechanism of generating magnetic anomalies, as proposed by
F. J. Vine and D. H. Matthews
(1963, Nature London 199:
947). [F. J. Vine in R. A. Phinney, 1968, The history of the
Earth's crust, Princeton Univ.
Press pp 73–89]

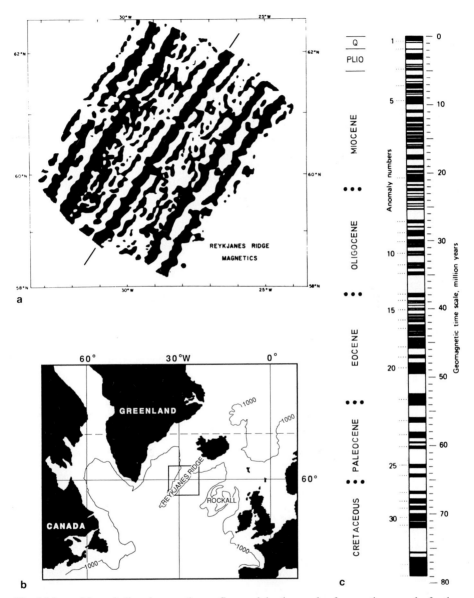

Fig. 1.16. a–c Magnetic lineations on the sea floor and the time scale of magnetic reversals, for the last 80 million years. Pattern on Reykjanes Ridge from J. R. Heirtzler et al. (1966, Deep-Sea Res. 13: 427). Time scale after Heirtzler et al. (1968) J Geophys Res 73: 2119). Biostratigraphic boundaries (**c**) modified. An up-to-date scale is given in Cande, S. C., and D. V. Kent (1992) A new geomagnetic polarity time scale for the Late Cretaceous and Cenozoic. J Geophys Res 97, 13 917.

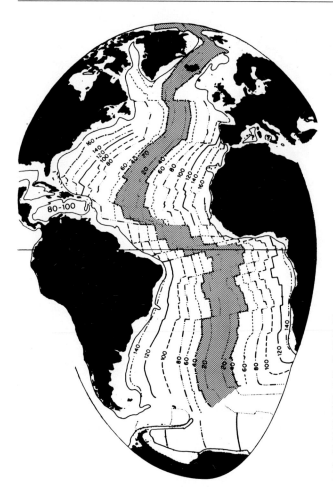

Fig. 1.17. Age of the sea floor in the Atlantic Ocean, based chiefly on the magnetic reversal scale. The gradual increase in size of the Atlantic is evident. It grows at the expense of the Pacific (whose age distributions are less well established. [W. H. Berger, E. L. Winterer, 1974, Int Assoc Sediment Spec Publ 1: 11]

If the age map is correct, we should then find that the oldest sediment lying on the basaltic substrate shows the same age progression. The deep-sea drilling ship *Glomar Challenger* set out in 1968 to test this prediction. It was first found to be correct during Leg 3 (1968/69). Since then, most magnetic and micropaleontological age determinations have coincided with an astonishing exactness (Fig. 1.18). In the end, this agreement had to be accepted as "proof" of the new theory.

1.9 Open Tasks and Questions

Plate tectonics brought a wealth of answers to fundamental questions – such as the origin of mountains! – that had long resisted the most painstaking inquiries. However, as any good theory should, it raised a host of new problems. Many of them simply concern refinement of existing concepts; for example, the details of *processes* active at the *plate margins* (Fig. 1.20).

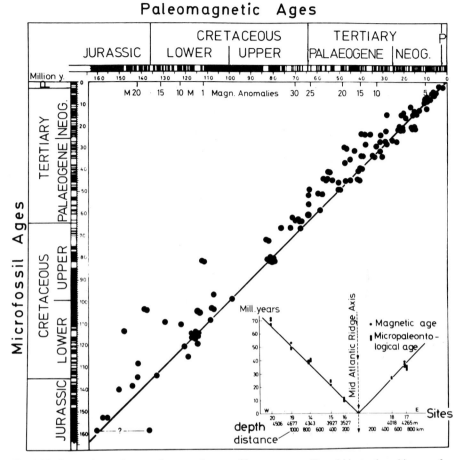

Fig. 1.18. Comparison of magnetic anomaly ages ("basement ages") and biostratigraphic ages for oldest sediment. *Glomar Challenger* Sites 1–417 (1968–1976). Compiled by M. Sarnthein, Kiel. Note that most of the sediment ages are slightly younger than the paleomagnetic basement ages as is expected. *Insert:* Age-distance plot of DSDP Leg 3 data, on a profile across the Mid-Atlantic Ridge off Brazil at 30° S. These data (published by A. E. Maxwell et al., 1970, Science 168: 1047) first demonstrated the agreement between paleomagnetic dating based on the hypothesis of sea floor spreading, and the dates derived from biostratigraphy.

Both the physics of motion and the chemistry of the fractionation of materials constitute the focus of such studies. How exactly are sulfide deposits formed near the Ridge crest? What is their fate in the subduction zones? Do they contribute to the formation of ore bodies there? What is the role of hydrothermal fluids in both environments? More fundamentally, how do these processes bear on the composition of seawater and the atmosphere? We shall return to some of these questions, especially in Chapter 10.

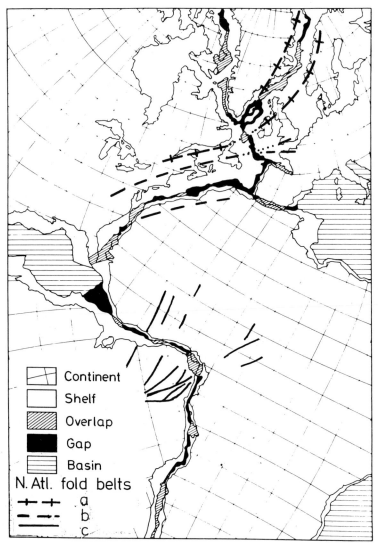

Fig. 1.19. The fit of the continents bordering the Atlantic, as proposed by E. C. Bullard et al. (in Blackett et al. 1965. A Symposium on Continental Drift. Philosophical Transactions, A258, Royal Society of London). The fit is based on matching the 500 fathom (900 m) contour. Note the Niger Delta overlap, which is expected, and the Bahamas overlap, which is unexplained. Note also the position of Gibraltar relative of Africa, and the "closure" of the Bay of Biscay. The alignment of the Paleozoic fold belts in North America and Europe and in South America and Africa (here added) is remarkably good. *a* Caledonian fold belt; *b* Hercynian fold belt; *c* Panafrican fold belt. [From various sources, Panafrican belt after C. J. Archanjo and J. L. Bouchez, Bull. Soc. Géol. France, 1991, 4: 638]

Other problems have to do not so much with processes, but with a better grip on *historical reconstruction.*

For the marine geologist interested in reconstructing the history of the oceans, reliable paleogeographic maps showing the distribution of land and water through geologic time are perhaps the most pressing need. E. C. Bullard et al., in 1965, demonstrated how to reassemble correctly the drifting continents and continental fragments in a sphere (Fig. 1.19). Bullard's co-workers and others since have greatly extended this type of work and have supplied a series of maps with ancient positions of continental masses. These, of course, are extremely useful for historical geology, both continental and marine. However, we know that mountain building and other processes active at the transitions between oceanic and continental crust have changed significant details of the continental configurations. Some of these changes are crucial in deciding, for example, whether there was a connection between one ocean basin and another. It will take many years of compilation, fieldwork, and detailed reassembly just to provide the kind of paleogeographic base maps necessary for explaining the distribution of ancient fossils, for instance.

These tasks are difficult enough for the time back to the Permian. For earlier periods the challenge takes on intimidating dimensions.

Continuous improvement of the *time scale of magnetic reversals,* for example, is an important task, because this scale forms the basis for discovering *rates* of change in Earth history, including continental drift or sedimentary processes. A more difficult task in this context is to determine the duration of magnetic reversals (less than 10 000 years mostly) and why the reversals occur in the first place. Or, the same question put differently, why reversals did not occur for considerable time spans, as, for example, in the middle Cretaceous (see Fig. 9.22). Was the release of *plumes* from the lower mantle (which apparently expressed itself millions of years later as out-pouring of basalts on the sea floor) in any way related to the cessation of magnetic reversals? If so, why should this be? Reversals, then, pose problems aplenty – and not just for migrating birds using the magnetic field to orient themselves.

Many fundamental questions have been raised (see Fig. 1.20). What determines sea-floor spreading rates? What is the reason for changing spreading directions and rates? What is the significance of the enormous clusters of islands and seamounts in the South Pacific (between the Bikini Atoll, some 2000 km southeast of the Marianas stretching parallel and inside the andesite line (Fig. A6.1) for about 8000 km to the Tuamotu Archipelago)? Presumably, they have to do with details of mantle processes which we are far from grasping as in other tectonic and volcanic intra-plate situations.

Clearly, all the major features on the thin skin of planet Earth, the crust, must ultimately owe much to such *mantle processes*. What does convection in the mantle actually look like? Are upper and lower mantle convection largely decoupled or not? What is the role of downgoing slabs in mixing the mantle? What about the rising plumes? How important are they within the convection scenario? Where is their source? How much is the material modified on the way up? How stable are the plumes through geologic time? Why do they have quite different expressions on the surface, ranging from oceanic plateaus to island groups?

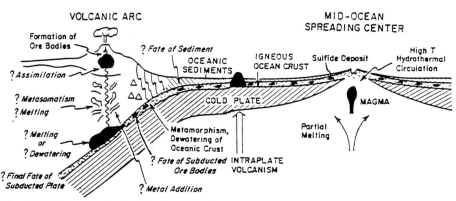

Fig. 1.20. Fluxes associated with plate tectonics and mountain-building at convergent zones. Processes within subduction zones are poorly understood. They include metasomatic activity, whereby existing rocks are partially or entirely replaced by reaction products of these rocks with invading matter from below. [European Science Foundation, 1988, Report on the Second Conference on Scientific Ocean Drilling, Strasbourg, Imprim. Reg.]

Exploration by *seismic tomography* (three-dimensional modeling of the mantle based on multidirectional propagation of seismic waves), and the systematic large-scale mapping of the chemistry of basalts are providing glimpses of what to expect. Apparently, the mantle is not just layered, but consists of a complicated patchwork of interlocking and mixing magma masses with different histories – and different physical and chemical properties. *Numerical modelling* – mathematical experiments based on current knowledge of the system – suggests the presence of quasi-cylindrical upwelling bodies, reminiscent of the updraft below cumulus clouds familiar to aviators. Of course, these updrafts in the mantle move extremely slowly on the human time scale. Downwelling, in the models, is represented by elongated sheets, perhaps derived from slabs descending in subduction zones. These slabs, being cold and therefore heavy, are part of the driving mechanism for the convection which we try to understand.

The various *messages from the mantle* – in the shape of large-scale topographic features (e. g., "superswell" in the South Pacific (McNutt et al. (1993) Science) and unusual isotopic composition of basalt – e. g. "Dupal" anomaly in the Indian Ocean: Dupré, B and CJ Allègre (1983) Pb-Sr isotope variation in Indian Ocean basalts and mixing phenomena. Nature, 303: 346. Hart, SR (1984). A large-scale isotopic anomaly in the Southern Hemisphere mantle. Nature 309: 137–144 – need to be mapped and coordinated in coherent models, before mantle processes are properly understood.

Further Reading

Cox A (ed) (1973) Plate tectonics and geomagnetics reversals. Freeman, San Francisco
LePichon X, Francheteau J, Bonnin J (1973) Plate tectonics. Elsevier, Amsterdam
Uyeda S (1978) The new view of the earth – moving continents and moving oceans. Freeman, San Francisco
Anderson RN (1986) Marine geology – a planet Earth perspective. Wiley, New York
Kearey P, Vine FJ (1990) Global tectonics. Blackwell Scientific, Oxford

2 Origin and Morphology of Ocean Margins

2.1 General Features of Continental Margins

The continents are very old: they contain rocks aged thousands of millions of years. Ultimately, they are derived from mantle materials by fractionation processes involving repeated mountain building and erosion – the "rock cycle". Because of low density the continents float on the mantle. The ocean floor, on the other hand, is geologically young, as we have seen. The basaltic rock which forms the ocean floor basement is rather close in composition to the mantle rock it came from. It is slightly heavier than continental rock (largely due to its high iron content, Appendix A6). The light-weight continental mass protrudes above the surrounding sea floor (Fig. 2.1a). Thick sediment piles accumulate at the boundary between continent and ocean,

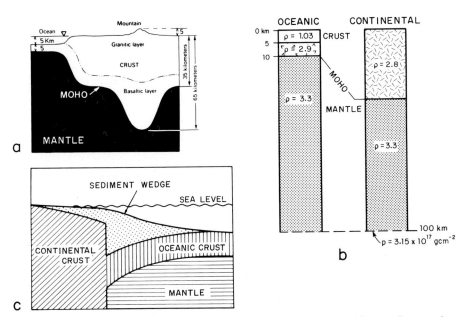

Fig. 2.1 a–c. Schematic isostatic block model for continent-ocean-transition. **a** Cross-section through continent "floating" on mantle (Uyeda, 1978). **b** Density profiles. **c** Sketch of general nature of continental margin

which build the actual margin (Fig. 2.1c). These sediments may be well-layered or strongly deformed, depending on the tectonic forces active at the margin.

The ocean margins, that is, the regions of transition between continent and deep ocean, differ greatly in their characteristics, depending on whether they occur in mid-plate areas (on the continent's trailing edge), or on the collision edge of a continent, or along a shear zone. The one thing most of the ocean margins have in common is the occurrence of large masses of sediment. Ocean margins are generally referred to as *continental margins* – a reflection of our landlubber point of view.

The importance of the continental margins in the overall geography of the ocean floor can best be illustrated with a few statistics (Table 2.1). The various numbers reflect in essence the efficiency of the exogenic and endogenic processes which provide the balance between the extent of continents and of ocean basins. This balance is produced by erosion of highlands, deposition around the continents, and the mountain-building processes briefly alluded to earlier.

The bulk of the *land area,* about 70 %, is within 1000 m above sea level. Continents wear down toward sea level because it is the baseline of erosion. Sea level is also the top level of deposition. Hence the sediments deposited offshore tend to build up to sea level. The great plains of the lower Mississippi Basin, and the entire Gulf Coast are prime examples of this tendency for large continental areas to be near sea level. These areas are underlain by sediments deposited close to sea level during times when the sea invaded the continent.

Large parts of the low-lying portions of continents are covered with marine deposits. Actually, such areas are part of the *shelf* of a continent, a part which is normally submerged in the course of geologic history. At the present time in our geologic period, an unusually large proportion of this shelf is exposed. If we put the outer end of the shelf at 200 m water depth (for convenience – it varies greatly), 28 million km^2, i. e., about 5 % of the Earth's surface consists of submerged shelves, that is, between 7 and 8 % of the ocean floor, or nearly the area of Africa.

Table 2.1. Statistics on continental margins. (After H. W. Menard and S. M. Smith, 1966, J Geophys Res 71 p 4305, and other sources)

	Shelf			Continental slope			Cont. rise	Trenches
World ocean without adjacent seas	Area (0–200 m) 10^6 km^2	Average width (km)	Average slope	Area (10^6 km^2)	Average width (km)	Average slope	Area (10^6 km^2)	Area (10^6 km^2)
Atlantic	6.080			6.578			5.381	0.447
(% of its area)	(7.9 %)	115	0°28'	(7.6 %)	260	1°19'	(6.2 %)	(0.5 %)
Indian Ocean	2.622			3.475			4.212	0.256
	(3.6 %)	91	0°23'	(4.7 %)	182	1°35'	(5.7 %)	(0.3 %)
Pacific	2.712			8.587			2.690	4.757
	(1.6 %)	52	0°49'	(5.2 %)	139	3°13'	(1.6 %)	(2.9 %)

The schematic drawing in Fig. 2.2a shows the relationships between shelf, continental slope, and continental rise, and introduces commonly used terms associated with continental margin environments. The physiographic diagram of the margin off the East Coast illustrates the various morphologic provinces (Fig. 2.2b).

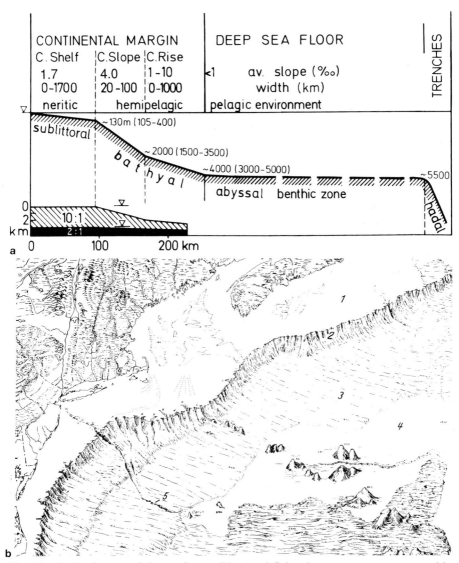

Fig. 2.2a, b. Depth zones of the sea floor. **a** Diagram defining the most common terms used in connection with sea-floor depth and distance from land. Profiles are usually strongly exaggerated; in reality slopes are gentle. The *inset* at the bottom shows this for a line off North West Africa. **b** Physiographic diagram of the continental margin of NE America. **a** *1* shelf; *2* continental slope; *3* continental rise; *4* abyssal plain; *5* submarine canyon. [B. C. Heezen et al., 1959, Geol. Soc. Am. Spec. Pap. 65.]

The terms *pelagic* and *neritic* refer to marine organisms as well as sediments, and mean *open ocean* and *coastal,* respectively. *Littoral,* etc. down to *hadal,* refer to water depths. *Littoral* is the same as *intertidal,* meaning between tides. *Supralittoral* refers to the spray zone, *sublittoral* to offshore from the tidal area.

2.2 Margins Are Sediment Traps

The continental margins are the dumping sites for the debris coming from the continents, the terrigenous sediments. The margins are also the most fertile parts of the ocean, where productivity is high. Thus, much organic matter becomes buried within the continental debris. If conditions are right, over millions of years, this organic material can eventually develop into petroleum. This happened, of course, in the Gulf Coast area, where oil is found buried under immense masses of sediment (see Chap. 10).

Which margins are likely to have thick sediment wedges? It is reasonable to expect that an ocean basin draining a large land area, relative to its size (Table 2.2), is going to have a thick accumulation of sediment. Indeed, the margins of the Atlantic Ocean have very thick wedges of sediment, up to 10 km and more. The Atlantic also has the largest proportions of slope and continental rise areas of the major ocean basins (Fig. 1.3). The reason for this is not only sediment supply, but also the fact that the Atlantic margins are old *"trailing edges"* and have not been disturbed by tectonic processes, other than sinking, for a long time.

Table 2.2. Land and ocean areas and drainage. (After H. W. Menard and S. M. Smith, 1966, J Geophys Res 71 p 4305, and other sources)

	Area (10^6 km^2)	% of Earth surface	Drained land area[b] (10^6 km^2)	Ocean area / land area	Average depth of water (km)
Asia	44.8	8.7			
Europe	10.4	2.1			
Africa	30.6	6.0			
North America	22.0	4.3			
South America	17.9	3.5			
Antarctic	15.6	3.1			
Australia	7.8	1.5			
Pacific	181.3[a] (166.2)	35.4	18	10:1	4.0 (4.2)
Atlantic	106.6[a] (86.8)	20.8	67	1.6:1	3.3 (3.8)
Indian Ocean	74.1[a] (73.4)	14.5	17	4.3:1	3.9 (3.9)

[a] Including adjacent seas (Black sea, Mediterranean and Arctic included with Atlantic). Numbers in parentheses: without adjacent seas.
[b] Excluding areas with interior drainage, and Antarctic.

2.3 Atlantic-Type (= Passive) Margins

Continental margins differ greatly, depending on their origin. As early as 1883, E. Suess (1831–1914) coined the terms *Atlantic margins* and *Pacific margins* to emphasize the major differences. Essentially, Atlantic-type margins are steadily sinking regions, accumulating thick sequences of sediment in layer-cake fashion. In contrast, Pacific-type margins are, on the whole, rising, and are associated with volcanism, folding, faulting, and other mountain-building processes. Atlantic-type margins also are called "passive", and Pacific-type margins "active", because of the difference in tectonic style and the occurrence of earthquakes and active volcanoes.

The origin of the margins must be understood in the context of sea floor spreading. In the Atlantic, the continental margins originated through a tearing apart of an ancient continent, along a line of weakness or great stress, and subsequently evolved through sinking and loading with sediment (Fig. 2.3).

The development can be illustrated taking the Red Sea as a model. Here, mantle material pushes up and tears the Arabian Peninsula from Africa.

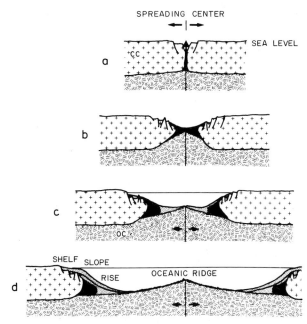

Fig. 2.3 a–d. Evolution of Atlantic-type continental margins. Uplift of Earth mantle material **a** expands the continental crust *(CC)* causing graben structures. Volcanism is common at this stage. The continental crust thins, subsides, and b splits apart. Coarse terrigenous sediments *(dotted)*, volcanogenic deposits *(black)* (and salt in some cases) accumulate. Rifting is followed by drifting, with further subsidence of continental margins. Mantle material forms new oceanic crust *(OC)* as shown in **c**. This stage resembles modern Red Sea conditions. **d** Sea-floor spreading widens newly formed oceanic crust area. Sediments cover older parts of sea floor, and build up margin.

The initial phase of this process can be studied in the East African Rift Valley, where the sea has not entered the rift. The pulling apart (and therefore thinning) of the continental crust opens a window for mantle material (Fig. 2.3 a, b) intruding from the asthenosphere. Heat flow increases further and there is a bulging upward from the rising mantle material, much as at the Mid-Ocean Ridge. However, between the separating edges of continental crust there is an increasing gap now. Pieces of the continental crust break off along "listric faults". This process continues to thin and strech the continental crust. ("Nonvolcanic" margin type as in the Northern Biscay or at Georges Bank, see Fig. 2.4).

Also, the central graben receives large amounts of sediment from the wall-like mountains surrounding it. These mountains – first uplifted by the rising mantle material – continue to rise as erosion unloads them. The central valley, which started as

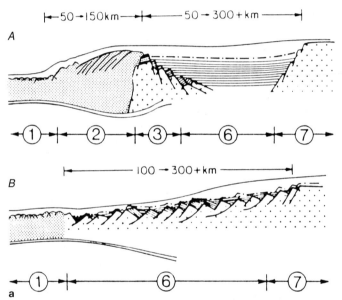

Fig. 2.4 a–c. a Comparison of typical structural elements of "volcanic" (*A*) and "nonvolcanic" (*B*) continental margins. *1* Normal thickness oceanic crust; *2* seaward dipping units (volcanic); *3* structural high in continental crust, often occurring adjacent to *2*; *6* thinned, subsided continental crust; *7* unstretched continental crust. *Parallel signatures* Sediments; *double line* Moho, Mantle underneath. [J. C. Mutter et al. 1987 and I. C. Sibuet and Z. Mascle 1978 in European Science Foundation, Cosod II Report, Strasbourg, 1987: 92.] **b** Rifted continental margins in the North Atlantic. "Volcanic" type (*a*, and black areas) and "nonvolcanic" type (*b*). Iceland as hot spot on the Mid-Atlantic Ridge. [R. S. White et al., 1987, Nature, 330: 439.] **c** Continuous seismic reflection profile across the "volcanic" type of a passive continental margin, with proposed drill sites for deep-sea drilling and the final drill hole 642 (central vertical line). (Voring plateau off Norway, see Fig. **b**). Dipping reflectors between horizons *E* and *K*. *E* Top of Lower Eocene basalt flows; *K* base of seaward-dipping reflector sequences; *M* and *O* Tertiary unconformities (*M* Middle/Upper Miocene, *O* Middle Oligocene). [Data from K. Hinz in O. Eldholm et al., Proc. ODP Initial Repots, 104, 12.]

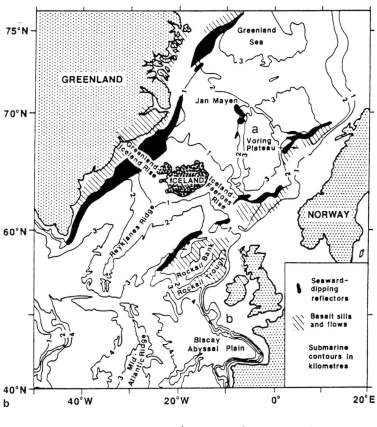

b

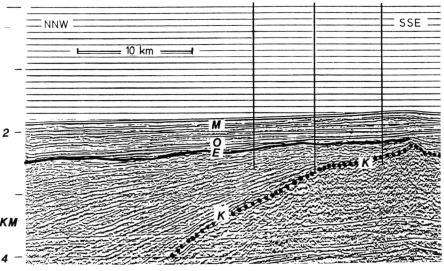

c

a gap, now sinks under the load of sediments and by cooling. As it widens, the sea invades and the magmatic accretion process asserts itself morphologically, as a young mid-ocean ridge with an axial valley.

The processes at depth, in the zone where continental crust meets the melts from the asthenosphere, are hidden from view, and are correspondingly less well understood. Magma intrudes between blocks of continental crust and partly remelts it and mixes with it. Some of the magma reaches the surface, producing large outpourings of basalt, as in the Afar desert of Ethiopia. Examples with volcanic units up to 3 or 5 km thickness from Atlantic rifting are volcanic margins as in East Greenland and off Norway (Voring plateau; see Fig. 2.4). Here deep-sea drilling confirmed the character of "dipping reflectors" as ancient lava flows.

Returning to the Red Sea (Fig. 2.3 c), where the submerged margins are sinking on the cooling lithosphere. Thick reef structures can grow on these sinking blocks, building up a carbonate shelf, and further depressing the crust with their weight. If the Red Sea were only slightly less open, salt deposits would form – indeed there is evidence from thick evaporite deposits that this happened in the past.

In sum, the receding margins sink and a rampart of reef carbonate may build up, and salt deposits also may form in the early phase of rifting, in low latitudes.

Ancient salt deposits and reef ramparts are just what we see along many of the Atlantic margins (Figs. 2.5, 2.6). Salt deposits, of course, also are well known from the Gulf of Mexico, where they push up as salt domes (diapirs), providing a path for petroleum migration (see Chap. 10). In the Atlantic proper, good evidence for salt deposits exists, especially off Angola. The salt in the South Atlantic was laid down, presumably, when this ocean was narrow, was closed to the north, and had restricted exchange to the south due to the Walvis Ridge-Rio Grande Barrier (now near 30° S). Large petroleum reserves may be associated with these salt deposits, because the South Atlantic also was the site for deposition of organic-rich sediments, during a long period in the middle Cretaceous.

The kind of material accumulating on sinking continental margins depends on the geologic setting of the region. In the tropics, and where no large rivers bring sediment or freshwater, reef carbonates can grow. Elsewhere, mixtures of lagoonal and riverine sediments may gradually be buried by offshore deposits – mainly hemipelagic mud, rich in the shells of *planktonic* (floating) and *benthic* (bottom-living) organisms. In places, the sediments can become extraordinarily thick: 10 to 15 km of sediment are reported from off the Niger, the Mississippi and from other large deltas (Fig. 2.6).

The end-result of rifting, then, are continental margins consisting of thick sediment stacks piled both on the sinking blocks of a continental edge and on the adjacent oceanic crust (Figs. 2.3, 2.6).

We can generalize these conclusions to all margins which originated by rifting and are riding passively on the moving plate (hence *passive* margins). Besides the Atlantic margins proper, there are the East African margin, the margins of India, much of the margin of Australia, and practically all of the Antarctic margin. In the Antarctic, of course, special conditions prevail with respect to erosion and deposition, at least since the formation of ice sheets. Thick ice sheets have been present there since the late Tertiary and possibly earlier.

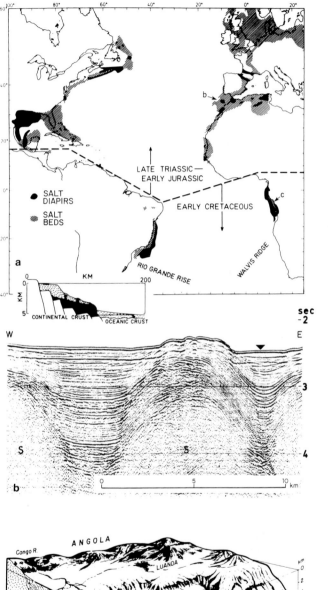

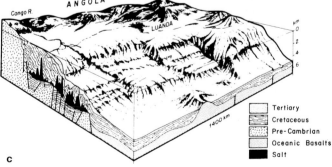

Fig. 2.5 a–c. Evaporite deposition in the early Atlantic. **a** Geographic distribution of Mesozoic evaporites. [K. O. Emery, 1977, AAPG Continuing Education Course Notes Ser 5: B-1.] **b** Salt diapir structures (**S**) as seen on air gun profile of *Meteor* Cruise 39, off Morocco (near 30° N). Water depth at triangle is approximately 1800 m. [E. Seibold et al., 1976.] **c** Relationship of salt diapirs to margin structure off Angola (SW Africa). The Aptian salt is underlain by nonmarine clastic deposits which fill graben-like depression within pre-Cambrian basement [R. H. Beck and P. Lehner, 1974, AAPG Bull. 58, 376.]

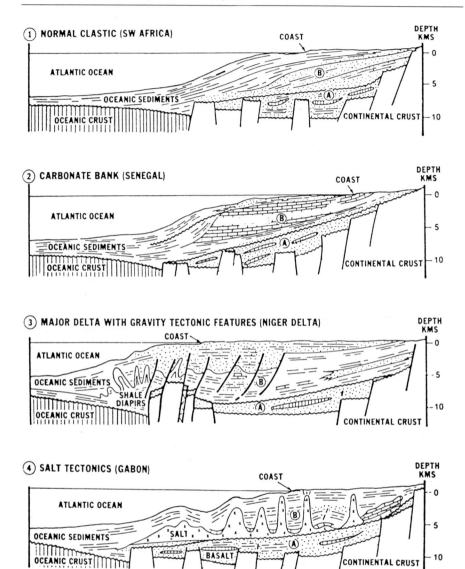

Fig. 2.6. Passive or Atlantic-type Continental Margins. Different types off Africa. *A* Nonmarine; *B* marine sediments. [K. T. Pickering et al., 1989: 252, Deep-marine Environments, Unwin Hyman, London.]

2.4 Unsolved Questions in the Study of Passive Margins

The unraveling of the exact origin and evolution of each stretch of passive margin poses its own problems. The commonly used analogy of the evolution of rifting – from the East African Rift to the Red Sea, to the Gulf of California, and finally the Atlantic – provides guidance as to which processes may be at work. Was there stretching before, and erosion of a crustal bulge at the site of future rifting? How wide was the original rift valley? How does the sinking of the outer parts of the continental edge affect landward crustal blocks? What are the rates of uplift and subsidence, and of erosion and sedimentation in time and space? With regard to the history of subsidence, what is the relative role of "floating in the mantle" (isostatic equilibrium) of the blocks, versus gravitational sliding? What are the forces provoking the very long lasting *uplift* of certain parts of the continental margin (as off South Africa), and the formation of long deep-seated *barrier ridges* along some margins? What is the significance of the lack of sediments of a certain age, in many margins? Was the lack caused by erosion? By nondeposition? By huge landslides?

One question of fundamental interest is whether and where the thick sediment stacks piling up on the passive margins will eventually be found in the geologic record on land. After all, the Atlantic cannot just go on rifting apart – sooner or later it runs out of space. One suggestion by J. T. Wilson is that a proto-Atlantic once was formed by rifting, *and then closed again,* running the previously passive margins into each other. The presumed product of this process: the chain of mountains from Norway through Scotland through Newfoundland and to the Appalachians. Check their positions on the "Bullard Fit" (Fig. 1.19) – Wilson's suggestion makes good sense (Wilson, J.T. (1966) Did the Atlantic close and then reopen?: Nature, 211:676–681).

If Wilson's hypothesis is correct, the passive margins would have turned into active margins when colliding with the trench that must have been there to make the proto-Atlantic disappear. What does such a collision margin look like? Would we be able to go to the mountains to check for the signs of collision?

To answer this question we must study the Pacific-type margins, that is, the collision margins.

2.5 Pacific-Type (= Active) Margins

We have previously alluded to the collision of a continent with a trench, focusing on the evidence for subduction (Sect. 1.5). Actually, there are at least three types of collision margins which we need to consider: those produced by continent-continent collision as in the Himalayas, by continent-ocean collision as at the Peru-Chile-Trench, with a shallowly dipping subduction zone (Fig. 2.7a) and those where the subduction takes place along island arcs, as along the Marianas, for example (Fig. 2.7b). This type has a deeply dipping subduction zone.

Perhaps the most important characteristics of collision margins are the folding and shearing of sediments, and especially the addition of volcanic and plutonic material,

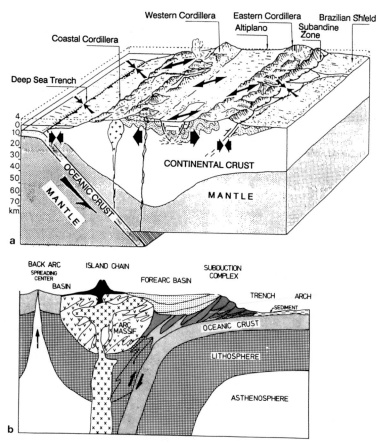

Fig. 2.7 a, b. Sketch of collision margins, in profile (not to scale). **a** Peru-type collision (ocean-continent). Slope sediments are being tectonically deformed. Igneous activity including volcanism derives from melts generated within the subduction zone. Complicated areal distribution of extension and compression. [J. Aubouin 1984, Bull. Géol. Soc. France, 3.] **b** Island-arc situation (ocean-ocean). Volcanic islands build up over subduction zone. Back-arc basin with spreading center. [Sources: J. R. Curray, D. G. Moore, in C. A. Burk and C. L. Drake 1974, ref. p. 250; and D. R. Seely, W. R. Dickinson 1977 Amer. Assoc. Petrol. Geol. Continuing Educ. Notes Ser. 5.]

derived from mobilizing matter from the downgoing lithosphere. The fractionation processes associated with partial melting on the descending slab, and with hydrothermal reactions, can lead to enrichment of melts with heavy metals – and hence to the formation of ore deposits, as in the Andes (Fig. 1.20).

The types of rocks which characterize the continental margins next to subduction zones are extremely varied, which comes as no surprise. The incoming lithosphere brings an assortment of basaltic rocks, serpentinite, gabbro, peridotite, which were derived from the mantle and altered by hydrothermal reactions under various conditions of pressure and temperature. In addition, various kinds of pelagic sediments may be added – deep-sea clay, shell carbonates, biogenous silica. When such rock assemblages are found on land, they are referred to as *ophiolites* and are mapped in the hope of finding ancient subduction zones. It is like hunting for lost oceans on land. Occasionally the reward is discovery of massive copper sulfides and other ores. The mechanism allowing these ophiolites to escape subduction by vertical displacement of several kilometers ("*obduction*") is a matter of speculation.

The steep slopes leading into the trench are favorable for large-scale gravitational transport of rock masses from the land side into the subduction zone. The jumbled masses (*mélange*) thus generated are then sheared, and metamorphosed (i. e., baked and cooked) under pressure (but at relatively low temperatures). Blue schists, and subsequently, amphibolites can form under these conditions.

In classic geologic literature, the sediments of trailing edges are known as *miogeosynclinal*, and those of collision edges as *eugeosynclinal*. The *geosyncline* part of terms, of course, stems from the observation that the Earth's crust must have subsided in order to accumulate the thick masses of sediment found in the mountains.

The nature of the active margins is subject of ongoing research and still holds many surprises. The early simple concept of an origin from scraped off material left by the downgoing slabs had to be modified. There is little transfer of material in places and, in fact, there is "tectonic erosion", whereby portions of the margin are swallowed by the subduction zone. This may be initiated by massive slumping into the trench (e. g., Japan Trench), which delivers materials for building continental roots, and for metasomatic processes within an island arc.

The role of fluids has received increased attention. Both tectonic motions (faulting, overthrusting) and chemical reactions within the accretionary prism are generally influenced by the presence and composition of such fluids expulsed by tectonic compaction and from dehydration reactions of commonly very high pore fluid pressures (Fig. 2.8). Gases are important, too. In the Caribbean Barbados Ridge Complex, for example, the low-angle fault between the accretionary wedge and the underthrusting oceanic crust is greased by methane bearing fluids, which keep the wedge detached from the downgoing slab.

Special complexity is added to the subduction system by the phenomenon of "back-arc spreading" (Fig. 2.7b). This is localized sea floor spreading, which occurs landward of volcanic arcs, as in the Philippines or west of Guam. More than 75 % of these marginal basins are concentrated in the Western Pacific. That extension (necessary to let magma rise) should be associated with collision is surprising. Are the island arcs drifting oceanward, pulled to the east by subduction?

Of course active, Pacific-type margins also are sediment traps. However, here sediments are piled up into chaotic mixtures of various types of rocks. In addition, one must keep in mind that enormous masses of material simply disappear deep into the mantle. The scale of the subduction activity is difficult to imagine – the lithospheric slab now entering the Japan Trench is more than 10 000 km long! At present rates, it will vanish in about 100 million years.

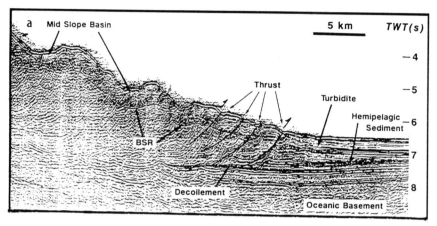

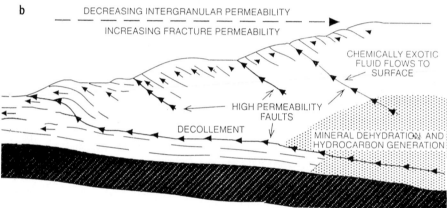

Fig. 2.8 a Reflection seismic profile from the subduction zone at the Nankai trough southeast of southern Japan. *TWS* Two-way travel time in seconds; *BSR* bottom simulating reflector. Note the downgoing oceanic crust of the Philippine plate with the decollement zone (in Miocene sediments). The accretionary prism above it consists of turbidites and hemipelagic sediments and is intensively deformed, thus opening paths for fluids. [A. Taira and Y. Ogawa, 1991, Episodes 14, 3: 209.] **b** Diagram showing paths for fluids in a sandy accretionary prism. [J. C. Moore et al., 1991, GSA Today, 1, 12: 269.] Where these paths reach the surface, seepage-related biological communities may occur, as observed by submersibles in the Nankai trough (see Chap. 6.9).

2.6 Shear Margins and Complex Margins

Passive or trailing continental margins normally are parallel to mid-oceanic ridges, as clearly seen in the North and South Atlantic. What about the nearly east-west-running margins of North Brazil and the African Guinea Coast? They are parallel to the many fracture zones near the Equator there and represent a third type, shear margins, with narrow shelves.

Not all margins can be readily classified. Some of these rim the marginal basins behind island arcs and elsewhere in adjacent seas. Some are originating from a combination of continent shearing and rift extension, with examples on both sides of the South Atlantic. Some have an extremely complex history, such as the California margin, which not so long ago used to be dominated by accretionary processes and is now characterized largely by tectonic shear (San Andreas Fault) and extension (Southern Californian Borderland). However, the task of the marine geologist is not so much to devise classifications which fit all possible instances, as to identify those processes which characterize various types of margins and set them apart from others.

2.7 The Shelf Areas

The submerged part of a continent is the *shelf*. We have seen earlier that emerged low-lying regions also can be argued to be part of the shelf. However, here we are concerned with the actual sea floor, covered by water.

Typically, shelves are flat and are not very deeply submerged. An average depth for the shelves rimming the Atlantic is near 130 m. Some shelves are quite wide, especially those on passive margins. These are, on the whole, depositional features; that is, they are built up by sediments. Narrow and rocky shelves, on the other hand, are common on active margins (see Fig. 2.7a). Here, erosional processes play an important role in shaping the shelf.

Some shelves extend deep into continents, and harbor *shelf seas* such as the Hudson Bay, the Baltic Sea, or the Persian Gulf. Most of the marine sediments found on land were originally deposited in shelf seas. To understand these sediments – which cover the greater portion of the continents – one needs to study sedimentary processes in modern shelf seas (see Chaps. 3 to 5).

Quite generally, present shelf environments and sediment types show considerable variety over short distances. In part, this variability stems from the fact that the sea level stood much lower only 15 000 years ago. Conditions were entirely different then, and many portions of the present shelf still reflect those conditions in topography and sediment cover. The reason for the low sea level, of course, was the presence of continental ice sheets which locked up enough water to make the ocean go down by about 130 m.

We see that the nature of shelves reflects tectonics (active versus passive) and the recent rise of sea level, on a grand scale. On a regional scale, climatic conditions and sediment supply are of prime importance. In low latitudes, buildup by reef forming organisms is (or was) important in many places. In high latitudes, ice has been an important agents for the last several million years.

In the northern North Atlantic, for example, shelves everywhere show the effects of ice. The growth of ice not only led to exposure of the shelves, it also dumped enormous amounts of debris in places, i.e., the *moraines*. This material still sits on the shelves, off Newfoundland, and in the North Sea. The ice moving far out onto the shelves, also actively carved deep ravines and depressions which have not yet been filled. The fjords of Norway, Greenland, and western Canada are witnesses to this powerful action of the ice.

Shelves formed by large deltas off river mouths (Amazon, Mississippi, etc.) can be very flat and monotonous, in striking contrast to both the rugged ice-carved shelf and the irregular coral reef shelf. The rich supply of fine sediment associated with delta environments allows redistribution and smoothing by waves and currents. Of course, waves and currents can also build up dunes, barriers, beach berms, and sand waves, depending on circumstances (see Chap. 7). Examples for these conditions of high terrigenous sediment supply include the North Sea, the shelf off the Siberian rivers, the shelf of the Yellow Sea. A prime example is the Senegal delta, where the shelf has less than 10 cm relief over several miles!

In any study of present shelves, we are chiefly faced with the question of how much of the observed morphology and sediment cover is inherited from the past and how much may be ascribed to the activity of the sea today. The task is complicated by the fact that "today" includes at least the last several hundred years. Within such time spans the sea can produce effects whose causes may not be obvious, especially if they include rare but powerful hurricanes and large earthquake-generated waves (*tsunamis*).

Tsunamis are produced chiefly in the trenches rimming the Pacific; they travel over thousands of miles in a few hours. They are very long waves, and so low in the open sea that they are not noticed on board a ship. When such a wave reaches a shelf area, however, it slows down and builds up, reaching enormous heights (tens of meters) where conditions are right. In this condition it can wreak havoc on the coasts of exposed shelves and also influence shelf morphology through the powerful bottom currents it produces.

2.8 The Shelf Break

The shelf break, where shelf joins upper slope, is a prominent morphological feature of continental margins. In principle, the break marks the depth below which the influence of sea level on erosion and deposition wanes rapidly (Chap. 5). The details are by no means understood, however.

The shelf break is commonly represented by a distinct increase in the slope at about 100 to 150 m. The global mean is near 130 m. In the Antarctic and Greenland it is very deep, down to 400 m. Here the break may mark the depression by the ice load and the maximum depth of carving by ice. However, it is equally deep off southwest Africa, where this explanation does not work. Generally, the break in slope is distinct (Fig. 2.9), but it can also be gentle in places.

The fact that the shelf break is commonly between 100 to 150 m deep strongly suggests that it marks the lowstand of glacial sea level. Apparently this lowstand was reached repeatedly in the maximum glaciations during the late Quaternary, thus exerting considerable control on shelf evolution. In any discussion of the shelf and the shelf break in a particular geographic area, the regional isostatic responses of the shelf to loading and unloading with water due to the changing sea level also must be

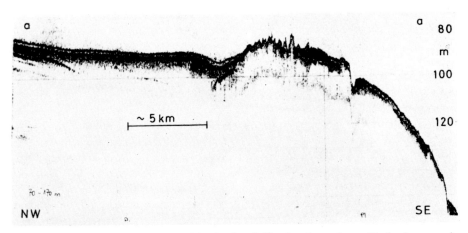

Fig. 2.9. Shelf break at the entrance of the Persian Gulf, subsurface echo profile by the research vessel *Meteor* (1965). Note the accumulation of soft, layered sediment behind the rugged reef structure. Upper slope collects reefal debris. The reef is dead. Shelf break is somewhat above 100 m depth. [E. Seibold, Der Meeresboden (1974), 15, Springer, Berlin, Heidelberg, New York.]

considered. However, such responses are difficult to separate from regional uplift and subsidence.

2.9 Continental Slope and Continental Rise

The classic profile of Atlantic-type continental margins shows a steepening of the slope at the shelf break and a gradually diminishing descent toward the deep sea floor (Figs. 2.2 and 2.6). The relatively steep part below the shelf break is the *continental slope;* the gently descending part leading into the deep sea is the *continental rise.* The boundary between slope and rise is ill-defined. Perhaps we can say that the slope is definitely part of the margin, whereas the rise is built on oceanic crust and is essentially part of the deep-sea environment.

Not all slopes and rises, of course, fit the textbook outline of the ideal Atlantic type – even in the Atlantic. Deep marginal ridges, as off Brazil, sheer walls of outcropping ancient sediments and deep-lying plateaus, such as the Blake Plateau off Florida, can interrupt the ideal sequence.

The collision margins off Peru and Chile (Fig. 2.7a) are characterized by a steep slope, without a rise – the trench swallows the material which would normally build the rise. A descent in a series of steps is typical for the collision slope.

Rather complicated conditions prevail off much of the western US coast. While the slope and rise off northern California can be described simply enough in terms of coalescing deep-sea fan deposits (Fig. 2.10 northern part), no such description is possible for the margin to the south. The Southern California Borderland looks much like an extension of the basin-and-range topography familiar from the Mojave Desert, into the sea. The tops of the ranges protrude as islands (Fig. 2.10 south-eastern part).

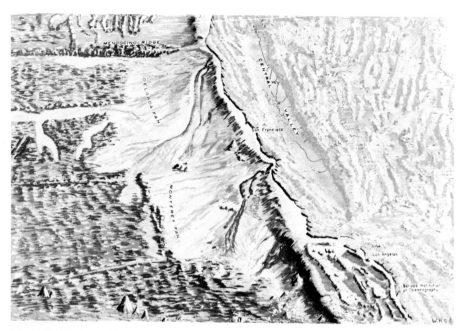

Fig. 2.10. Physiographic diagram of the ocean margin off California. Note the narrow shelves bounded by sea cliffs toward the land (uplift!). The Continental Borderland in the south is a submerged basin-and-range province. The continental slope essentially consists of enormous coalescing fans which transgress over the abyssal hills province. [Sketch based on physiographic diagram of H. W. Menard, 1964.]

The various physiographic diagrams commonly depicting the rapid descent from shelf edge to deep sea are somewhat misleading in that the slopes are, in reality, quite gentle: between 1° and 6° (see Table 2.1). A 1° slope, of course, would appear as a plain if one were standing on it.

From the variety of types of slopes, it is clear that a number of different forces are at work in shaping them. We have referred to some of the endogenic forces when comparing Atlantic-type and Pacific-type margins. Changes in the thickness of continental crust, through remelting and assimilation into the mantle (subcrustal erosion), have been proposed as one mechanism influencing the evolution of margins. Such processes may be active, for example, in the Southern California Borderland.

Basically, most slopes are the surfaces of thick accumulations of sediment washed off the continents and mixed with marine materials of biologic derivation. The deep sea receives a rather miserly share of the continent's products: the bulk accumulates on the margins. The high rate of accumulation on many slopes leads to a precarious balance in places, especially off deltas. When there is not enough time to de-water and solidify the sediment, immense landslides can result from rather small disturbances, even in very gentle slopes. The slides tend to move on surfaces defined by clayey layers with high water content. Water pressure may be unusually high in such horizons, thus "floating" the overlying sediment stack. Slides can be coherent over large areas, preserving much of the original stratigraphy, or they can be chaotic (called slumps), producing a jumbled mess. Earthquakes can trigger slides, especially in the active margins, but also elsewhere.

Off Cape Hatteras, a tongue-shaped mass of displaced sediment on the Upper Continental Rise is 60 km wide and over 190 km long, and has a hummocky relief of up to 300 m in places. During and following slope failure, deformation can span a number of processes, from rigid-block motion to turbulent flow. Probably most of the slope failures occurred during Pleistocene low sea level stands (J. S. Schlee and J. M. Robb, Geol. Soc. Amer. Bull., 1098, 1991).

An example of a large-scale slide is shown in Fig. 2.11; the jumbled masses have come to rest at the foot of the slope and now form part of the continental rise. The seismic profile was obtained by echo-sounding *into the sea floor* with powerful "booms" of sound, rather than the "pings" used to define the surface of the floor. (The method is called *continuous seismic profiling*).

The arrangement of the sedimentary layers below the sea floor in Fig. 2.11 shows that erosion has taken place in earlier times, alternating with periods of deposition. Erosion can be produced both by slides and by the action of strong deep currents flowing horizontally along the slope. Such currents have been dubbed *contour currents,* to distinguish them from the *turbidity currents* which are mud-laden bodies of

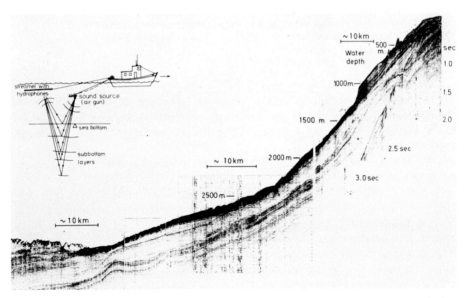

Fig. 2.11. Submarine mass movements off Dakar (NW Africa). Air gun record of *Meteor* Cruise 25/1971. Shelf edge *upper right*. Slide starts at 1050 m depth (return time for outgoing sound pulse: 1.4s). Thickness of slide ~ 200 m. Material came to rest below about 2300 m depth (= 3.6 s) at the foot of the continental slope. *Insert left* Air gun system with air guns as sound source. Acoustic signals are reflected by the sea floor and by subbottom layers and are recorded by hydrophones in the streamer. [E. Seibold, Der Meeresboden (1974), 17, Springer, Berlin, Heidelberg, New York.]

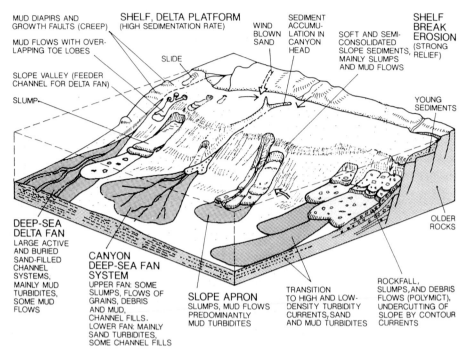

Fig. 2.12. Compilation of exogenic processes shaping (passive) continental margins. [G. Einsele et al. (eds.). Cycles and events in stratigraphy, Springer, Heidelberg, 1991: 318.]

water flowing downhill (Sect. 2.11). It is now thought that much of the sediment on the continental rises is carried there originally by turbidity currents and by slides starting somewhere near the shelf break, and that subsequently it is redistributed by contour currents.

A compilation of different exogenic processes shaping the continental margin is given in Fig. 2.12.

2.10 Submarine Canyons

The continental slope is commonly cut by various types of incisions, ravines, and valleys, the most spectacular of which are the *submarine canyons*. The origin of these impressive features (Fig. 2.13) has long puzzled marine geologists and is still a matter of debate.

Many large submarine canyons are very much like their counterparts on land: tributary systems in the upper parts, with meandering *thalwegs* (similar to river beds), and steep sides in places (20° to 25°, even 45°). Overhanging walls also occur (Fig. 2.14). Walls may consist of hard rocks, even granite in some places. As in river

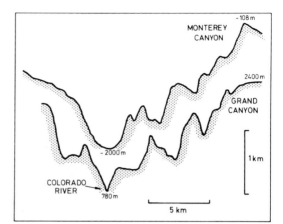

Fig. 2.13. Profile of Monterey Canyon compared with that of the Grand Canyon of Arizona. [F. P. Shepard and R. F. Dill, 1966.]. The resemblance is coincidental, but illustrates the enormous size of the Monterey Canyon (see Fig. 2.10.)

Fig. 2.14. Illustration of morphological similarities of subaerial canyon (Grand Canyon *left) and submarine canyon (La Jolla Canyon right).* Note steepness and overhang in both cases [Photo *left,* E. S.; underwater photo courtesy R. F. Dill.]

canyons, there is a continuous descent of the deepest point of the valley, with maximum slopes of 15 % near the coast and gentle slopes of about 1 % farther out to sea.

Quite commonly, canyons cut right through the shelf and may line up with valleys on land. Ancient fishing villages are sometimes located at the heads of submarine canyons: the greater water depth disperses high waves away from the coast and provides some safety for boats and houses on shore. Examples are Nazare in Portugal, Monterey Canyon of Central California, and Kayar in Senegal. Some canyons, however, start only at the edge of the shelf. The seaward extension of submarine canyons can reach to the end of the continental rise and beyond, and run for hundreds, even thousands of kilometers across the deep-sea floor (Fig. 2.2b, Hudson Canyon; Fig. 1.3. Mid-Ocean Canyon, off Eastern Canada).

Canyons and various other types of slope valleys can be extremely abundant in places, such that the slope is dissected like the remnants of a mesa on land. Nor do those canyons necessarily run straight downhill; they may be cut at an angle to the slope. In other places, submarine canyons seem to be absent, presumably by reason of low sediment supply (insufficient for producing downhill currents) or because of gentle slopes, or both. The best-known canyons are associated with the mouths of large rivers – the Congo, Indus, Ganges, and the Hudson. The fan valleys which continue these canyons down-slope are bordered by levees (Fig. 2.17).

A large number of hypotheses have been put forward over the years to account for the origin of submarine canyons. In fact, different types of canyons must have different origins. For example, it has been demonstrated through deep-sea drilling that the Mediterranean became isolated from the world ocean some 5 to 6 million years ago and dried up during periods of evaporation. At those times, deep canyons could have been carved by the familiar action of rain and run-off, into the margins of the Mediterranean. Indeed, the true floor of the Nile Valley is very deep, supporting the idea of canyon-cutting during desiccation. At present the river runs on top of a thick pile of sediment, which has since filled the canyon. However, we can hardly invoke such drastic falls of sea level everywhere in the world ocean. Thus, there must be a way to make submarine canyons by cutting them under water. The fact that so many canyons are off river valleys suggests a mechanism: some sort of submarine river flowing on the sea floor. This submarine flow cannot be an extension of the river entering the sea: its water is fresh and therefore less dense than seawater. River water floats on seawater. But water with a high mud content is heavy enough to flow downhill on the ocean floor.

One way to stir mud into the water is to have slides start in the soft sediments deposited off river mouths. Large amounts of mud are brought down the river during floods, and such mud is rather unstable. Hurricanes and wave action may produce large amounts of heavy, muddy water through stirring up of the sediment. Earthquakes may be agents for starting mudslides which turn into muddy downhill flow, or *turbidity currents*.

During glacial periods, when sea level was greatly lowered by the buildup of ice caps, the exposed shelves could not act as mud traps for the rich sediment load coming from land. Also, wave attack must have been strong, and its effects may have reached much greater depths than today. Storms and storm waves were probably more frequent then, so that sediment on the outer shelf and on the upper slope could be resuspended periodically, providing for mud-laden, heavy water bodies, which could then move downslope and initiate powerful turbidity currents.

The realization that such currents play an extremely important role in present-day marine processes as well as in the geologic record, came only in the 1950's, largely through the work of Ph. H. Kuenen (Fig. 2.15a).

Kuenen established, by experiment, that such heavy downhill flowing currents can exist in nature and that they would deposit layers which show the *grading* familiar from previously unexplained sediment-series *(flysch)* in the Alps and other mountain ranges (Fig. 2.15b).

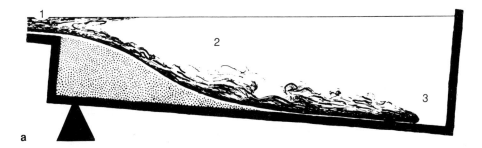

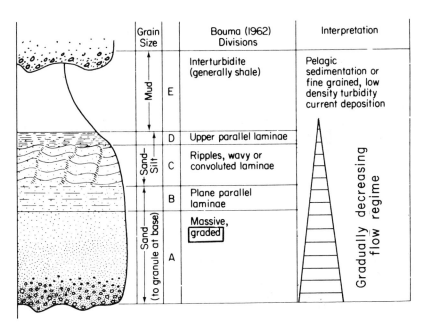

Fig. 2.15 a, b. Origin of graded layers. *a* Experiment of Ph. H. Kuenen. *1* Turbid, sediment-laden water is introduced into the tank; *2* water in tank remains still and clear over the bottom, where the denser muddy water rushes downslope; *3* turbulent front of the turbidity current. Depending on its strength, a turbidity current can erode and redeposit enormous amounts of sediment. [J. Gilluly et al., 1968, Principles of geology. W. H. Freeman, San Francisco, after photos by H. S. Bell, Cal. Tech.] **b** Standard sequence of divisions in a turbidite layer, as proposed by A. H. Bouma. The *lower part* is the graded bed, produced by a turbidity current. The *upper part* results from "normal" sedimentation; it contains almost all the geologic time represented. Sudden loading of these pelagic clays may produce "load casts". High velocities of the currents are indicated by drag and flute marks at the base of the turbidite. [G. V. Middleton, M. A. Hampton. 1976, in D. J. Stanley, D. J. P. Swift, Marine sediment transport and environmental management, John Wiley, New York.]

Before the *turbidite hypothesis* became a standard tool in the interpretation of sediments, the "desolate thousand-fold alternation of sandstone, limestone, and shale" which faced the alpine geologist in the flysch was a complete mystery. Now these alternations are generally regarded as a record of pelagic sedimentation with periodic

influx of mud-laden bottom-hugging water bodies. These muddy waters, upon de-
celerating, drop their load within a short time as *graded layers*. Such layers are found
within the fan deposits of continental slopes and in the sediments building the abyssal
plains (Fig. 2.17).

Kuenen's work inspired marine geologists to look for direct evidence of the action
of turbidity currents in the present ocean. A well-known example of this search is the
investigation of the succession of telegraph cable breaks down the continental slope,
which occurred after the 1929 Grand Banks Earthquake off Newfoundland. From the
timing of the breaks (as recorded by the telegraph companies involved), B. C. Heezen
and M. Ewing concluded (1952) that the quake had set off turbidity currents which
moved downhill at high speed and snapped the cables (Fig. 2.16). The speeds they
estimated were on the order of 25 to 50 miles per hour (10 to 20 m/s) – velocities of
powerful super-currents. (For comparison, the fastest normal ocean currents run at

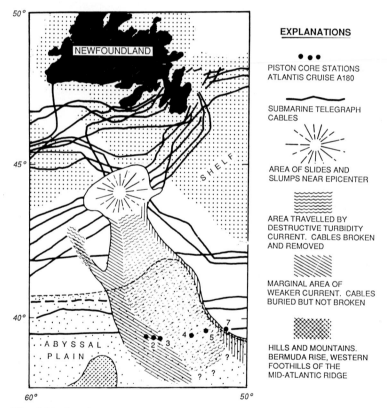

EXPLANATIONS

• • •
PISTON CORE STATIONS
ATLANTIS CRUISE A180

SUBMARINE TELEGRAPH
CABLES

AREA OF SLIDES AND
SLUMPS NEAR EPICENTER

AREA TRAVELLED BY
DESTRUCTIVE TURBIDITY
CURRENT. CABLES BROKEN
AND REMOVED

MARGINAL AREA OF
WEAKER CURRENT. CABLES
BURIED BUT NOT BROKEN

HILLS AND MOUNTAINS.
BERMUDA RISE, WESTERN
FOOTHILLS OF THE
MID-ATLANTIC RIDGE

Fig. 2.16. Grand Banks 1929 earthquake. The cable break sequence (later combined with the
stratigraphic record in the cores) was interpreted by B. C. Heezen and M. Ewing (1952. Am J Sci
250: 849) as evidence for high velocity turbidity currents. [B. C. Heezen, in M. N. Hill, 1963, The
Sea, 3: 744.]

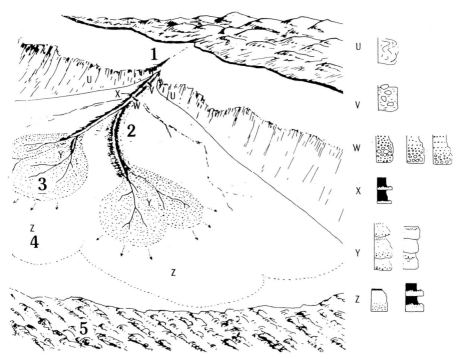

Fig. 2.17. Build op of deep-sea fans. *1* Canyon-cutting through shelf and upper slope traps and funnels sediment to the fan; *2* upper fan valley, walls with slump features (*U*), bottom with debris flows (*V*) and mostly graded coarse grained beds, forming conglomerates in fossil examples. Levees with thin-bedded turbidities (*X*). Levees can be breached (note dead channels); *3* active suprafan with distributary channel, filled with pebbly or massive sands (*Y*); *4* outer fan with classical turbidites (*Z*); *5* abyssal hill region beyond fan. Valleys between hills may have distal fan material. [Based on a sketch by W. R. Normark, 1970, Am Assoc Pet Geol Bull 54: 2170 and on R. Walker, 1978, Am. Assoc. Petrol Geol. Bull. 62: 932.]

about 2 m/s, or 5 miles per hour.) While their calculations were not generally accepted, their studies did raise the possibility that such currents exist.

There are but fewe direct observations of turbidity currents in the marine realm. These currents are rare events, and when they do occur, instruments for measuring velocities or suspended matter are in much danger of being lost. We do not know, therefore, what a "typical" turbidity current might look like. In Bute Inlet, in British Columbia, where direct observations were made in the 1980s, floods from two rivers entering the inlet produced muddy flows with velocities of about 3 m/s. The thickness of the currents was more than 30 m, and coarse sand occurs at least up to 7 m above the sea floor. Fine sand was transported far offshore; it was seen out to 50 km, and at a depth of 620 m. The transport channel has bottom slopes of generally less than 1° (D. P. Prior et al., Science 237, 1330, 1987).

Returning to the submarine canyons: Normally, nothing much is happening in them except for the gentle back-and-forth motion of the tides and internal waves.

But from time to time, perhaps once in a century or a millenium, a great turbidity current rushes down-canyon, moving enormous masses of sediment. Additionally, undercutting of canyon walls causes mass wasting similar to land slides. Such currents, with high velocities and inertia, and carrying sand or even pebbles and boulders, are certainly able to deepen and widen the canyons by erosion, even in hard rock. As the power of the current wanes downslope, particles settle out, generally first the coarse and later the finer ones. These deposits form huge semi-conal sediment bodies, the deep-sea fans.

2.11 Deep-Sea Fans

Deep-sea fans consist of overlapping tongues of sediment, variously dissected by channels, which are in turn re-filled when abandoned (Fig. 2.17). Details of the fan landscape, with its distributary system (including channels, levees formed by spill-over, slumps, and slides), have become available through side-scan surveys. Meanders of several km width have been mapped off the Amazon River, and elsewhere. Fans are of interest from an economic point of view, as potential reservoirs for hydrocarbons, with their huge dimensions, and their many meters-thick sand bodies offering high porosity and permeability.

Much or most of the sediment in deep-sea fans consists of *turbidites,* which is what the deposits of turbidity currents are called. Turbidites are also common both within slope sediments and in the *abyssal plains* (Fig. 2.18). Depending on their magnitude, turbidity currents rushing down a fan valley leave the distributary channels at various points, and build up turbidites as their velocity slows (Fig. 2.17).

Most turbidites are thin and are soon destroyed through reworking by bottom-living organisms and by bottom currents. Thick layers, of course, can survive this process and are then recognizable in the sediment sequence.

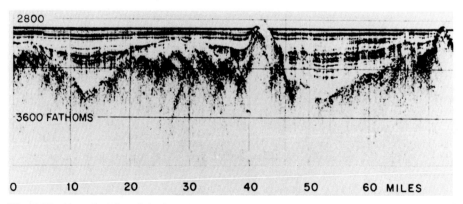

Fig. 2.18. Abyssal plains. Seismic echo profile across a stretch of abyssal plain. (Courtesy C. D. Hollister). Note that the sediment surface is perfectly horizontal regardless of the underlying basement topography. (2800 fathoms = 5100 m; 3600 fathoms = 6600 m)

The very distribution of abyssal plains supports the idea that turbidity currents run along on the floor over extremely long distances. This can be observed in turbidites off river mouths such as off the St. Laurence, Hudson, Mississippi, and Amazon Rivers and more than 3000 km to the south of the Ganges-Brahmaputra Delta.

For such currents, trenches would form an absolute obstacle, and we should not find abyssal plains beyond them. Once a trench is filled, however, plains can develop on the other side. This is exactly the distribution seen in the Gulf of Alaska. As the supply of turbidity deposits runs out, far in the deep sea, the abyssal ocean floor starts showing its typical bumpy morphology: the familiar abyssal hills, which are covered by pelagic oozes and clay.

It reflects the youth of the science of marine geology that major processes of *sediment erosion, transport, and deposition*, which are responsible for much of ocean margin morphology, were essentially unknown until the 1950s. It took yet another decade for the *tectonics* of ocean margins to come into focus, through the theory of seafloor spreading. The history of ocean margin development is still, in the whole, poorly understood because of the lack of samples from within the sediment wedges: recovery by deep drilling is now possible, but is extremely expensive.

While turbidity currents and turbidites are of great importance in helping shape the morphology of continental margins, they are but one aspect of the large-scale transfer of material from the continent to the ocean. To appreciate the scope of this subject, we next discuss marine sediments in general.

Further Reading

Shepard FP, Dill RF (1966) Submarine canyons and other sea valleys. Rand McNally, Chicago

Burk CA, Drake CL (eds) (1974) The geology of continental margins. Springer, Berlin Heidelberg New York

Dickinson WR, Yarborough H (1981) Plate tectonics and hydrocarbon accumulation. Am Assoc Petrol Geol, Continuing Education Ser 1, revised edn. Tulsa, Okla

Kuenen Ph H, Migliorini CI (1950) Turbidity currents as a cause of graded bedding. J Geol 58: 91–127

Watkins JS, Drake CL (eds) (1983) Studies in Continental Margin Geology. AAPG Mem, 34. Am Assoc Petrol Geol, Tulsa, Okla

Bally AW (1981) Geology of passive and continental margins: history, structure and sedimentological record. Am Assoc Petrol Geol, Education Course Note Ser 19

3 Sources and Composition of Marine Sediments

3.1 The Sediment Cycle

Marine sediments show great variety: there is the debris from the wearing down of continents, the shells and organic matter derived from organisms, salts precipitated from seawater, and volcanic products such as ash and pumice (Fig. 3.1).

Most sediments ultimately derive from the *weathering* of rocks on land. The action of ice and water, of hot and cold, breaks up the rocks into small particles and leaches them, destroying the more soluble minerals. A special type of "weathering" occurs at the ridge crests and other young volcanic features in the sea: the reactions of heated seawater with basalt. Such reactions may contribute considerable amounts of matter to seawater. Just how much is not known, however.

The sea itself takes a toll from the continents: waves and tides can eat into land, taking debris offshore. The material may stay on the shelf, or it may bypass the shelf to accumulate on the slope or below.

Sediment which stays on the shelf, of course, essentially stays on the continent. The other sediment is eventually carried back to the continents or disappears into the mantle by mountain building and subduction processes, as discussed earlier (Sect. 2.5).

3.2 Sources of Sediment

3.2.1 River Input. The dissolved and particulate load of rivers constitutes the main source of sediment: the coalescing fans which build the continental slope off California (see Fig. 2.10), for example, are largely riverine mud, with an admixture of shell and organic matter. These precipitates (calcium carbonate, $CaCO_3$; and opal, $SiO_2 \times nH_2O$) also derive to a great extent from river input, namely from the dissolved load of rivers entering the ocean. We can make a rough estimate of how much material enters the sea. Deep-sea sedimentation rates are between 1 and 20 mm per 1000 years. Slope sediments accumulate at a rate of up to 100 mm per 1000 years, approximately (Fig. 3.13).

Let us take the value of 100 mm/1000 years for 10 % of the ocean, and a value of 5 mm/1000 years for the rest: the average is near 15 mm/1000 years ($0.1 \times 100 + 0.9 \times 5$). The ocean floor covers more than twice the area of the land. Hence, the continents – if they are the ultimate source of all sediment – must wear down at a rate of near 30 mm/1000 years. Not a bad guess, actually. Recent estimates of denudation

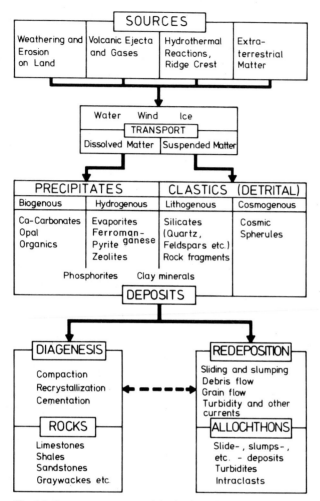

Fig. 3.1. Sources, transport, and destination of marine sediments

of the US portion of North America suggest 65 mm/1000 years. Of course, man's agricultural activities have greatly increased erosion rates, although we may not yet be able to define this effect unequivocally.

Another rough calculation may be based on an estimated sediment supply from rivers of 12 km^3/year. If we distribute this amount on the 362 million km^2 of sea floor, we have a sedimentation rate of about 30 mm/1000 years, and a corresponding erosion rate of 60 mm/1000 years for the continents. These calculations, of course, neglect the dissolved influx of silicate and calcium from reactions of seawater with the oceanic crust, a process that has turned out to be of considerable importance (see Sect. 10.44).

In general, mechanical weathering dominates in high latitudes, where water acts mostly as ice, and in deserts, where it is subdued as an agent. Chemical weathering (leaching) is favored by high rainfall and high temperatures and dominates in tropical areas. In extrapolating these and other present-day patterns into the past, we must always remember that we live in a highly unusual period. The growth of mountain ranges and the powerful abrasive action of the continental ice masses have greatly increased mechanical erosion for the last several million years.

3.2.2 Glacier Input. The great importance of ice in delivering sediment to the sea in high latitudes is readily appreciated when contemplating the immense masses of outwash material they bring to the shores, to be reworked on the shelves. Less important as concerns sediment mass, but very interesting for paleoclimate reconstruction, is the fact that calving glaciers can transport both fine and very coarse material far out to sea. When the icebergs melt, they drop their load. Around Antarctica this type of transport reaches to about 40° S. The *"drop stones"* record not just the movement of ice, but also the geology of the hinterland. Each glacier carries samples from the mountains where it originates; this is very helpful in obtaining materials from ice-covered Antarctica. In the North Atlantic, the drop stone limit roughly follows the present boundary between very cold and temperate waters (Fig. 3.2). During the last ice age, this limit extended much further south, to a line between New York and Portugal (Sect. 7.2.3). At present, about 20 % of the sea floor receives at least some ice-transported sediments.

3.2.3 Input from Wind. In contrast to ice transport, wind can move only the fine material. Medieval Arabian scientists had noted the dust coming out of the Sahara into the "dark sea" of the Atlantic Ocean (Fig. 3.3). In the 1800s, Charles Darwin assumed (correctly) that the dust must build up the ocean floor. As the dust blows out to sea, the larger particles settle out first and the grain size becomes continuously finer. Particles from a large Saharan dust storm, in 1901, had an average size of about 0.012 mm in Palermo, and 0.006 mm in Hamburg. (This would be classified as extremely fine silt, see Appendix A5). During this same storm, up to 11 g of dust per m^3 were measured over the Mediterranean. Occasionally, rather large particles can be transported over great distances. Recently, for example, particles larger than 0.075 mm were found in the air over the North Pacific, 10 000 km from their sources in China, where their journey started in a huge dust storm (P. R. Bener, 1988, Nature, 336: 568).

The rate of dust-fall from the air can best be measured in snowfields and ice cores with annual layering. Even in the Antarctic and in Greenland – far from desert sources – the rate is quite appreciable: 0.1 to 1 mm dust in 1000 years. Exactly how much dust is falling on the sea floor is not known. Some estimates suggest that much or most of the deep sea clay is derived from wind input. Such clay accumulates at between 1 mm/1000 years in the North Pacific, and 2.5 mm/1000 years in the Atlantic.

The rates undoubtedly vary considerably through time, being high during glacials, during periods of loess input (see Sect. 9.3.4).

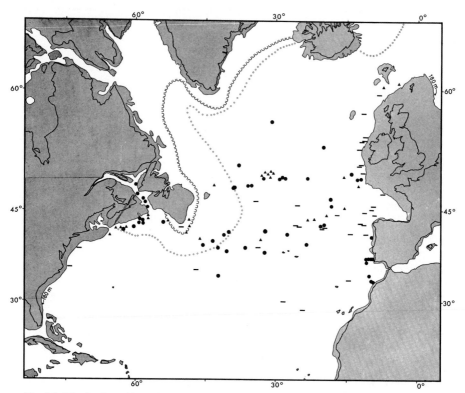

Fig. 3.2. Distribution of ice-rafted materials in the North Atlantic. Present limits of drift ice (normal and extreme) run from around Newfoundland toward Greenland and Iceland. During the last ice age, this limit went from New York straight across to Portugal. *Triangles* Surface samples; *dashes* dredge samples; *circles* core samples. [H. R. Kudrass, 1973, Meteor Forschungserg Reihe C 13: 1]

3.2.4 Volcanic Input. A substantial amount of material is delivered by volcanoes, especially those associated with active oceanic margins. While much of the "*volcanic ash*" carried by winds is finely dispersed, several-cm-thick layers are not uncommon in deep-sea sediments. Some of these correlate over great distances, and mark periods of major volcanic eruptions. A well-known example is the Toba super-eruption in Sumatra (Indonesia) 73 500 years ago, which produced a layer 500 to 3500 km in extent. The consequence of volcanic aerosols in the stratosphere is a strong cooling, which in this case could have lasted for several years. If so, this cooling could have contributed to a rapid reglaciation at the end of the last interglacial. Near island arcs, volcanic ash layers ("*tephras*") build aprons that are several thousand meters thick.

In times past, volcanism was a dominant source of sediment over large regions of the ocean. The composition of deep-sea clay (Chap. 8) suggests that in the time before 10 million years ago, before vigorous mountain-building and glaciation drastically changed conditions, the main source of deep-sea clay in the Pacific was the decomposition of volcanic ash.

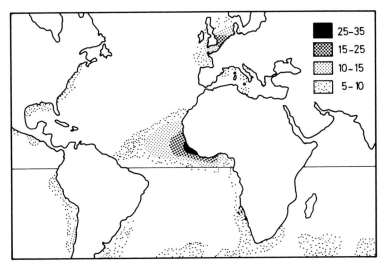

Fig. 3.3. Abundance of haze from dust, over the Atlantic Ocean. Numbers are % of observations. [G. O. S. Arrhenius, 1963, in M. N. Hill, Sea 3: 695]

Not all volcanogenic sediment is brought by wind: pumice, being highly porous, can float and move with ocean currents for long distances, even transporting attached organisms.

Of course, volcanic material is also eroded from land and brought into the sea as terrigenous sediment. Differences in color, mineral composition, glass properties, and subtle differences in chemistry, "chemical fingerprints", hold the clue as to which is the source for a given volcanogenic deposit.

Geologically speaking, volcanoes have a short life, and single eruptions are flash-like events. Therefore ash layers can be used for regional stratigraphic purposes *(tephrochronology),* for example, in the Mediterranean (Fig. 3.4) or around Iceland. Volcanic activity also adds gases and hydrothermal solutions to the ocean. These fluxes have an important bearing on the evolution of the chemistry of ocean and atmosphere and are the subject of intense study at present (Sect. 10.4.4).

3.3 Sediments and Seawater Composition

3.3.1 Acid-Base Titration. To a first approximation, seawater is a solution of sodium chloride (NaCl), that is, table salt. Sodium and chloride make up 86 % of the ions present, by weight (Table 3.1). The other major ions are magnesium, calcium, and potassium (alkaline and earth alkaline metals), and the acid radicals sulfate and bicarbonate. Since the major cations form strong bases, but bicarbonate forms a weak acid, the ocean is slightly alkaline, with a pH of near 8, rather than the point of neutrality (pH = 7). (pH is a measure of the acidity of a solution.) On the whole, the salty ocean may be understood as the product of emission of acid gases from volcanoes (hydrochloric, sulfuric, and carbonic acid) and the leaching of common silicate

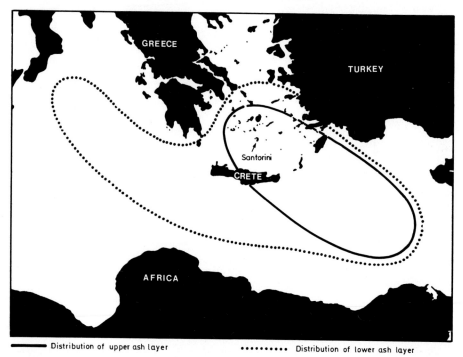

Distribution of upper ash layer   •••••••• Distribution of lower ash layer

Fig. 3.4. Distribution of volcanic ash produced by two large volcanic explosions in the Aegean Sea, presumably of Santorini. The lower ash layer marks a prehistoric event (> 25 000 yrs.). The upper layer is less than 5000 yrs old; the volcanic explosion creating it may be the one which brought catastrophe to the Minoan culture, some 3600 years ago. [D. Ninkovich, B. C. Heezen, Nature London 213: 1967, 582, in K. K. Turekian, 1968, Oceans, Prentice-Hall, New Jersey]

rocks whose minerals have the form [Me Si_a Al_b O_c], where Me stands for the metals Na, K, Mg, and Ca, and the remainder makes insoluble silica-aluminium oxides, that is, clay minerals.

How stable was the composition of seawater through geologic time? If we use the above concept of acid-base titration, and assume that seawater is a solution in equilibrium with the sediments on the sea floor, the result is that the composition should be rather stable. Also, we can obtain samples of seawater salt in the ancient salt deposits. Their composition indicates that sea salt did not vary much over the last 600 million years, probably by less than a factor of 2 for any one major component. Paleontologic evidence certainly agrees with this assessment. Already in the early Paleozoic there are organisms whose closest modern relatives have rather narrow salt tolerances: radiolarians, corals, brachiopods, cephalopods, echinoderms. Of course, adaptation of the organisms to an increasing or changing salt content cannot be ruled out.

On comparing the average composition of river water with that of seawater, one notes drastic differences. Essentially, river water is a very dilute solution of calcium

Table 3.1. Comparison between seawater and river water

Ion	Seawater ppm[b]	Weight %	Rank	Average river water ppm	Weight %	Rank	Residence times[c] in seawater in millions of years
Cl^-	18 980	55.0	(1)	7.8	6.4	(5)	>200
Na^+	10 561	30.6	(2)	6.3	5.2	(6)	210
SO_4^{2+}	2 649	7.7	(3)	11.2	9.3	(4)	–
Mg^{2+}	1 272	3.7	(4)	4.1	3.4	(7)	22
Ca^{2+}	400	1.1	(5)	15.0	12.4	(2)	1
K^+	380	0.4	(6)	2.3	1.9	(8)	10
HCO_3^-, CO_3^{2-}	140	0.2	(7)	58.8[a]	48.6	(1)	–
Br^-	65	0.1	(8)	0.02	–		–
H_3BO_3	23	–		0.1–0.01	–	–	
Sr^{2+}	23	–		0.09	–		–
F^-	1.4	–		0.09	–		–
H_4SiO_4	1	–		13.1	10.8	(3)	(Si =) 0.04
Fe^{2+}, Fe^{3+}	0.01	–		0.67	0.5		–
$Al(OH)_4^-$	0.01	–		0.24	0.2		–
Sum	34 479	= 100 %		120.8	= 100 %		

[a] From D. A. Livingstone (1963) U.S Geol Surv Profess Paper 440 G ("Hard" water contains roughly twice the average total, "soft" water one half).
[b] ppm = parts per million (g per ton).
[c] The residence time is the ratio between the mass within the reservoir and the mass introduced per year. (Data mainly from E. D. Goldberg 1965, in Chemical Oceanography vol. 1, 163–196, Academic Press, New York)

bicarbonate and silicic acid, with a small admixture of the familiar salts – a substantial proportion of which are recycled marine salts. We see from this discrepancy in the composition of seawater and river water that the river influx per se is irrelevant to the makeup of sea salt. What counts in the *solubility* of the salts. In the simplest terms, the soluble salts are abundant in seawater and the others are not.

3.3.2 Interstitial Water and Diagenesis. Fine-grained sediments (clays and silts) have porosities of 70 to 90–% by volume when first deposited on the sea floor, while sands have around 50 %. This pore space is initially filled with trapped seawater. As the sediments are buried below new material, the load increases, pore space is reduced by compression, and part of the pore water leaves, mostly by escaping upward. The water thus lost does not necessarily have the same composition as the water originally trapped. Chemical reactions with the surrounding sediments cause changes both in the interstitial waters and in the composition of the solids. Compaction and chemical reactions involving pore fluids (or solids only, as in recrystallization) constitute "*diagenesis*", which is the process transforming loose sediments into rocks.

Typically, diagenesis is most active in the uppermost meter of freshly deposited sediments. *"Redox"* reactions usually dominate, and these reactions are the more intense the more organic carbon is present. The oxidation of organic carbon leads to removal of dissolved oxygen from pore waters. Additional oxygen demand is satisfied by stripping oxygen from dissolved nitrate and from solid iron oxides and hydroxides (e. g., coatings on grains). If the demand is strong enough, sulfate is stripped of its oxygen also. These reactions are all mediated by bacteria. In the process, gases are produced (carbon dioxide, ammonia, hydrogen sulfide), as well as iron sulfide. Thus, the escaping pore water is enriched in these gases and impoverished in nitrate and sulfate, and the remaining sediments are enriched in sulfide (e. g.,pyrite). Fermentation of remaining organic carbon compounds can subsequently lead to formation of *methane,* which can react with water (given low temperature and high pressure) to make *clathrates.* Such clathrates are thought to be widespread in sediments below coastal upwelling regions, where they produce prominent seismic reflectors *(bottom simulating reflectors).* The massive escape of gases from organic-rich muds can produce *"mud volcanoes"* (see Fig. 3.5) or smaller *"pockmarks"*.

Besides redox reactions, the dissolution of carbonate and opal is of prime importance in diagenesis. During *"early diagenesis"*, much of the dissolved matter can leave with the escaping pore water, or depart simply by diffusion out of the sediment into the overlying waters. Once the sediment is deeply buried, concentrations within the interstitial waters can increase to the point where reprecipitation occurs. This leads to cementation of remaining grains by carbonate and silica cements. These

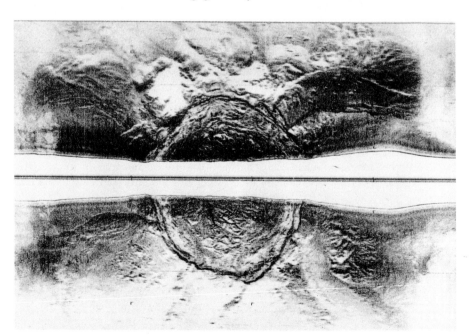

Fig. 3.5. Mud volcano. Side scan record from the Black Sea bottom. (Side scan principle see Fig. 4.15). [Courtesy Dr. Glunow, Moscow, from UNESCO-IMS-Newsletter 61, Paris]

various processes, including recrystallization (crystal growth within pre-existing solids), can be reconstructed by studying the distribution of the elements and compounds involved, within the solids and the interstitial waters. The distribution of certain isotopes (oxygen, carbon, strontium) is of special interest in this context, because dissolution, migration, and reprecipitation under various conditions result in altered ratios of the isotopes.

Of course, diagenetic processes are extremely important for hydrocarbon source rocks, migration and reservoir rock porosities, and permeabilities (Fig. 10.2).

3.3.3 Residence Time. For steady-state conditions, output must equal input. Thus the seawater has to rid itself of all new salts coming in, in the same proportions as they are added. Where are the "sinks" for this material? The quantitative assessment of sinks is a major geochemical problem. For calcium carbonate the sink is calcareous skeletons built by organisms; for silica it is opaline skeletons. The metals presumably leave the ocean in newly formed minerals such as authigenic clay, oxides, and sulfides, and in zeolites, as well as in alteration products resulting from reactions between hot basalt and seawater, at the ridge crest. The sulfur is precipitated in heavy metal sulfides in anaerobic sediments near the land. Some salt leaves with the pore waters in the sediments. Under the assumption that the ocean does not change its composition, we can calculate the average time a seawater component remains in the water, before going out as sediment. This time is called *residence time.*

Calculating the residence time is analogous to figuring out how long people will stay in a museum: count the people present, and the number entering per unit time. The ratio is the average viewing time. Similarly,

$$t = A/r, \qquad\qquad\qquad (3.1)$$

where A is the amount present, and r is the input. Some residence times are given in Table 3.1. Clearly, sodium and chloride have a long residence time, and silica has a very short one. The residence time is, in essence, a measure of the geochemical solubility (or inversely, reactivity) of the substance in question. Equation (3.1) was once used to calculate a "salt age" for the ocean, under the assumption that the ocean had started out fresh and retained all sodium since. This salt age came out near 100 million years. In our museum analogy, calculating the salt age corresponds to figuring the time the exhibit opened, from the number of people present and the rate at which they are coming in. The salt age was once useful as a minimum estimate for the scale of geologic time.

3.4 Major Sediment Types

There are essentially three types of sediments: those that come into the ocean as particles, are dispersed, and settle onto the sea floor; those that are precipitated out of solution directly; and those that are made by organisms. For convenience we may call

Fig. 3.6 a, b. Familiar examples of two major types of beach sand **a** medium sand, lithogenous, La Jolla **b** coarse sand, biogenous, Hawaii. (Photos: W. H. B.)

the first type *lithogenous,* the second *hydrogenous,* and the third *biogenous* (Fig. 3.6; Box 3.1). Around the ocean margins, lithogenous sediments are predominant, although there are also salt deposits and biogenic sediments. In the deep sea, biogenous sediments predominate, especially *calcareous ooze.*

3.5 Lithogenous Sediments

3.5.1 Grain Size. The bulk of sediment around the continents consists of debris washed off the continents. This debris resulted from the mechanical breakup, with or without leaching, of continental igneous and sedimentary rocks. The product is fragments of rocks and minerals. One property that is extremely important for assessing source and transport processes is *grain size* (Chap. 4.1.1). One distinguishes *gravel, sand, silt,* and *clay.* Thus *sand,* for example, is all material of a size between 0.063 mm and 2 mm, regardless of its composition or origin. *Silt* has sizes between 0.063 and 0.004 mm (see Appendix A5).

Likewise, *clay* is everything below a particle size of 0.004 mm (or 4 µm). (In some scales 0.002 mm is used as limit.) This terminology is sometimes confusing, because "clay minerals" are minerals of a certain type, abundant in clay-type deposits, but they are not invariably of "clay size".

Overall, there is a gradation of grain sizes from source to place of deposition, such that the bigger, heavier particles remain closer to the source while the clay-size particles can be carried over large distances. *Gravel* (2–256 mm) and boulder-size material (> 256 mm) does not commonly travel far, except when taken along by ice (which does not care much about the size of its load). Since gravel consists of pieces of rock which can be identified and matched with the source area, it is commonly possible to determine the origin of a gravel assemblage. Except on sea floor supplied by glaciers or by coral reef, gravel is not an important constituent of marine sediments.

Box 3.1 Classification of Marine Sediment Types

Lithogenous Sediments. Detrital products of disintegration of pre-existing rocks (igneous, metamorphic, or sedimentary, see Appendix A6) and of volcanic ejecta (ash, pumice; ash is < 2 mm). Transport by rivers, glaciers, winds. Redistribution through waves and currents. Nomenclature based on *grain size* (gravel, sand, silt, clay). Additional qualifiers are derived from the *lithologic components* (terrigenous, bioclastic, calcareous, volcanogenic, etc.), and from the *structure* and *color* of the deposits.

Typical examples are as follows (environment in parentheses).
– Organic-rich *clayey silt,* with root fragments (marsh).
– Finely laminated *sandy silt,* with small shells (delta-top).
– Laminated quartzose *sand,* well-sorted (beach).
– Olive-green homogeneous *mud* rich in diatom debris (upper cont. slope) (*mud* is the same as terrigenous clayey silt or silty clay).

Fine-grained lithogenous sediments (which become *shale* upon aging and hardening) are the most abundant by volume of all marine sediments (about 70 %). This is largely due to the great thickness of continental margin sediments.

Biogenous Sediments. Remains of organisms, mainly carbonate (calcite, aragonite), opal (hydrated silica), and calcium phosphate (teeth, bones, crustacean carapaces) (see Table 3.3). *Organic sediments,* while strictly speaking biogenous, are commonly treated separately. Arrival at the site of deposition by in situ precipitation (benthic organisms living there), or through settling via water column (pelagic organisms; coarse shells fall singly, small ones commonly arrive as aggregates). Redeposition by waves or currents. Redissolution common, either on sea floor or within sediment. Nomenclature is based on *type of organism* and also on *chemical composition.* Additional qualifiers from *structure, color, size, accessory matter.* Typical examples:

– Oyster bank (lagoon or embayment)
– Shell sand (tropical beach)
– Coral reef breccia (slope below coral reef) (*breccia* is coarse broken-up material, in this case reefal debris)
– Oolite sand, well sorted (strand zone, Bahamas)
– Light gray calcareous ooze, bioturbated (deep-sea floor)
– Greenish gray siliceous ooze (deep-sea floor)

Biogenous sediments are widespread on the sea floor, covering about one half of the shelves and more than one half of the deep ocean bottom, for a total of 55 %. About 30 % of the volume of marine sediments being deposited at the present time may be labeled biogenous, although they may have considerable lithogenous admixture.

Hydrogenous Sediments. Precipitates from seawater or from interstitial water. Also products of alteration during early chemical reactions within freshly deposited sediment. Redissolution common. Nomenclature based on *origin* ("evaporites") and on *chemical composition.* Additional qualifiers from *structure, color, accessories.* Typical examples:

– Laminated translucent halite (salt flat).
– Finely bedded anhydrite (Mediterranean basin, subsurface).
– Nodular grayish white anhydrite (ditto).
– Manganese nodule, black, mammilated, 5 cm diam. (deep Pacific).
– Phosphatic concretion, irregular slab, 15 cm diam., 5 cm thick, light brown to greenish, granular (upwelling area).

Hydrogenous sediments, while widespread (Ferromanganese deposits and metalsulfides around deep-sea vents), are not important by volume at present. At times in the past, when thick salt deposits were laid down in the newly opening Atlantic, in the Mesozoic, and much later in a dried-up Mediterranean (end of the Miocene) the volume of hydrogenous sediments produced was considerable. The salinity of the ocean may have been appreciably lowered during those times of evaporite deposition.

3.5.2 Sand is typical of beach and shelf deposits (Fig. 3.6) Like gravel, it may consists of rock fragments, a striking example being provided by black beach sands on Hawaii, which are made of volcanic rock. Commonly, however, sand consists of fragments of the minerals quartz, feldspar, mica, and others. Except for quartz, these are compounds of the type [Na, K, Mg, Ca]$_a$ [Si, Al]$_b$ O$_c$, that is, alumino-silicates. Quartz (SiO$_2$) is the mineral most resistant to abrasion and to leaching, and sands which have been extensively reworked are therefore greatly enriched in quartz (mature sands). The first minerals to be destroyed during weathering, transport, and reworking are the iron-rich minerals. Feldspar also is not very resistant to chemical destruction.

In many tropical beaches, sand may consist entirely of fragments of calcareous skeletons, of mollusks, corals, and algae. One might argue whether a deposit made up of such "bioclastic" material should be counted as "lithogenous", as derived from pre-existing rocks or as "biogenous". Strictly speaking, there is a cycle of mechanical breakup and abrasion involved (hence lithogenous), but for geochemical balance calculations we might include this material with chemical deposits. Carbonate particles are easily abraded and chemically eroded. Thus, in a mixture of carbonate and quart sand, the quartz will soon dominate when the sand is reworked.

The source areas and dispersal history of sands can also be explored by noting the compositional types of "heavy minerals" (densities greater than 2.8 g/cm^3; examples: the silicates hornblende, pyroxene, olivine, and also magnetite, ilmenite, rutile) (Appendix A4). The *heavy mineral association* allows the mapping of depositional provinces, which, in turn, provides clues to the action of shelf currents and other factors (Fig. 3.7).

The *shape* of sand grains, especially quartz grains, can give clues to their origin. For example, grains in glacial deposits have sharp edges, whereas reworking by waves leads to rounding. Dunes especially collect well-rounded grains, and such grains may be "frosted". One problem with a straightforward application of these concepts is that quartz grains are so resistant that they can be recycled many times by erosion and redeposition of old sedimentary deposits *(polycyclic sands)*. Another problem is that etching of grains can take place *after* deposition, within the sediment, obliterating and confusing the surface markings.

3.5.3 Silt-Sized sediment is very characteristic of continental slope and rise, but may occur at any place on the shelf where conditions are quiet, so that it is not washed out by wave or current action. The composition of the silt is much like that of sand in the coarser end of the range, and like that of clay in the finer end. Mica is especially abundant in terrigenous silt.

Sand is commonly studied with a binocular microscope, and the composition of clay is investigated by X-ray diffraction. The study of silt traditionally fell into the crack between these two methods and has had a rather low popularity rating. More recently, the scanning electron microscope (SEM) has made it possible to investigate this size fraction in more detail (Fig. 3.8). The composition of the silts is usually closely related to that of the associated fine sand fractions.

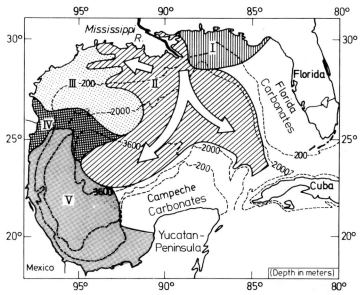

Fig. 3.7. Heavy mineral provinces of the Gulf of Mexico, based on typical mineral associations. *I* East Gulf; *II* Mississippi; *III* Central Texas; *IV* Rio Grande; *V* Mexico. Carbonate particles are dominant off Mexico and Yucatán. The heavy mineral patterns contain clues about the sources and transport paths of terrigenous sediments. [D. K. Davies, W. R. Moore, 1970, J Sediment Petrol 40: 339]

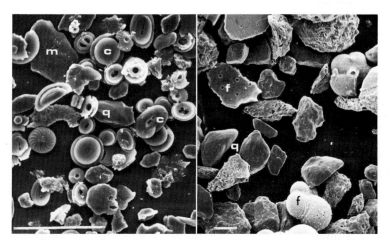

Fig. 3.8. Recent hemipelagic sediment from Continental Rise off NW Africa (Cape Verde Rise). SEM photos, *bar* 20 μm. *Left* Fine-silt fraction (2–6 μm) predominantly composed of coccoliths (*c*) and of some detrital mica (*m*) and quartz (*q*); *Right* coarse silt fraction, mainly detrital grains (*q* quartz) and foraminifera tests and fragments (*f*). [Photo courtesy D. Fütterer]

3.5.4 Clay-Sized sediment is ubiquitous on shelves, slopes, and in the deep sea. Like the silt, it indicates *low energy environments* where abundant, because it is easily picked up and washed away. Exceptions are clays in those high energy environments where clay supply is extremely high, as off some tropical rivers. Much of the clay, as mentioned, consists of clay minerals, products of weathering that are brought in by rivers and winds and are dispersed by waves and currents.

The common clay minerals, or clay mineral groups, are montmorillonite (or smectite), illite, chlorite, and kaolinite (Appendix A4). Their distribution and paleoclimatic significance will be discussed in more detail in Chapter 8.

The high surface areas of clay particles give clayey sediments special chemical properties. For example, they readily adsorb a large variety of substances, and react with the ions in seawater and in interstitial waters. Thus, during *diagenesis,* within buried sediments, new clay minerals can form slowly. Ultimately, this has important implications for the chemistry of seawater and for geochemistry in general. Also, clays can readily form aggregates, in seawater.

Clayey sediments in regions of high sedimentation rate are commonly rich in organic matter, partly because organics cling to the clay during deposition, and partly because where conditions are quiet enough for the deposition of clay they also are favorable for the deposition of the fluffy organic particles. Much of the clayey matter may actually be brought to the floor within fecal pellets of organisms filtering the water. Such clay is associated with food particles. Recent experiments with sediment-trapping equipment in the Baltic Sea, in the open ocean off California, and elsewhere, suggest that this *"fecal pellet transport"* mechanism is a significant agent of sedimentation. The association between clay and organic matter is important in petroleum exploration in the search for source rocks.

3.6 Biogenous Sediments

3.6.1 Types of Components. Organisms produce sediments in the form of shells and other skeletal materials and organic matter. The label "biogenous" generally refers only to calcareous, siliceous, and phosphatic hard parts (see Table 3.3).

Virtually all of the calcium carbonate deposited on the sea floor belongs in the category of biogenous sediment. On the shelf it is chiefly the shells and skeletons of benthic organisms, while on the slope there is an increasing proportion of planktonic remains away from the coast. In the deep sea, finally, plankton remains (foraminifers, coccoliths; Appendix A9) are entirely dominant. Opaline skeletons also are important – again mainly benthic remains (sponges) on certain shelves, and planktonic ones (diatoms, radiolarians) in slope sediments and in the deep sea. In addition to carbonate and silica, phosphatic particles are produced by organisms. The sedimentation of such particles plays an important role in the phosphorus budget of the biosphere, and is geochemically of great interest (see Sect. 10.3.1). the various other types of hard parts – strontium-sulfate, manganese-, iron-, and aluminum compounds – are intriguing from an evolutionary standpoint, but do not noticeably contribute to sediment patterns. Organic matter (organogenic sediment) will be taken up in Chapters 6 and 10, when discussing productivity and hydrocarbons, respectively.

Table 3.3. Inorganic constituents and carbonate minerals in hard parts of marine organisms (for systematics see Appendix A9)

Bacteria	Aragonite ($CaCO_3$), iron oxide, manganese hydroxide
Diatoms	Opal ($SiO_2 \cdot nH_2O$)
Coccolithophores	Calcite ($CaCO3$)
Chlorophyta	Aragonite
Rhodophyta	Aragonite, Mg-calcite
Phaeophyta	Aragonite
Foraminifera	Calcite, Mg-calcite, aragonite (rare)
Radiolarians	Opal, celestite ($SrSO_4$)
Sponges	Mg-calcite, aragonite, opal; rare: iron oxides, celestite (rare)
Corals	Aragonite, Mg-calcite
Bryozoans	Aragonite, Mg-calcite + aragonite
Brachiopods	Calcium-carbonate-phosphate, calcite
Echinoderms	Mg-calcite; rare: amorphous phosphates, Fluorite, opal
Mollusks:	
Gastropods	Aragonite, aragonite + calcite
Pelecypods	Aragonite, aragonite +calcite, calcite
Cephalopods	Aragonite
(Various phosphate minerals and iron-oxides also have been detected in certain mollusks)	
Annelid worms	Aragonite, arag. + Mg-calcite, Mg-calcite rare: phosphates, opal, iron oxides
Arthropods	
Decapods	Mg-calcite, amorphous Ca-phosphate
Ostracods	Calcite, aragonite (rare)
Barnacles	Calcite, aragonite (rare)
Vertebrates	Ca-phosphates

3.6.2 Benthic Organisms are responsible for building up enormous masses of shelf material off certain coasts. We have earlier referred to the Great Barrier Reef, which is composed of calcareous shells and skeletons of corals, algae, mollusks, and foraminifera. Much of Florida sits on top of shelf carbonate built by benthos. Both aragonite and calcite are secreted by various types of organisms (see Table 3.3). The content of *magnesium* varies between groups of organisms, but also within the same group. There is a tendency for higher magnesium values in those skeletons precipitated in warmer water.

The great importance of carbonate-secreting benthos in producing sediment is apparent all through the Phanerozoic record, that is, the last 600 million years or so.

Early on, phosphatic skeletons were actually more important than calcareous ones. The reason for the switch is still unknown. Mesozoic and Cenozoic Platform carbonates were deposited all around the *Tethys,* the seaway which once linked the western Pacific and the Atlantic through Asia and Europe. The oil fields in Arabia and in Mexico are associated with cavernous Mesozoic reef limestones, built up within the Tethys seaway. In the Paleozoic, shelf carbonates built the Permian reefs in the Rockies, which contain a rich fauna of ancient corals, brachiopods, and mollusks. The carbonates of the Caledonian mountains running from Norway to Scotland and

Wales into Newfoundland and Appalachia mark an ancient Atlantic seaway (Fig. 1.19).

The older the carbonate deposits, the more recrystallized they tend to be. Aragonite is replaced by calcite in this process, and many original structures of fossils are destroyed through *diagenesis.* "Dolomitization" is an extreme form of diagenetic alteration, whereby every second Ca-ion is eventually replaced by Mg.

The shallow water limestones of the geologic record, produced in the shelf environments of ancient oceans, commonly contain an admixture of siliceous rocks, as layers or nodular masses of *chert* arranged along horizons parallel to the bedding.

Mineralogically, this material is, in the main, finely crystalline quartz; in Tertiary sediments it may be *cryptocrystalline,* or it may be still amorphous in parts. The origin of these flinty masses has long been a mystery (although this did not prevent their being used for tool-making in the Stone Age, of course). In the Miocene Monterey Formation of California, the source of the silica is shells of planktonic diatoms. However, in shelf limestones the source may well have been siliceous sponges. Such sponges are abundant on shelves and upper slopes in the present ocean wherever there are silica-rich waters, that is, in high latitudes and in upwelling areas.

Why is there no evidence for incipient chert formation in modern shelf carbonate areas?

The reason is that silicate concentrations are very low in present tropical waters. Diatoms and sponges precipitate silica, but their skeletons are delicate and are quickly redissolved in the highly undersaturated seawater, which is stripped of its silicate by diatoms in upwelling regions. Could ancient oceans have had higher concentrations of silicate in the water? This is quite possible, if such oceans were much less fertile than the present ones. Also, it has been suggested that a high level of volcanic activity favored the formation of siliceous sedimentary rocks by supplying large amounts of silicate to seawater. For the Eocene, this idea has been offered to explain the abundance of cherts in deep-sea carbonates. These cherts have caused considerable difficulties for deep-sea drilling, since they quickly destroy the drill bit, and impede recovery of sediment cores.

3.6.3 Plankton Organisms contribute a significant amount of material to slope sediments (Fig. 3.8). Deep-sea carbonates are made almost entirely of planktonic remains (see Chap. 8). Shelf sediments, especially those of the geologic past, can also contain considerable amounts of planktonic remains. For example, the English chalk is made largely of the remains of planktonic calcareous algae, the *coccolithophorids* (which produce the little platelets called *coccoliths*). The role of diatoms in supplying opal to margin sediments has been mentioned.

In general, the remains of planktonic organisms increase relatively to those of benthic organisms in the offshore direction. A well-known application of this principle is the determination of the ratio of planktonic to benthic foraminifera in the mapping of marine facies. In the deep sea, the ratio is greater than 10:1. On the edge of the shelf it is close to 50:50. In the Persian Gulf, a somewhat restricted shelf sea, the plankton-benthos ratio is even lower: about 3:7 near the entrance, and less than 1:10 in the interior. Thus, the planktonic remains are typical for the open ocean.

3.7 Nonskeletal Carbonates

3.7.1 Carbonate Saturation. In the present ocean, carbonate precipitation is either within organisms (shells, skeletons) or is associated with their metabolic activity (algal crusts). This need not always have been so: a certain proportion of the limestones and associated carbonates in the geologic record may have been precipitated inorganically.

Where would one look for such inorganic precipitation today? To guide us in the search for likely places, we need to consider briefly the chemistry of carbonate precipitation and dissolution.

Seawater which spontaneously precipitates a mineral – for example, aragonite, $CaCO_3$ – is said to be *supersaturated* with this mineral phase. Seawater which dissolves the mineral is *undersaturated.* When precipitation just equals dissolution, *saturation* obtains, that is, the solution is *in equilibrium* with the solid. The degree of saturation is expressed as the ratio of the ionic product of reactants present to the product necessary for saturation (brackets indicate concentrations):

$$D_{sat} = [Ca^{2+}] [CO_3^{2-}]_{observed}/[Ca^{2+}] [CO_3^{2-}]_{equilibrium}. \tag{3.2}$$

Clearly, D_{sat} equals 1 for saturation, less than 1 for undersaturation, and is greater than 1 for supersaturation.

Whenever D_{sat} is greater than 1, we expect spontaneous precipitation. However, even though D_{sat} is indeed greater than 1 in the surface waters of tropical oceans, inorganic precipitation is negligible. It is commonly assumed that the presence of magnesium interferes with the expected reaction. Thus we need to find an unusually high D_{sat}, if we are to see inorganic precipitation. Additionally, the presence of appropriate crystal nuclei should be favorable.

High temperature and a low CO_2 content in the water increase the product $[Ca^{2+}]$ $[CO_3^{2-}]$ by increasing the concentration of carbonate ion. The effect of adding or subtracting CO_2 is readily seen from the following equations:

$$CO_2 + H_2O = H^+ + HCO_3^-, \tag{3.3}$$
$$HCO_3^- = H^+ + CO_3^=. \tag{3.4}$$

Almost all the inorganic carbon in the ocean is in the form of bicarbonate, HCO_3^-. Removal of CO_2 drives the reaction of Eq. (3.3) to the left, subtracting hydrogen ions. In turn, this drives the reaction of Eq. (3.4) to the right, opposing the change in hydrogen ions. Increasing the temperature lowers the solubility of CO_2. Also, CO_2 is taken up by algae during photosynthesis – hence the precipitation of $CaCO_3$ on the surface of many shallow-water tropical algae.

The following equation summarizes the process surrounding carbonate precipitation and dissolution in a simplified manner:

$$CaCO_3 + H_2O + CO_2 \rightleftharpoons Ca^{2+} + 2 HCO_3^-. \tag{3.5}$$

Dissolution proceeds from left to right, precipitation from right to left. Note that CO_2 is used up during dissolution of carbonate (to help make bicarbonate), and is released

materials during high stands, while yielding little during low stands. When platforms are flooded, the growth of algae, foraminifers, mollusks and corals produces a large amount of calcareous debris which is then available for transport down the flanks (and also for making storm ramparts, in cases). In contrast, during low stands, the platforms fall dry and, while soils or karst from carbonate dissolution mediated by runoff may develop, there is little production of loose sediment.

3.7.3 Dolomite. Calcite may contain magnesium in different concentrations. "Mg-Calcites" are, in fact, the rule rather than the exception in the shelf environment. Dolomite [Ca, Mg (CO$_3$)$_2$], however, is a different story, and poses many unsolved problems.

One might expect that dolomite would precipitate from seawater because it is much less soluble than aragonite. There is, after all, plenty of magnesium in seawater (Table 3.1). However, such precipitation has not been observed in today's ocean. Dolomite apparently forms *within* calcareous sediments, either through partial replacement of Ca with Mg in preformed carbonate or possibly by precipitation from pore waters. As mentioned, these processes are part of "diagenesis". A classic area for the study of incipient dolomite formation is the southern margin of the Persian/Arabian Gulf (Fig. 3.10). Here, the intertidal flats lie behind barrier islands, within a lagoon with warm and very saline water. In the pore waters of sediments above low tide, magnesium is greatly enriched with respect to calcium (and sulfate is reduced) due to the precipitation of gypsum (calcium sulfate) in adjoining evaporite pans (i. e. the *"sabkha"*).

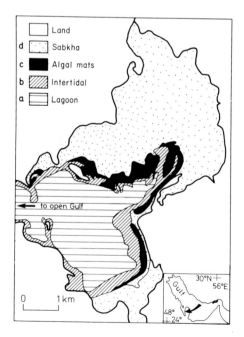

Fig. 3.10. Present-day dolomite formation. Lagoon on the southern coast of the Persian/Arabian Gulf (*a*), floor covered with calcareous mud. Intertidal zone (*b*) and algal mats of uppermost litoral (*c*) rim the lagoon. Intermittently flooded evaporite flats (*d*, sabkhas) follow landward [After L.V. Illing et al 1965, Soc Econ Paleontol Mineral Spec Publ 13: 89, Simplified]

Obviously, in this instance of dolomite formation, the dolomite is closely associated with evaporation flats, tidal conditions, and unusually warm and saline water – an environment unfriendly to higher organisms but favorable for recrystallization. Thus, we should not expect many macrofossils in sediments rich in this kind of dolomite. Stromatolithic textures, however, may be preserved.

Experiments on dolomite formation by P.A. Baker and M. Kastner (1981, Science 213:214) suggest that the removal of the sulfate ion from the interstitial waters (by reduction and precipitation as iron sulfide) is the crucial step allowing dolomite formation. If so, the importance of the precipitation of gypsum in evaporite lagoons may lie in the decrease of the sulfate concentration rather than in the increase of the Mg/Ca ratio.

Microbial sulfate reduction in the uppermost few tens of meters of continental margin sediments also seems to favor dolomitization. According to investigations with deep-sea drilling cores, this occurs if accumulation rates are less than 500 m/million years and organic carbon contents are greater than 0.5 weight %.

Dolomite has also been found in deep-sea sediments deposited under *anaerobic* conditions. This has been explained by bacterial sulfate reduction, removing this obstacle to precipitation. Sulfate reduction also increases the alkalinity, providing additional carbonate ions for precipitation. In general, diagenetic signals are strong in hemipelagic sediments with suboxic processes. Under certain conditions, dolomitization of preformed carbonates may occur by dilution of interstitial seawater by freshwater, penetrating from nearby highs on tidal flats, as happens on the Bahaman Islands, and on some coral reef islands.

3.8 Hydrogenous Sediments

With the question of dolomite formation, we have entered the evaporite environment. Within continental margins, the bulk of hydrogenous sediments are evaporitic salt. Strictly speaking, calcareous shells and skeletons are also hydrogenous, since they originate in the water. However, we have called the minerals precipitated by organisms biogenous, and set them apart.

3.8.1 Marine Evaporites are those sediments which form when seawater evaporates. Restriction of exchange with the open ocean, in a semi-enclosed basin, is necessary to drive the salt content high enough for precipitation to begin. Such restricted bodies of ware are (1) coastal lagoons; (2) salt seas on the shelves; or (3) early rift oceans in the deep sea. A special case is the Mediterranean, which was partially isolated in the latest Miocene, 6 to 5 million years ago (see Sect. 9.5.2).

How much salt can be produced by evaporating a 1000-m- high column of seawater? Salt constitutes 3.5 % (or 35 ‰) of the weight of the column; its density is about 2.5 times that of water. Thus, we would obtain about 14 m salt. Most of this would be table salt (halite) (see Table 3.1). The least soluble salts precipitate first: calcium carbonate (aragonite), and calcium sulfate (gypsum). To precipitate halite, the brine needs to be concentrated about tenfold. Many evaporites only contain carbonate and gypsum (or anhydrite), others have thick deposits of halite or, rarely, of

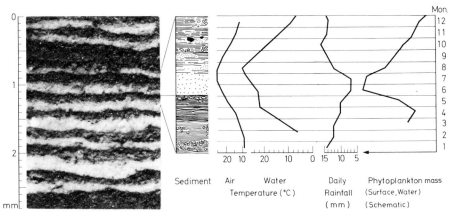

Fig. 3.14. Annual layers (varves) in the sediment of a bay in the Adriatic Sea (Mljet Island). The photo shows light and dark laminae. One light-dark pair corresponds to one year. The lower boundary of light layers are generally sharp; they are due to precipitation of carbonate by phytoplanktom, which bloom in early summer (record to the *right*) as temperature rises. In fall and winter, rains bring terrigenous matter which – together with organic detritus – provides for dark colors. The interpretation of the varves is a complicated matter; recent studies use statistical procedures to reconstruct climatic conditions in detail. (Photo E. S.)

sediments of Santa Barbara Basin off California using diatoms, dinoflagellates, and other microfossils (Sect. 7.7.2). Unfortunately, such ideal calendar pages are extremely rare in the marine realm. They can only be preserved where burrowing organisms cannot live, that is, in the absence of free oxygen above the sediment-water interface.

It is somewhat hazardous to apply sedimentation rates found for a given type of sediment to a geologic sequence of such sediments. Episodic or periodic events, especially erosional events, can change overall rates considerably. In many cases, sediments of a certain depositional regime are underlain by erosional features of the same regime: ice-polished rocks under moraines, wind-polished pebble pavements under dunes, channels under river gravels, lopped-off sea floor beneath graded storm deposits (shallow water), or turbidites (deep water). Erosion, of course means loss of information to the geologists. Missing sections – *hiatuses* – are erased history, lost time as it were, whose duration can hardly be estimated. Therefore *accumulation rates* averaging long time spans and including these losses are smaller than instantaneous sedimentation rates. Of course, the presence of hiatuses suggests we must look for information about the processes producing them. Also, hiatuses extending over large regions can be used for the correlation of sediment sequences, especially in seismic stratigraphy.

In concluding this chapter about sources and composition of marine sediments, some more esoteric components may be mentioned. So-called *"cosmic spherules"* were first described by Murray of the *Challenger* Expedition. Black magnetic spherical objects up to 0.2 mm in diameter, commonly rich in Fe and Ni, can be isolated from pelagic sediments, typically several per gram pelagic clay. Because it was thought that they represent a steady influx of matter from space, their abundance was used occasionally as an indicator for accumulation rates.

Glassy objects, normally up to 1 mm in diameter, may be produced by meteorite impacts on terrestrial rocks and are called "*microtectites*". They were found in deep-sea sediments in strewn fields around Australasia (0.7 million years old), off the Ivory coast (1.1 million years), and in the Caribbean (33–35 million years), for example. Where abundant, they can be used as a time marker and serve for correlation by "*event stratigraphy*".

Further Reading

Selley RC (1976) An introduction to sedimentology. Academic Press, London

Bouma AG, Normark WR, Barnes NE (eds) (1985) Submarine fans and related turbidite systems. Springer, Berlin Heidelberg New York

Reading HG (ed) (1986) Sedimentary environments and facies. Blackwell Scientific, Oxford

Tucker ME, Wright VP (1990) Carbonate sedimentology. Blackwell Scientific, Oxford

Friedman GM, Sanders JE, Kopaska-Merkel DC (1992) Principles of sedimentary deposits. Stratigraphy and sedimentology. Macmillan, New York

4 Effects of Waves and Currents

We have seen that the morphology of the ocean margins is largely determined by tectonics and sediment supply (Chap. 2), and we have reviewed the various types of sediments which are involved (Chap. 3). We now turn to the all-important role of water motion in determining the distribution of sediments on the sea floor. A first impression of this role can be gained from contemplating Fig. 4.1 which illustrates the redistribution of sediment supplied by a river in a setting typical for southern California. The water motions indicated are not the only ones that need to be considered, as we shall see.

Fig. 4.1. Redistribution of sediment on the continental margin, by water motion. The drawing reflects a West Coast geographic setting; note river input, longshore transport (powered by wave energy), interception by submarine canyon. Fine sediment bypasses the shelf, and comes to rest on slope or deeper. Note starved beaches and rocky shelf beyond canyon. [Based on a drawing in D. G. Moore, 1969, Geol Soc Am Spec Pap 107: 142]

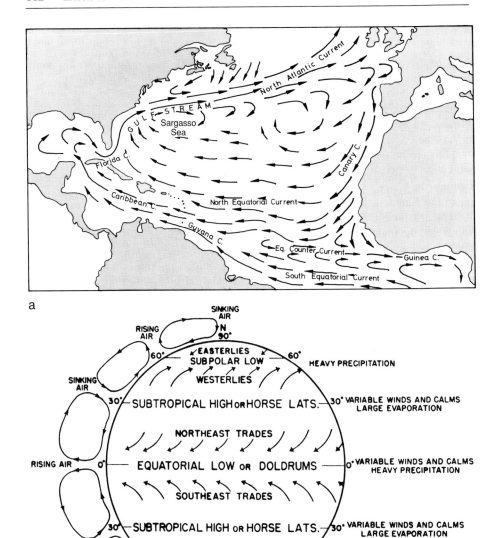

Fig. 4.14 a–b. Gulf Stream, and wind-driven ocean circulation. **a** The Gulf Stream carries warm surface waters toward the seas of NW Europe. It is part of the giant subtropical gyre of the North Atlantic whose center is the Sargasso Sea. [Mainly after G. Neumann, W. J. Pierson, 1966, Principles of physical oceanography. Prentice-Hall, Englewood Cliffs]. **b** The global wind system, schematically. [R. H. Fleming, 1957, Geol Soc Am Mem 67: 87]

WINDS   CURRENTS

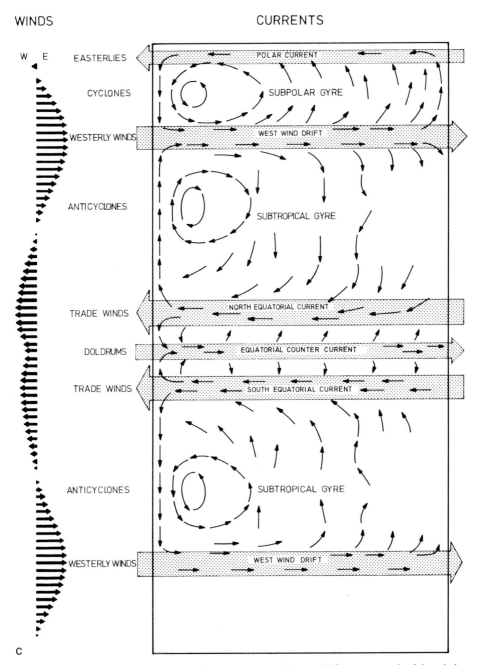

Fig. 4.14. c Idealized rectangular ocean with the currents which would form as a result of the wind forces shown on the left. The asymmetry of the gyres is caused by the Earth's rotation. (From W. Munk, 1955, Sci Am 193 (3) 6; modified)

Surface currents and associated temperature distributions control not only plankton patterns, but also the growth of benthic assemblages, especially coral reefs (see Sect. 7.4). The western boundary currents bring warm water to high latitudes, the eastern ones bring cool water to low latitudes. Thus, the belt of tropical coral reefs is much wider in the west than in the east, within each ocean basin.

The drift of icebergs and hence the transport paths of their load of rocks is controlled by currents, as are the paths of various other drifting materials such as trees – some of which are thought to be responsible for the dispersal of land animals on islands. The drifting larvae of benthic organisms *(meroplankton)* reach their uncertain destinations with the aid of currents, and disperse throughout the ocean basins by island-hopping if necessary.

Finally, surface currents can affect the sea floor directly, through erosion on shelves and even on upper slopes. The deep-reaching Gulf Stream sweeps fine material off the Blake Plateau east of Florida (see Fig. 1.3), keeping the manganese pavements there clean.

4.3.2 Current Markers. How can we detect the action of currents by studying the sea floor? The winnowing of fines from remaining coarser material, that is, the "sorting" of sediment, was discussed earlier (Sect. 4.1.2). Usually, it is necessary to make grain size determinations to show that sorting occurred. However, sometimes effects of winnowing can be seen in bottom photography, which shows pavements of cobbles, nodules, shells, or other winnowed layers of coarse residual material. Also, scour marks behind obstacles are common indicators of currents.

The use of side-scan sonar, on the shelf and more recently in the deep sea, has greatly increased our knowledge about the activity of near-bottom currents, from their effects on sediment patterns and morphology of the sea floor (Fig. 4.15). Streaks of coarse material in fine sediments, indicating current action, were discovered in the entrance to the Baltic Sea by such acoustic sensing (Fig. 4.16).

The *direction* of a current can be readily deduced from photography and acoustic scanning. Current *strength* is more difficult to deduce. Direct observations on the effects of current strength are possible in tidal flats. The German and Dutch *Wadden* flats of the North Sea have long been a natural laboratory for these kinds of observations (Fig. 4.17).

We have seen some features which *parallel* currents. Some are *at right angles* also: the current ripples. Much like wind dunes, ripples have a gentle up-current side and a steep down-current slope (Fig. 4.18). This type of ripple differs from the oscillation type mentioned earlier (Sect. 4.2.1). The internal laminae of the current ripples consist of the down-current slopes made by material falling over the edge at the crest. Thus, fossil ripple marks *(cross-bedding)* yield clues to ancient current systems in contact with the sea floor (Fig. 5.3c).

The relationship between current velocities and ripple formation has been studied in the laboratory in some detail. Ripple characteristics depend both on the type of sediment present, and on velocity distributions in a complex manner. The formation of ripples of dimensions of centimeters to decimeters starts at current velocities of about 25 to 100 cm/s. The origin and architecture of *giant ripples* and *underwater*

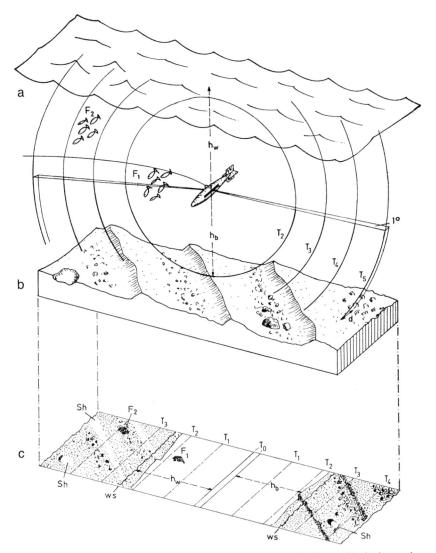

Fig. 4.15. Principle of side-scanning echo sounder. **a** Water surface; **b** Sea floor with ripples, rocks, and a small depression *(left);* **c** a section of the acoustic recording (sonograph). T_o Outgoing pulse of the towed submerged sound source (the "fish"); T_1, T_2, etc., time markers (= distance to sound reflecting object); *Sh* acoustic shadow; F_1, F_2 schools of fish, and their acoustic image; h_w distance "fish" to water surface *(WS);* h_b distance "fish" to sea floor. [R. S. Newton et al., 1973, Meteor Forschungsergeb. Reihe C 15: 55]

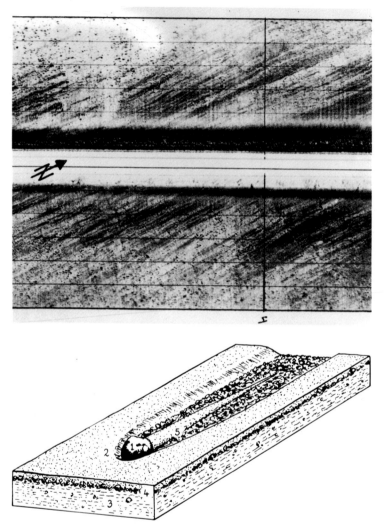

Fig. 4.16. Comet marks in the channel Store Belt, Baltic Sea, between the Danish islands of Langeland and Lolland. *Upper* Sonograph of the sea floor. Length of record about 2 km, width ~ 150 m, water depth about 12 m. *Dark spots* are boulders (morainal debris). Some boulders have a "tail", toward left and down. These "comet marks" indicate currents from the north. [Sonography by F. Werner, Kiel.] *Lower* Morphology of a typical comet mark, as seen by divers. *1* Obstruction (boulder); *2* crescent scour; *3* till; *4* residual pebble layer (fossil erosional pavement); *5* "comet tail" of fine sand, built into erosional depression produced by turbulence behind obstacle. [F. Werner et al., 1980, Sediment Geol 26:233]

Fig. 4.17 a, b. Current action in tidal flats. **a** *Mya arenaria* in life position uncovered by erosion of about 20–30 cm of mud, by lateral migration of tidal channel. Note the residual bivalve shells which collect as a "pavement" on the erosional surface. **b** Large current-produced ripples, covered by small ripple marks. German Bight north of Weser estuary. Ripples migrate ocean-ward *(to the left),* hence they are produced by currents of the outgoing tide. [Photos E. S.]

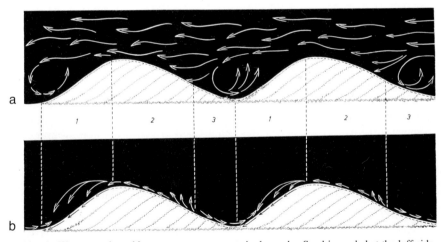

Fig. 4.18 a, b. Water **a** and sand **b** movement at current ripple marks. Sand is eroded at the luff side *(sector 2)* and migrates to the ripple crest. From there sand is transported on the bottom or in suspension *(white arrows)* downward, forming lee laminae. Horizontal bottom vortex in the ripple valley removes fine grains, concentrates coarse material, including shells. Therefore ripples move over coarser-grained base laminae. [H. E. Reineck, 1961, Senckenbergiana Lethaea, 42:51]

dunes is less well known. These features may be several meters high, and tens, even hundreds, of meters apart. Some resemble the barchan dunes of the great deserts. What kind of currents are necesary to produce such dunes? Do they only form where certain types of sediment are available? Are the dunes relics from periods of unusual activity on the sea floor? These questions are still open at the present time.

In coastal areas with a sufficiently large tidal range, alternation between high and low tide produces *tidal currents*. Considerable erosion can take place where tidal currents are confined because of the high velocities attained and the large volumes of water which are commonly involved. In a number of harbors in estuaries, the tidal currents help flush out the entrance and prevent it from silting up. The inlets of lagoons on the East Coast are kept open by fast tidal currents. As these currents slow on either side of the inlet, *tidal deltas* form (see Fig. 3.9).

4.3.3 Upwelling. So far, we have contemplated the effects of horizontal currents on the sea floor. *Vertical water motion* also affects the sea floor, in more than one way. The *overall mixing rate* of the ocean is ultimately tied to vertical motion, and this rate is closely tied to the productivity of the ocean. In turn, the sea's productivity influences what types of biogenous sediments (carbonate, silica, phosphates) will end up on what part of the sea floor.

A more obvious example for the effects of vertical motion is *upwelling*. Off coasts with eastern boundary currents, surface water has a tendency to move seaward, due to deflection by the *Coriolis Force* which results from the rotation of the Earth. Deflection is to the right in the nothern hemisphere, and to the left on the southern one (Fig. 4.19a). The seaward motion is reinforced when winds blow off-shore. Outward-bound surface waters are replaced by cold, nutrient-rich waters from within the thermocline (100 to 200 m depth) (see Fig. 4.19b). Hence the low water temperatures off Northwest and Southwest Africa, California, Peru, and Chile, and hence also the high production of algal plankton, which feeds zooplankton, fish, and even birds up the food chain. The intensity of upwelling changes seasonally and also from year to year. For example, during *"El Niño"*-years, upwelling is greatly reduced both off California and off Peru (Sect. 7.7.3). Precipitation tends to be significantly increased during such periods; along the North Central Coast of Peru large floods can occur and have been documented in river flood deposits for the last several thousand years. The upwelling variations are being followed by satellite sensing of temperature and chlorophyll abundance at the sea surface.

The sediments below areas with strong upwelling are typically rich in organic matter. For example, sediments in the Walvis Bay area, Southwest Africa, have up to 20 % of organic carbon (C_{org}). Opal also usually is abundant: off Walvis Bay one finds up to 70 % opal from the frustules of diatoms. Fish debris and other vertebrate remains also are increased in abundance, presumably delivering part of the phosphate necessary for phosphorite formation (see Fig. 10.7).

The sediment contains many detailed clues to the intensity of upwelling. The planktonic species (foraminifera, diatoms) tend to indicate cold water. The high supply of organic matter depresses the oxygen content, due to decay in the deep water and on the sea floor. In extreme cases anaerobic sediments with annual layers

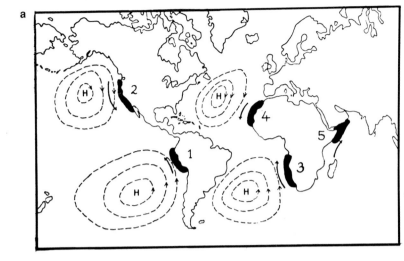

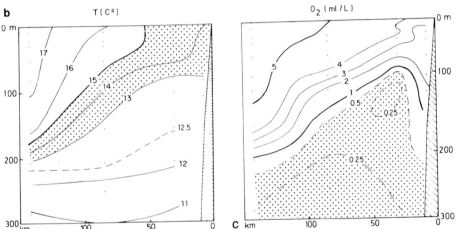

Fig. 4.19. Upwelling currents. *Upper* Global distribution of main upwelling areas. Note position on the east side of subtropical gyres, at edge of "eastern boundary currents". Surface water moves offshore due to Coriolis Force, and is replaced from below. [Science, 1980, 208: 39]. *Lower* Temperature and oxygen distributions in an upwelling area off San Juan, Peru, Sept. 1968. [S. Zuta et al., in R. Boje, M. Tomczak, 1978. Upwelling ecosystems. Springer, Heidelberg] Note the water layer near 200 m depth in the ocean extends to the surface at the continent. This water is oxygen-poor and nutrient-rich. The high nutrient supply stimulates algal production (that is, growth of dinoflagellates and diatoms).

(varves) may develop. Using the clues described, it is possible to follow changes in position and strength of upwelling, and hence the migration and change of nature of the subtropic climatic belts in the geologic record. This is a field of active research involving time scales from centuries to millions of years.

Upwelling is not restricted to coastal areas. The divergence of surface waters anywhere in the open ocean causes replacement from below. The great equatorial divergences in the Pacific and Atlantic Oceans cause the well-known *equatorial upwelling* which we shall encounter when studying carbonate distributions (Chap. 8). The high productivity along these belts not only raises the sedimentation rates of calcerous and iliceous ooze, but also ultimately may be responsible – through concentration of trace metals in organisms and transport to the sea floor in organic debris – for the high values of copper, nickel, zinc, and other metals in Pacific manganese nodules (Sect. 10.4).

4.3.4 Deep Ocean Currents and Benthic Storms. There was a time, not so long ago, when the deep ocean was thought to be a quiet, calm environment whose currents are weak and relatively unimportant as far as shaping the sea floor. The first indication that this is not necessarily so came from the calculations of Georg Wüst (1890–1977) and Albert Defant (1884–1974) in the 1930s. They showed that the temperature-salinity distributions seen in the closely spaced profiles of the *Meteor* Expedition in the central Atlantic implied strong bottom-near flows driven by density differences. Such currents hug the slopes of ocean margins and the flanks of mid-ocean ridges, and they follow density surfaces which are practically horizontal. Hence, the currents are parallel to bottom contours, and marine geologists refer to them as *contour currents*. For physical oceanographers they are *deep geostrophic currents*. Their effects are clearly visible on deep-sea photographs (Fig. 4.22). Such currents can erode, especially where confined to passages. Erosion occurs, for

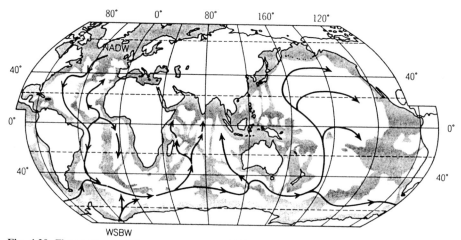

Fig. 4.20. Flow patterns of deep-sea currents at 4000 m depth. The main source areas are in the Northern Atlantic (North Atlantic Deep Water, NADW), and in the Weddell Sea (Wed. Sea Bottom Water, WSBW). Dissipation of these deep currents is by general upward movement. [E. Seibold et al., 1986, The sea floor, Japanese edition, after Broecker and Peng, 1982, Tracers in the sea. Eldigio Press, New York]

ATLANTIC

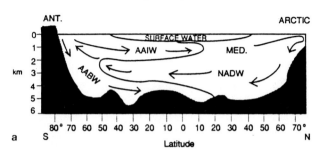

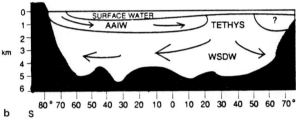

Fig. 4.21 a, b. Evolution of deep and intermediate water circulation during the Cenozoic as inferred from isotopic results. The ocean of the Eocene (**b**) was dominated by halothermal circulation, while the modern ocean (**a**) is dominated by thermohaline circulation. *AABW* Antarctic Bottom Water; *AAIW* Antarctic Intermediate Water; *NADW* North Atlantic Deep Water; *WSDW* Warm Saline Deep Water; *MED* Mediterranean water. [J. P. Kennett and L. D. Stott, 1990; in P. F. Barker and J. P. Kennett et al., Proc. Ocean Drilling Program, Sci. Res. 113, 875]

example, where the abyssal Antarctic Bottom Water passes through the Vema Channel off Argentina, or through the Samoan Passage in the eastern South Pacific.

The *Antarctic Bottom Water* (AABW) fills the deepest parts of essentially all major ocean basins, through such passages (Fig. 4.20, WSBW; Fig. 4.21a). In the western South Atlantic and in the South Pacific, the AABW causes dissolution of carbonate on the sea floor – a process that profoundly affects sedimentation patterns and the nature of the substrate for benthic organisms. The AABW originates on the shelf of the Antarctic, mainly in the Weddell Sea (WSBW in Fig. 4.20) by cooling of polar waters with admixtures of saline *North Atlantic Deep Water* (NADW) (Fig. 4.21a). With a salinity near 34.7‰, when reaching a temperature of near -0.4°C, the Antarctic surface water becomes heavier than the underlying water and sinks through it. Since it is also heavier than almost any water in the deep ocean, it sinks to the abyss as AABW and fills it from below. Its spreading depends on a deep access to a basin. For example, the Angola Basin can only be reached through the Romanche Deep, at the Equator. Hence the AABW makes a long detour to get there (Fig. 4.20).

Besides chemical and mechanical erosion, AABW can produce particle sorting and small-scale features such as scour and ripple marks, and (presumably) large-scale ones, such as the giant dunes in the Argentine Basin.

a

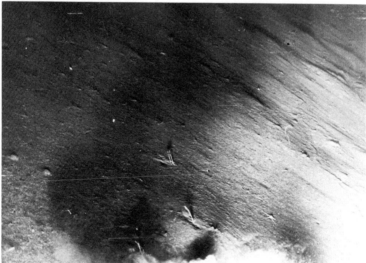

b

Fig. 4.22a, b. Effects of deep geostrophic currents. **a** Continental rise off the US. East Coast, 2.5 km depth. Little or no evidence of current action. (Organism is a "sea whip", pennatulid coral). **b** Same area, but 5 km depth. Current action is obvious from scour marks and lineations. Current flow is *from lower right to upper left,* about 20–30 cm/s. [Photos courtesy C. D. Hollister; see A. H. Bouma, C. D. Hollister, 1973, SEPM Pacific Section, Short Course, Turbidites and deep water sedimentation, Anaheim]

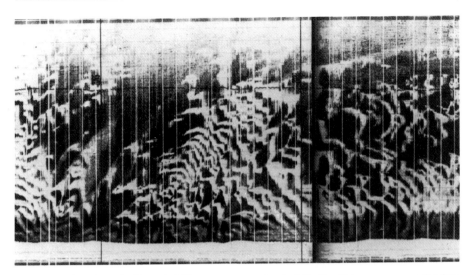

Fig. 4.23. Giant dunes in Carnegie Ridge area, eastern tropical Pacific. Side-scan sonograph taken at 2.4 km depth, covering 1600 x 450 m. Sediment is calcareous ooze. [Recording courtesy P. Lonsdale and B. T. Malfait, S. I. O.]

The deep-towed side-looking sonar developed by F. Spiess and co-workers has revealed an abundance of dunes, sediment ridges, and erosional ravines in many places, even where deep currents had not been suspected (Fig. 4.23).

Of course, the possibility always exists that (a) calculations on current distributions are in error, and (b) the features seen have nothing to do with the present water motion, but are the witnesses of conditions long past.

The NADW sinks in the Norwegian Sea and falls over shallow passages over the Greenland-Faroe (Scotland) Ridge. West of Iceland, a huge undersea cataract starts at a sill depth of about 650 m, reaching velocities of more than 1 m/s. Some 5 million m^3/s are added to the NADW, which moves to the south. The NADW then fills much of the North Atlantic Basin, and rides up the AABW on its way south, where it eventually joins the Circumpolar Current (Fig. 4.20). Deflection by Coriolis Force results in the *Western Boundary Current,* whose effects are seen in water depths of about 4900 m at the base of the Continental Rise off New York, for example. Coarsening of silts and aligned magnetic grains, mud waves, ripples, and scour marks around morphological obstacles are indicators for its action, forming *"contourites".*

Deep *contour currents* were investigated in detail during the "HEBBLE" Experiment, about 75 km east of Woods Hole in some 4700 m water depth (A.R.M. Nowell et al., Marine Geol., Spec. Issue, 99, 3/4, 275, 1991). A surprising result was the discovery of strongly pulsed *"abyssal* or *benthic storms"* lasting between 2 and 20 days and occurring several times a year. Currents of such "storms" can reach velocities of 15–40 cm/s, measured 10 m above bottom, and transport sediment concentrations of 3.5–12 mg/l, with particle diameters of 0.004 mm on average, several meters above bottom. During these storms, erosion may remove the uppermost 5–10 cm of sediment, leaving a smooth bottom, but after a few weeks, bioturbation re-establishes a rough microtopography. This activity by benthic organisms facilitates erosion during the next storm.

Drift-like storm deposits hundreds of meters high, as well as hiatuses in deep-sea sediments, may result from these processes. Obviously, the existence and nature of these storms are of great interest in connection with plans to dump noxious waste into the deep sea (Sect. 10.5.5).

At this point, it seems that these dramatic abyssal storms are restricted to the neighborhood of the axes of very cold bottom-water flows, as on the west side of the North and South Atlantic Basins, around South Africa, in Circum-Antarctic waters. Also, benthic storms occur in regions with strong large eddies in surface waters.

If confirmed by further investigations, these results raise the exciting possibility of a direct link between surface and deep-sea currents.

During the last Ice Age the production of NADW was greatly reduced, for at least two reasons. One, there was less evaporation in the North Atlantic and less export of water vapor to the North Pacific. Thus, the necessary high salinities to make the water heavy enough to displace existing bottom water were not achieved.

Two, the Norwegian Sea probably was largely covered with pack ice which blanketed the water surface and prevented rapid cooling. Thus, the deep circulation in the Atlantic (and probably also in the Pacific) was entirely different during glacials from what it is today.

On top of the cold, deep waters there is another layer before we reach the warm surface waters. This is the *intermediate water,* which sinks in the subarctic and subantarctic open ocean convergences, and forms the bottom for the subtropical gyres. Along the ocean margins, where the intermediate waters intersect the continental slope, the flow tends to be in a direction opposite to that of the surface waters.

In general, the frictional interaction between bottom currents and the sea floor results in a *"benthic boundary layer"* which is on the order of a few hundred meters thick. Compared with the overlying deep water, it has a high content of suspended matter, and is characterized by turbulent motion. Also, its chemistry is distinct. This layer, and the bioturbated mixed layer at the sea floor, form a system within which seawater/sea-floor interaction is a dominant process controlling sedimentation patterns.

4.3.5 Exchange Currents. Geologically highly significant are the currents which provide for the exchange of waters between marginal, semi-enclosed seas, and the open ocean. The exchange entirely dominates the chemistry and fertility of the marginal basin and hence its sedimentation (see Sect. 7.6 and Fig. 7.12).

In arid zones, excess of evaporation over precipitation produces heavy water in the marginal sea; as it flows out over the sill, it is replaced by surface water from the open ocean. This is the case in the Mediterranean, and this circulation is called *anti-estuarine.* The salty Mediterranean waters are found at 1500 m depth over the entire central Atlantic, and profoundly affect the development of abyssal and deep water masses (Fig. 4.21a). A similar situation obtains, on a smaller scale, for the Red

Sea and the Persian Gulf. Here the saline waters sink to intermediate depths below the thermocline, and affect the development of the oxygen minimum in the Arabian Sea. Basins with *anti-estuarine circulation* tend to collect *carbonate* on the sea floor, and tend to discriminate against organic carbon, opaline silica, and phosphatic deposits.

The influx of Atlantic surface waters into the Mediterranean is a strong current and a huge river: 50 times the Mississippi at flood stage. It is said that the Battle of Trafalgar (1805) owed its outcome to the Gibraltar Current. The reason given is that the French and Spanish fleet had to wait for favorable winds to buck the Current, while Nelson was able to assemble his fleet and poise it for strike. In modern times, the submarines invading the Mediterranean had to struggle against the deep outflow. Already in 1820, the British Admiral W. H. Smyth observed the in- and outflow through the Gibraltar Straits, and gave the correct explanation. It was not, however, generally accepted for another 50 years of discussion.

The inverse situation – shallow current out of, deep current into the basin – is typical for estuaries and fjords, and occurs on a large scale in the Black Sea and the Baltic Sea. Excess precipitation over evaporation is necessary to develop this circulation. At the Bosporus, the connection between the Black Sea and the Mediterranean, fishermen have known since time immemorial that their nets would pull toward the Mediterranean at the surface and toward the Black Sea at the bottom. L. F. Marsili (1681), pioneer in marine geology, confirmed the effect by measurement and correctly explained the currents as being due to density differences. The light excess water flows out of the Black Sea at the top; the heavy saline water from the Mediterranean falls toward the bottom of the Black Sea, displacing the less saline water above. Ultimately, anaerobic conditions result from this circulation, called *estuarine,* and organic-rich deposits accumulate on the sea floor. The hows and whys are discussed in Section 7.6.

4.3.6 Turbidity Currents. We have already discussed the activity of mud-laden, gravity-driven currents in connection with the origin of continental slopes, submarine canyons and deep-sea fans (Chap. 2.10). These currents differ from the others here listed in several important ways: (1) they are strictly episodic, (2) they derive their power from excess density due to suspended sediment, and (3) they run downhill, transporting large amounts of material.

Further Reading

McCave IN (ed) (1976) The benthic boundary layer. Plenum, New York
Komar PD (1976) Beach processes and sedimentation. Prentice-Hall, Englewood Cliffs NJ
Allen JRL (1982) Sedimentary structures, their character and physical basis. Elsevier, Amsterdam
Pickard L, Emery WJ (1982) Descriptive physical oceanography – an introduction, 4th ed. Pergamon Press, Oxford
Hollister CD, Nowell ARM (eds) (1985) Deep ocean sediment transport, vol 1. Elsevier, Amsterdam
Hollister CD, McCave IN, Nowell ARM (eds) (1988) Deep ocean sediment transport, vol 2. Elsevier, Amsterdam

5 Sea Level Processes and Effects of Sea Level Change

5.1 Importance of Sea Level Positions

When studying sedimentary rocks on land, the first question a geologist will ask is whether the sediment was laid down above or below sea level, that is, whether or not it is of marine origin. For marine sediments, the next question usually is about the depth of deposition, that is, about the *position of sea level* relative to the sedimentary environment. On the present sea floor, depth of deposition rather dominates the major facies patterns of the material accumulating on it: the size distributions of clastic sediments, the chemistry of biogenous and authigenic matter, the distribution of benthic organisms. For the past, *sea level fluctuations,* on scales between thousands and millions of years, dominate the calendar of geologic history (Fig. 5.1).

Sea level fluctuations are of two kinds: *global* and *regional.* Global fluctuations produce contemporaneous transgressions and regressions on the shelves of all continents. These changes in sea level are called *eustatic,* they originate from changes in the volume of ocean water or in the average depth of the ocean basin. Examples are changes in ice volume, or changes in sea-floor spreading rates, as we shall see. Regional fluctuations consist in transgressions and regressions on one particular shelf; they are produced by regional sinking or uplift of the shelf. Hence, such sea level fluctuations are called *tectonic,* with the understanding that the tectonics are of regional importance only.

Where the sea level intersects the continental margin, physical, chemical, and biological processes are of high intensity. Waves, tides, and currents show maximum activity. The productivity of tidal lands and of the littoral is exceptionally great, and the sediment is intimately associated with rapid nutrient cycling, gas exchange, and life processes in general. Furthermore, sea level is the baseline of erosion and deposition: exposed areas erode, submerged areas build up. The erosional and depositional processes at and near sea level to a large extent determine the *coastal morphology* that we see. They also leave their distinct imprint in the record; they are *sea level indicators.*

On a larger scale, the position of sea level with respect to the global hypsographic curve is of great importance. It determines the degree to which shelves are submerged. Flooded shelves absorb more sunlight than exposed ones, adding heat to the global budget.

Also, submerged land experiences practically no chemical weathering, which normally keeps down CO_2 levels in the atmosphere. Thus, CO_2 rises and climate warms

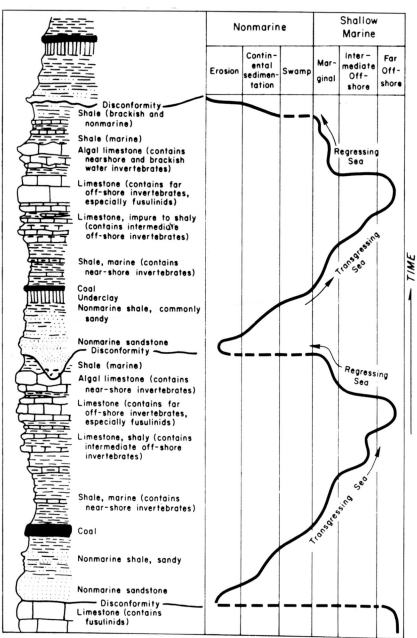

Fig. 5.1. Sea level fluctuations: calendar of geologic history. The diagram shows a section of the late Paleozoic cyclothem successions in Kansas, which consist of alternating marine and nonmarine sediments. Description of facies to the *left*; interpretation in terms of depositional environment to the *right*. The interpretation uses "Walther's Rule" introduced a century ago, which states that only those sediment types that are neighbors geographically will be neighbors in stratigraphic succession. [R. C. Moore, as drawn by J. C. Crowell, 1978, Am J Sci 278:1345]

when sea level rises globally. In a very simplified way, one might say that climate is mild during times of high sea level, and harsh during times of retreat of the oceans into its basins. Sea level fluctuations, then, are closely tied to *paleoclimatic evolution.*

Global changes in sea level also have important bearing on the production of hydrocarbons and coal. Sediments rich in organic matter, which have the potential to deliver petroleum when heated under pressure (Sect. 10.2) accumulate especially during periods when the continents are extensively flooded. The black shales of the lower Jurassic and the middle Cretaceous derive from such widespread transgressions. Coal deposits typically form during more regressive periods, when low-lying coastal plains are available for growth of vegetation in swamps and moors. Thus, sea level fluctuations are of interest in *economic geology.*

Quite generally, the rates of erosion of the continents and the sites of deposition in the ocean depend much on continental relief and position of sea level with regard to the shelves. How sediment is transported to the deep ocean is also controlled by sea level. During high stands, the transport of sediment by turbidity currents will be reduced. Turbidity currents depend on supply of mud to the outer edge of the shelf. This supply is highest when sea level is low, and is greatly reduced during high stands when submerged shelves and estuaries trap the material delivered by rivers. Thus, the types of *sediment bodies* found at the base of the *ocean margins* should depend on the history of sea level fluctuation.

Changes of sea level positions through time, then, are of fundamental importance in the geologic sciences. The dynamics of coastal morphology, the mapping of sea-level indicators, the interaction of sea level and climate, and the origin of sediment bodies in continental margins are topics which must be considered. However, sea level is also of the utmost interest to people living at the coast: their livelihood and their very survival depend on various manifestations of sea level change. This is true for both short-term fluctuation (tides, storm waves, tsunamis) and long-term trends, such as the *isostatic uplift* in Scandinavia or Canada, resulting from the removal of the ice caps between 15 000 and 8000 years ago, or the steady sinking of major delta regions (Mississippi, Nile, Ganges, Rhine). Also, the current general rise in sea level due to *global warming* over the last 100 years which results in expansion of seawater, is of concern. The rise is of the order of 2 cm per decade (Fig. 5.22).

We shall next review the evidence for sea level processes and fluctuations, starting with the short time scale and the obvious, and proceeding to the Pleistocene and pre-Pleistocene changes. At the end we append a story on Venice: the old city is slowly but surely sinking below the sea.

5.2 Sea Level Processes and Indicators

5.2.1 Wave Action. The most conspicuous indicators of sea level are those associated with wave action. Familiar examples are wave-cut terraces, and all kinds of beach deposits. We have already discussed some of the pertinent processes (Sects. 4.2.1 and 4.2.2).

Wave action leaves both erosional and depositional witnesses in the record. Many coastlines are marked by a steep escarpment where the land ends and the sea begins.

Fig. 5.2. Wave-cut terrace at Enoshima, Pacific coast of central Japan. [Photo E. S.]

Such escarpments or *sea cliffs* commonly are the result of regional uplift and wave attack. The wave-cut platform in front of a cliff is the most conspicuous lower boundary of wave erosion. It may be thought of as the floor of a series of notches and caves which are cut into the retreating cliff (Fig. 5.2). The rate of growth of such a platform varies greatly; it depends on the force of the waves, the resistance of the cliff material, and the time available for cutting. In Southern California, rates of cliff retreat are between 1 and 100 feet per century, for example. Rates of sea cliff retreat on the order of 1 m/yr are common on certain coasts of the North Sea in the Atlantic, where stormy winter seas eat their way into unconsolidated glacial deposits. In England, many coastal villages have been lost through the centuries, due to erosion.

Waves transport and rework the sediment, changing its texture by sorting and influencing its structure, for example, by producing ripple marks (Fig. 5.3). Caution is indicated, however, since ripple marks also occur on the deep-sea floor, far below sea level (Fig. 5.3b). Thick enrichments of heavy minerals, so-called *placers*, depend on a process which only exists on a beach (Sect. 10.3.3). Well-sorted skeletal remains – shell pavements or *coquina* – also are indicators of wave action. Waves usually influence the sea floor only down to about 10 to 20 m. Even great storm waves have a wave base rarely deeper than 30 m, although wave effects on sediments have been reported for depths greater than that (Sect. 4.2).

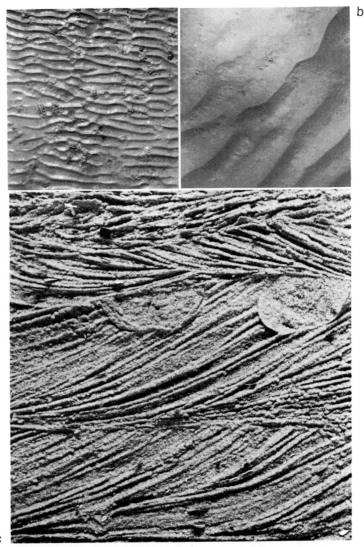

Fig. 5.3 a–c. Ripple marks as wave indicators. **a** Oscillation ripples, wave length about 5 cm, with *Arenicola* fecal strings: tidal flats. North Sea. **b** Oscillation ripples, wave length 30 cm. Top of Sylvania Sea Mount near Bikini Atoll, depth 1500 m, calcareous ooze. **c** Internal structure of current ripples, sandy tidal flats. German North Sea Coast. Frame is 22 cm wide. [Photos E. S. **a,** H. W. Menard **b,** H. E. Reineck **c**]

Waves can also be responsible for *producing* certain types of sediments, mechanically as well as chemically. Polished beach shingles and calcareous oolites are examples. Calcareous oolites are characteristic for unusually warm and saline environments (Chap. 3.7.2). Apparently they grow only within the zone agitated by waves and tides.

5.2.2 Tides and Storm Action: the Intertidal Zone. The tides, those long waves made by the Moon's daily passage, raise and lower the sea level along the coasts by a few centimeters or by many meters every day. The tides originate in a complicated interplay between the rotation of the Earth-Moon system about its common center, the system's rotation about the sun, the Earth's spin, and the topography of the floor of oceans and seas. The tides which we witness at the coast are large rotational waves radiating from points of no motion situated in the central areas of the ocean basins. The amplitudes of the waves are highest close to the coast (Fig. 5.4).

Thus, tides are strictly tied to astronomical forcing and are shaped by basin morphology. If the distribution of their frequencies and amplitudes can be reconstructed for the geologic past, we can obtain important clues about changes in the Earth-Moon system, and about basinal morphology.

Here, we are concerned with the "indicators" of tidal action in a general sense. Of course the classic sea level indicator facies is represented by the whole of the deposits of the intertidal zone, which is alternately submerged and exposed by the tides. We have already introduced one such type of broad intertidal, the *wadden* (Fig. 4.17). Biological features will be discussed in Sections 6.2–6.6.

Tidal flats can be associated with river deltas as in the Bay of Bengal and at the mouth of the Amazon River, or with estuaries, as in the Gulf of Korea or around the North Sea. Characteristic small-scale features of these sediments are flaser- or lenticular bedding, bimodal ("herringbone") cross-bedding, a variety of erosional and bioturbational markings, channel lag deposits, and graded laminae.

Rain imprints and desiccation cracks may also occur. On a larger scale, a landward decrease of grain sizes due to decreasing current activities and time of water cover marks a contrast to normal offshore conditions.

Intertidal flats and adjacent areas are subject to rapid changes in the conditions of sedimentation. Peak winter storms wreak havoc along the coastal lowlands of the North Sea from time to time. In 1362, an enormous storm, the "Great Man Drowning", raised the sea level by about 6 m along the Friesian Coast, where Germany borders with Denmark. The storm flood created new access for the tides into the salt marshes and moors lying behind natural barriers, inland from the intertidal flats. Soon the tides brought mud which accumulated on top of the marshes and the peat, producing a *storm layer*. To the people living there, this geologically common event meant destruction of homes and pastures. In 1634, another fierce storm produced widespread flooding, destroying villages and killing thousands. The present coastal geography of the area is essentially a product of those two storms. A faint indication of plowed fields can be seen to this day on certain intertidal flats of the region. Some of the land has been reclaimed since, aided by the influx of sediment from the wadden, and by the building of dikes.

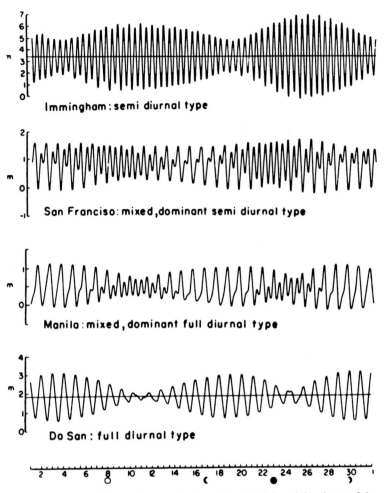

Fig. 5.4. Tidal records from harbor cities. At *bottom* the dates (March 1936) and the phases of the moon. [A. Defant, 1966, in G. Neumann, W. J. Pierson, 1966. Principles of physical oceanography. Prentice-Hall, New Jersey]

In areas such as these lowlands, then, slight changes in sea level and the action of heavy storms bring marked changes in the type of sediment deposited. Typically, the record which is preserved on the slowly sinking floor consists of an intercalation of peat, salt marsh deposits, marine muds, and the sands and shells of the beach. Storm deposits are common within the intercalation. Similar sequences are quite familiar from the geologic record (see Fig. 5.1). They are characteristic sea level deposits in areas with high terrigenous sediment supply. The sea level controls the groundwater level by damming up and floating the freshwater. The ensuing lack of drainage can produce swamps from which peat bogs develop. In the geologic record these appear as coal beds between marine sediments.

a   b

Fig. 5.5 a, b. Desiccation cracks in algal mats, in a lagoon on the west coast of Baja California near San Quintin. **a** View from a sand dune, toward inner lagoon. Marsh vegetation, stressed by high salinity from evaporation, gives way to algal mats toward the intertidal. **b** Detail of algal mats (section 2 m wide). The mats separate into polygonal pieces and curl upon drying. Upturned rims collect evaporites. [Photos E. S.]

In muddy intertidal flats of the subtropics, similar bedding features can be observed. High evaporation rates can produce widespread desiccation cracks (Fig. 5.5). Other clues to an intertidal environment of deposition are rain-drop impressions, *pseudomorphs* of cubic halite crystals (that is, infilling of crystal forms by sand or mud), precipitation of gypsum and tracks of land animals on sediments with marine organisms.

The sequences of algal mats constructed by *cyanobacteria* (but also in subtidal, normal marine settings) are known back into the Precambrian, as *stromatolites*. They are preserved as irregular layers, finely laminated mounds, or stubby columns. Trapped sediments consist mostly of carbonate particles and eolian dust, but additional microbacterial mineralization cannot be excluded.

Cyclically laminated and thinly bedded rhythmites in the Late Proterozoic of South Australia indicate that 650 million years ago the year had some 400 solar days (as compared with 365 at present) and 13.1 lunar months (12.4). The day had 21.9 hours (24). Similar clues to a slowing of the *Earth rotation* – due to tidal friction – have been found in stromatolites, bivalves, and corals in younger formations.

5.2.3 Photosynthesis. A large number of different benthic organisms can indicate shallow water by their presence, but none as convincingly as those dependent on light. Photosynthesis can only proceed when sufficient light is available. The light intensity drops to 1 % of the surface value at anywhere between 10 and 200 m water depth – depending on the clearness of the water. Sessile plants, such as the geologically important calcareous algae and algal mats, generally occur no deeper than 100 m or so. Animals living in symbiosis with algae also indicate shallow water. These animals include large foraminifera, stone corals, even certain mollusks (Sect. 6.1.3).

5.3 Coastal Morphology and the Recent Rise in Sea Level

5.3.1 General Effects of Recent Sea Level Rise. The coastal regions of uplifted margins are largely shaped by the interplay between tectonic forces and marine processes, especially wave erosion. Raised marine terraces are a characteristic morphologic feature of such uplifted margings; the West Coast delivers many fine examples. In contrast, the coastal morphology of slowly sinking margins is entirely dominated by the processes associated with the sea level itself. In North America, the Gulf of Mexico and the East Coast provide prime examples of this type of morphology, in Europe the "Wadden Sea" mentioned earlier.

In order to understand the coastal landscape, we must be constantly aware of one central fact: the recent rapid rise in sea level which began about 15 000 years ago and lasted until about 7000 years ago (Fig. 5.6). The sea level rose in response to the melting of glacial ice, chiefly the Laurentian and the Scandinavian Ice Sheets (Antarctic melting is thought to have contributed a comparatively minor amount). The maximum addition of water to the ocean occurred some time between 14 000 and 9000 years ago. The sea level changed by approximately 120–130 m on the whole; that is, it rose from the average depth of the shelf edge to the present positions.

As mentioned earlier, sea level is rising right now, after a long period of stability during the Holocene. During the last few decades it has risen by some 1–2 mm per year, presumably as the result of thermal expansion of the upper ocean layers, due to a general warming since the turn of the century (Figs. 5.22 and 7.16).

The effects of the deglacial transgression on coastal morphology and sedimentation were varied and profound. On gently sloping shores the sea advanced very rapidly. In the upper Persian Gulf, for example, the coastline must have retreated by some 100 m per year, during the maximum rise of sea level. Whoever inhabited the Shatt-al-Arab at that time must have been very aware of the invasion of sea.

In temperate humid regions, coastal peat bogs grew upward with the rising groundwater level; finally they became flooded by salt water and covered by marine sediments. Dunes were eroded by the approaching surf, except where cemented by lime derived from shell material. Resistant matter left from the erosional process collected as a *transgression conglomerate* below the march of the waves, the *basal conglomerate* typical of many marine transgression sequences in the geologic record.

The exact time sequence of the rise of sea level during the melting of the northern glaciers has long been a matter of controversy, as the various interpretations in Fig. 5.6a indicate. Was it an even, rapid change? Did it occur in pulses, as indicated by terraces in many shelves and coral islands?

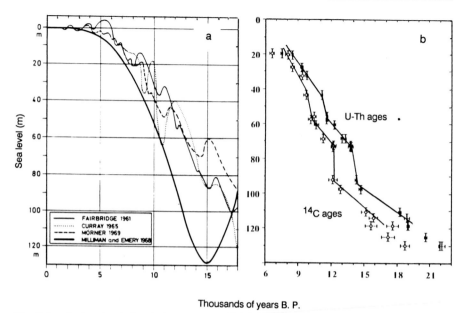

Fig. 5.6 a, b. Sea level rise during deglaciation. **a** Various hypotheses of how the sea level is supposed to have risen during deglaciation (about 15 000 to 9000 years ago), and afterwards during the Holocene. Dates are based on radiocarbon determinations. [E. Seibold, 1974, R. Brinkmann (ed). Lehrbuch der allgemeinen Geologie, vol 1, 2nd ed. F. Enke, Stuttgart], **b** Results from thorium-uranium dating on corals (*Acropora palmata* off Barbados) suggest that these radiocarbon dates are too young, by about 10 % for 10 000 years, and 20 % for 15 000 years. Two major pulses of sea level rise are indicated. [E. Bard et al., Nature, 1990 345:405]

Results of research during the past decade indicate that there were two major and several minor *pulses* of meltwater addition to the ocean. The pulses are reflected in rapid changes in oxygen isotope composition of the oceans (meltwater has a low $^{18}O/^{16}O$ ratio), and this change, in turn, is recorded in the shells of foraminifera and corals (see Sect. 7.3.2). The two major pulses are dated near 14 000 and 11 000 years on the absolute age scale (12 000 and 10 000 years by radiocarbon; see Fig. 5.6b).

We are much closer now than we were even a decade ago to going from sea level curves to the rate of *mass of water added per time* to the ocean. This will make it possible to attack questions related to regional tectonics. What is the local imprint of vertical coastal motion in each sea level curve? In northern latitudes, what exactly was the regional uplift of the crust in response to the removal of ice? How much was the crust depressed by the addition of water on the shelf? How did the Earth's rotation respond to the shift of weight from Canada and Scandinavia to the open ocean? How did this shift influence the *geoid,* the shape of the Earth, and its sea level? The answers will profoundly affect our understanding of coastal morphology.

An outstanding example for the effects of the rise of sea level is the general morphology of the East Coast of the USA. It is a coast of drowned rivers (Fig. 5.7a).

Fig. 5.7 a, b. Sea level rise and East Coast morphology. **a** Drowned river valleys, Cape Cod. Barrier beaches migrate into the entrance of the estuary, as spits, and restrict access to the open sea. **b** Close-up of a migrating spit, Buzzard's Bay. Migration is to the *left*. Note forced detouring of the tidal channel. [Photo courtesy D. L. Eicher]

Longshore sand transport (Sect. 4.2.2) tends to close off the estuaries which fill with sediment to make marshlands (Fig. 5.7b).

Drowned rivers can make good harbors: the East Coast is dotted by ports with excellent deep-water access. Drowned rivers are sediment traps; they tend to catch in their estuaries whatever material is brought from the continent. On the whole, there-fore, the shelf off the East Coast is starved of sediment. What we find on the sea floor there is largely relict material from the last ice age. The distribution of such *relict sediments,* naturally, is not very indicative of the present-day processes active on the shelf, except through the evidence for reworking.

With this general background, let us now consider the geologic-morphologic pro-cesses at sea level in more detail, first at river mouths, then at lowland coasts between the rivers.

5.3.2 River Mouths. A glance at a world map shows that river mouths can be indentations or protrusions in the coastline – that is, estuaries or deltas (Fig. 5.8). Tidal motion can reach far into the estuaries. Commonly, salt water intrudes along the bottom. This intrusion of the marine realm brings marine sediments upriver. Such sediments are easily recognized within the river deposits: echinoderm remains, for-aminifera, marine ostracods.

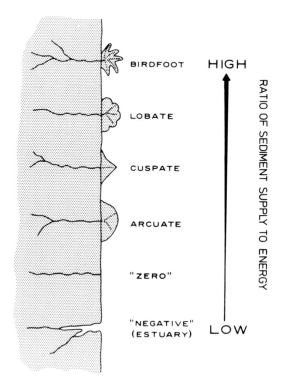

Fig. 5.8. Subaerial delta shapes as a function of sediment supply and dis-tributional energy (waves, currents). [J. R. Curray 1975, in A. G. Fischer, S. Judson, eds, Petroleum and global tectonics. Princeton Univ Press, New Jersey]

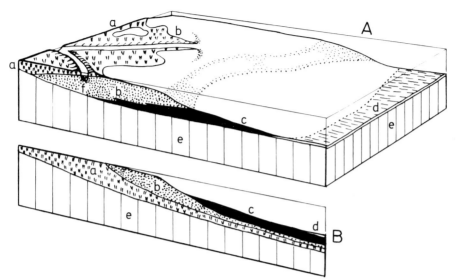

Fig. 5.9 a–f. Schematic cross-section through a birdfoot delta. **a** Inter-levee lowlands, marshes (levees are ridges confining the distributary channels); **b** delta front; **c** prodelta; **d** open shelf floor; **e** older base (may or may not be deltaic); **f** distributary channel with fill and levees. *A* Delta builds up and out, the sea retreats, the sedimentary sequence is "regressive" (*a* across *b*, *b* across *c*, etc); *B* The sea level rises and the delta retreats, the sedimentary sequence is "transgressive" (*a* below *b*, *b* below *c*, etc). Note that the regressive or transgressive nature of the sequence cannot be seen on the surface, but only by studying the sequence of sediments within the delta.

Thus, both the river and the sea bring sediment. Why then are the estuaries not filled in? River floods, tidal action, and especially the young age of estuaries are the cause. The drowned river valleys produced by the transgression have not yet come to equilibrium with the sediment load. Where the river valley runs across the shelf and into a submarine canyon on the slope, the filling-in of the river mouth is slowed, because sediment can travel out of the estuary onto the deep-sea floor. The Congo and the Hudson are examples. The best harbors, therefore, are those with a canyon offshore like New York, and those where strong tidal action keeps the outer river channel open, as in London, Bordeaux, or Hamburg.

For the geologist, who must learn to read the record, the river mouth of deposition, the delta, is of special interest (Figs. 5.9 and 5.10). Why a delta forms and not an estuary depends on many factors: for instance, low tidal activity as in land-sur-rounded seas (Mississippi and Nile deltas); high sediment load due to strong seasonal rains, and high erosion rates in the mountainous hinterland (the Indus, Ganges, Ir-rawadi or Huangho deltas).

When coast and sea level are stable, the delta builds out to sea, the marine facies retreats, the record shows *regression*. A borehole in such a delta shows a *regressive sequence:* shallow deposits over deep ones (Fig. 5.9A). If sea level falls, of course, the regression is accelerated. Conversely, a rise of sea level or a sinking coastline will

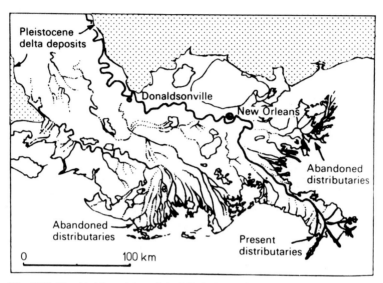

Fig. 5.10. The bird-foot delta of the Mississippi. The subsidence of the delta allows the sea to re-invade the areas of abandoned distributaries. [J. Gilluly et al., 1968, Principles of geology. W. H. Freeman, San Francisco, after H. N. Fisk]

ideally produce a *transgressive sequence,* which is the reverse order of the regressive one.

Deltas are economically interesting because of their *oil and gas potential.* High organic productivity and enhanced preservation of organic matter by high accumulation rates may form source rocks. Sands in distributary channel fills or offshore bars at the delta front can become reservoir rocks. General subsidence, aided by sediment loading, results in multiple stacks of deltaic sequences, caused by repeated progradation and abdandonment of delta lobes, each of them typically several tens of meters thick.

5.3.3 Lagoons and Barriers. Next to deltas we usually find low-lying coasts showing a facies zonation parallel to the coastline. Offshore bars or barrier islands commonly occur together with a sand beach (Fig. 5.11). The beach may be backed by dunes, which the wind piled up using sand from the beach. In turn, this beach-dune complex may form a barrier for interior lagoons, as is the case along much of the Gulf of Mexico and the East Coast. The narrow barriers can be tens of miles long. Rivers emptying into the laggons may cut one or several channels through such barriers, especially when flooding, chopping them up into a series of barrier islands. From the seaward side, storm waves may break through a barrier, forming *overwash fans* on the lagoon beach (Fig. 5.12). Tidal action keeps such channels open, building deltas on both sides.

How does barrier-and-lagoon morphology reflect changes in sea level? Again, the balance between the rise of sea level relative to the land and the supply of sediment

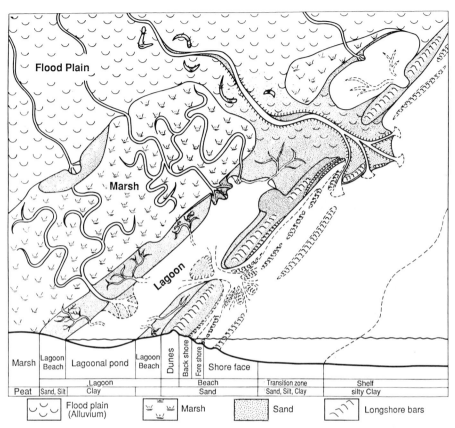

Fig. 5.11. Typical geomorphic features of a barrier-type coast. [H. E. Reineck, I. B. Singh, 1973, Depositional sedimentary environments. Springer, Heidelberg, based on a diagram by C. D. Masters, 1965]

to the coast is the crucial factor to consider. Since the sea level stabilized, about 6000 years ago, supply of material has been all-important. Off Galveston Bay, for example, the large supply of sand allowed a rapid building-out of the offshore barrier (Fig. 5.13). On the whole, however, barriers and lagoons must have migrated landward over the last 15 000 years.

The barrier-type coast is very abundant: of 244 000 km coastline, its share is about 32 000 km, that is 13 %. North America and Africa each have about 18 %, Europe only about 5 %. A stable broad shelf and a high supply of sediment would appear to be favorable for the development of barrier coasts, as illustrated by the barrier island coast of the Southern North Sea. Coastal uplift, narrow, dissected, and starved shelves are unfavorable.

5.3.4 Mangrove Swamps. During the recent rise of sea level, generally speaking, the various facies zones paralleling the coast migrated landward. In the tropics, one of the

Fig. 5.12. Overwash fans on St. Joseph's Island, Texas, produced by storm waves. Open Gulf to the *left,* lagoon to the *right.* [Photo courtesy D. L. Eicher]

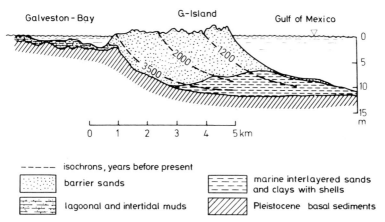

Fig. 5.13. Architecture of Galveston Island. Regressive sequence below the barrier beach near Galveston, Gulf of Mexico. Advance of barrier island is due to high supply of sand and relatively stable sea level. The dates (1200, 2000, 3500), are based on the ^{14}C content of shells. [J. R. Curray, 1969, in D. J. Stanley, ed, The new concept of continental margin sedimentation. Am Geol Inst, Washington DC]

a   b

Fig. 5.14 a, b. Mangrove vegetation on Bimini (Bahamas). **a** *Rhizopora* at low tide. The roots are covered at high tide. Plant about 2 m wide. **b** *Avicennia* with air roots. Width of view at proximal edge 2.5 m. [E. Seibold, 1964, Neues Jahrb Geol Paleontol Abh 120:233]

most impressive migrations of this type must have been the retreat – and landward invasion – of the mangrove swamps. Mangrove growth dominates many intertidal zones in the tropics (Fig. 5.14). Mangroves need year-round temperatures of 20 °C or higher. In equatorial regions with high annual rainfall, for example, off Cameroon and Guyana, mangrove forests form a broad belt offshore. Here, the mangrove forests blend into the tropical rainforests at high tide level. The expansion (and successive burial) of mangrove swamps during deglaciation due to a rising sea level, must have produced an organic-rich layer on many tropical shelves, much as peat growth did on temperates shelves. Such layers, when buried, will coalify. The coaly layers in ancient cyclothems (Fig. 5.1) apparently originated in similar fashion. Rapid burial of carbon extracts CO_2 from the atmosphere. The question of possible CO_2 changes due to changes in sea level is a topic of active research in marine chemistry and geology (see Sect. 8.5.6).

Mangrove swamps and other coastal habitats are quite sensitive to disturbance by human activities. Changes in the coastal environment, whether by nature or by human impact, can now be readily monitored using *remote sensing,* from satellites (such as GEOSAT 3, which circled the globe in 1986–1989). This type of information will greatly increase our appreciation of the rates of change in tidal flats, mangrove swamps, deltas, lagoons, barrier islands, and coral reefs, and also provide clues regarding the processes at work, including human impact through land use and pollution.

5.4 Ice-Driven Sea Level Fluctuations

5.4.1 The Würm Low Stand. In the last section we referred to the rapid rise of sea level, between 15 000 and 7000 years ago, which was caused by the melting of glacial ice (see Fig. 5.6) This rise is but one phase in the long series of sea level fluctuations of the Pleistocene. Sea level has been constantly changing over the last several hundred thousand years as a result of the waxing and waning of large continental ice masses. The rise of the last deglaciation, which was witnessed by our ancestors, was one of the biggest and fastest sea level changes ever, as far as is known.

It was the result of a maximum change of climate: from a peak cold period to a peak warm period.

About 20 000 years ago, during the last major ice age (the *Würm* in Europe, the *Wisconsin* in North America), enough water was tied up in the continental glaciers to depress sea level by some 120 m. Large shelf areas fell dry as a result of the ice-caused regression of the sea. Rivers crossed the shelves and entered the sea at the shelf edge, cutting backward into the shelf. Their sediment load was dropped in a narrow zone at the upper slope, became unstable there and slid, starting turbidity flows which rushed down the undersea canyons. In front of the ice, outwash plains (or moraine ridges next to the terminal ice tongues) developed on the exposed shelf. Dune fields evolved on shelves where the climate was favorable.

Land animals spread over the emerged shelves and used them as land bridges in some cases, for example, between Northeast Siberia and Alaska. Mammoths roamed over the large area covered by the North Sea today. Their remains are found on the sea bottom now, together with the tools of prehistoric hunters. Could these people of the late Stone Age have left a memory of the effects of rapid sea level rise in the ubiquitous legends of a Great Flood?

5.4.2 Pleistocene Fluctuations. There is an indication that, throughout the late Pleistocene, major transgressions were much more rapid than any of the regressions. In general, apparently, the buildup of ice (and hence the sea level drop) proceeded at a slower rate than the melting of ice. The evidence for this concept rests on the oxygen isotope record of pelagic foraminifera (Fig. 5.15).

How can we measure sea level changes in the chemistry of foraminiferal shells?

The principle of this method was introduced by C. Emiliani in 1955. (Pleistocene temperatures. J Geol 63, 538–578.)

The shells of the foraminifera consist of calcium carbonate, $CaCO_3$. Hence, they contain oxygen. The water within which the shells grow, H_2O, also contains oxygen. There are three kinds of oxygen, the normal one, with atomic weight 16, and two rare ones, with atomic weights 17 and 18. The oxygen-17 is very rare and we need pay no further attention to it. Through mass spectrometry, one can determine the ratio between oxygen-16 and oxygen-18 (commonly written ^{16}O and ^{18}O). The ratio of these two isotopes in the shells is in equilibrium with the ratio of the isotopes in the water in which the shells grew. In other words, if the $^{18}O/^{16}O$ ratio in the water changes, it will change similarly in the shell.

This fact relates to sea level change as follows. Every time the sea level drops, the $^{18}O/^{16}O$ ratio of the water increases, because the glacial ice is made of water which is impoverished in the isotope ^{18}O. Hence, seawater is enriched in ^{18}O during glacials (Fig. 5.15b). The $^{18}O/^{16}O$ ratio of the carbonate shells reflects this change in the chemistry of seawater. Temperature also influences the $^{18}O/^{16}O$ ratio of the shells. In

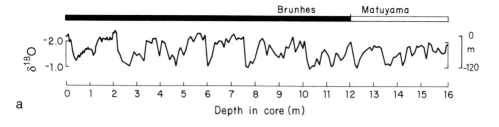

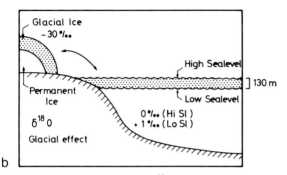

Fig. 5.15 a, b. Fluctuation of δ ^{18}O in the planktonic foraminifer *Globigerinoides sacculifer,* western equatorial Pacific. **a** *"Brunhes"*: present normal magnetic epoch, since 790 000 years ago. *"Matuyama"*: previous epoch, during which the Earth's magnetic field was reversed. Eight isotope cycles are visible within the Brunhes; the average duration is therefore 100 000 years. The temperature of surface waters is nearly constant through time, in the area where the core was taken. Hence, the isotope fluctuations are produced largely by the build-up and decay of northern ice masses, as shown in **b**. [Core data from N. J. Shackleton, N. D. Opdyke; 1973, Quat Res 3: 39]

Fig. 5.15a, however, the long sediment core on which the measurements were made comes from an area where the effect of temperature on the isotopic signal can be neglected in the present context.

Note that the oxygen isotope curve (Fig. 5.15a) indicates that sea level fluctuated between rather well-defined limits. Sea level was never much higher than right now, and it was never much lower than during the last glacial. There must be a climatic factor or factors which prevent a buildup of ice beyond a certain limit, and other factors which prevent additional melting once a certain amount of ice has been melted. We shall return to this puzzling observation in Chapter 9.

5.4.3 Effects on Reef Growth. The sea level fluctuations of the Pleistocene left their imprint also in the shallow water carbonates, notably in the tropical reefs. Each highstand of the sea level resulted in a buildup of reef carbonates, while the lowstands resulted in erosion. In fact, the origin of atolls, those ring-shaped islands dotting the central Pacific, has been contemplated under this aspect (Fig. 5.16). Thus, while Darwin's hypothesis of submergence is correct in a general sense (see Sect. 7.4.3, Fig. 7.8), the influence of the fluctuating sea level must not be forgotten.

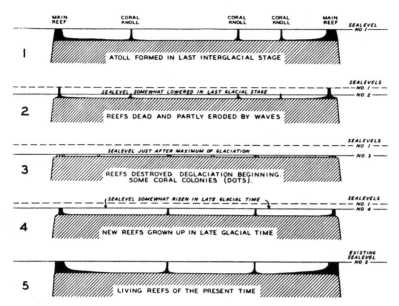

Fig. 5.16. Daly's Glacial Control Theory of coral reefs. The time sequence is from *1* (124 000 years ago) to *5* (the present). The ring shape of atolls is due to the more favorable situation for coral growth at the edges of the island (due to cleanness of water, and high food supply). The knolls in the lagoon grow up on slightly elevated, mud-free ground. [R. A. Daly, 1934, The changing world of the ice age. Yale Univ. Press, New Haven]

On rising shores with reef belts, such as Barbados in the Caribbean, the fluctuating sea level translates into raised reef terraces. The terraces correspond to the high stands in sea level. They have been dated, by measuring the concentration of radioactive uranium within individual coral heads, and the product of uranium decay, the element thorium. Since at the time of growth of the coral there is essentially no thorium present, one can tell how much decay of uranium took place from the ratio of these radioactive isotopes (Fig. 5.6b). Results show that the uplifted corals grew during several high sea level stands, namely 124 000 and 103 000 years ago – very much as expected from the oxygen isotope curve shown in Fig. 5.15.

Ice-driven sea level fluctuations seem to have been less pronounced in the earlier Pleistocene than in the later part of the period. However, even back to the Miocene there were such fluctuations, since large-scale ice buildup apparently started on Antarctica some 15 million years ago. As far as we can tell from the oxygen isotope record (see Chap. 9), the strength of the sea level fluctuations increased at 6 million years ago (in the latest Miocene) and again about 3 million years ago, when northern glaciations set in.

5.5 Tectonically Driven Sea Level Fluctuations

5.5.1 Sea Level and Sediment Bodies. Sea level fluctuated considerably all through the Phanerozoic, even during periods when apparently no ice was present. As to the present sea floor, fluctuations since the Jurassic are of special interest: essentially they determined the sequence of sedimentary layers within the continental margins.

The thick sediment stacks in the "passive" continental margins have been much studied for economic reasons. When a margin sinks more or less continuously, coastal sediment bodies must reach great thickness, provided the sediment supply keeps pace with subsidence, and deposition remains locked into the sea level. Tertiary sandy beach deposits can attain up to 1500 m thickness in the northwestern Gulf of Mexico, for example, also widths of up to 40 km and lengths of several 100 km (Chap. 5.3.2).

The significance of the sand bodies in economic geology lies in their porosity and permeability. Thus, they can retain (and deliver) great quantities of water, petroleum, or gas. It is for this reason that the origin, dimensions, and properties of coastal sand bodies have received much attention from marine geologists as well as oil geologists. Both drilling and seismic exploration help define their extent in the coastal areas of interest.

To interpret the sequences of marine sediments on land, in the margin, and on the deep-sea floor, and for economic reasons as well, we would like to know how sea level fluctuated over the last 150 million years. However, in as much as the sea level variations within this geologic period were not driven by the growth and decay of ice caps, they were not reflected in the isotopic composition of seawater. We cannot, therefore, find them in the isotopic composition of the foraminifera in the manner indicated earlier.

How then can we measure these fluctuations?

5.5.2 Reconstruction of Sea Level Changes. The intensive world-wide exploration of continental margins by seismic profiling has recently led to the realization that the sediment-stacking patterns in margins of different ocean basins are quite similar – hence, it is assumed, they must be due to global sea level variation. Using this hypothesis, P. Vail, R. M. Mitchum, B. U. Haq, and their associates developed a method to derive sea-level fluctuations from the geometry of sediment layers, as recognized on seismic reflection records. For the time since the beginning of the Triassic they have found more than 100 major global sea-level changes, about one for every 2 million years, on average. The basic idea is this: during a relative rise of sea level (transgression), sediment layers expand into shallower water, and they become wider as they build up. During a fall of sea level (regression) the reverse occurs, and erosion sets in on the shelf. Erosion, of course, produces a *hiatus:* a surface which joins older and younger sediments in a discontinuous way. Hence, the course of regression is poorly documented, and it looks as though it happened rapidly. It is much like cutting a section out of a movie: the change between "before" and "after" becomes very sudden.

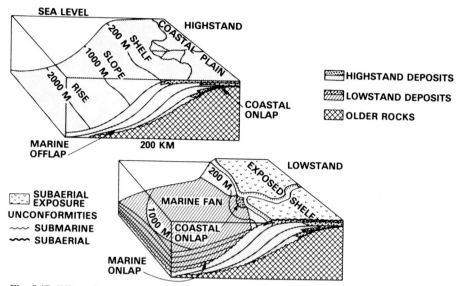

Fig. 5.17. Effect of sea level position on depositional pattern at the continental margin. Highstand and lowstand of sea level. [P. R. Vail et al., 1977, Am Assoc Pet Geol Mem 26: 49]

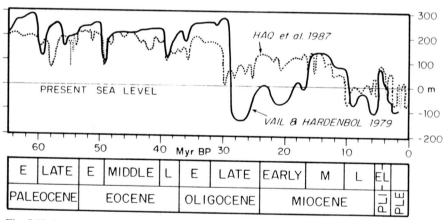

Fig. 5.18. Relative changes of sea level as deduced from the geometry of margin sediment bodies. The underlying model *(solid line)* is that given in Fig. 5.17. [P. R. Vail and J. Hardenbol, 1979, Oceanus 22:71]; *Dotted line* [Haq et al. 1987, Science 235:1156.] [As drawn by W. H. Berger and L. A. Mayer 1987, Paleoceanography 2,6,620]

The depositional patterns during a highstand and a lowstand of sea level are illustrated in Fig. 5.17. In essence, the sediment pile moves into deeper waters during the lowstand, and its geometry changes correspondingly. The construction of the apparent sea level cycles (see Fig. 5.18) is based on this changing geometry of sediment stacks (Fig. 5.19). The falls in sea level appear as instantaneous events,

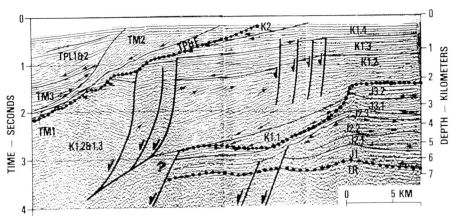

Fig. 5.19. Reflection seismic line offshore Northwest Africa. The Triassic *(TR)*, Jurassic *(J)*, and Cretaceous *(K)* Supersequences are separated by *dotted lines. Arrows* indicate marine on- and offlaps. During the Jurassic six seismic sequences are separated by minor unconformities. With the beginning of the Cretaceous, carbonate sedimentation including reef building at the shelf margin (e. g. *J3.1*) ends. The subsiding continental margin after the opening of the Atlantic has been covered afterwards by thick sequences of deep-water shales *(K1.2)* and deltaic sandstones. They were truncated by erosion with the beginning of Tertiary *(T)*. Complicated Tertiary sequences. Faults marked by *half arrows.* The fundamental pattern is similar to that observed in the Gulf of Mexico. [Based on H. Füchtbauer (Ed), Sedimente und Sedimentgesteine, 1988, Schweizerbart, Stuttgart, 859; after P. Vail et al. 1977, Amer Assoc Petrol Geol Mem 26]

due to the presence of erosional (or nondepositional) hiatuses. While sea level might recede quickly, it cannot do so instantaneously, of course.

The *"Vail sea level curve"* is a useful tool for the correlation of seismic stratigraphies of continental margin sediments. To what extent this curve reflects the true global sea level variations is not known. One problem is that the rate of sediment supply (which is a factor more or less independent of sea level) must play an important role in controlling the geometry of the sediment bodies. Another is that the rate of sinking of the margins must be considered. Quite possibly, sea level fell continuously since the latest Cretaceous and only varied its *rate* of falling. When it fell slowly, the sinking passive margins overtook it, and accumulated transgressive sequences. When sea level fell quickly, the passive margins could not keep up: the result was regression.

Other difficulties also arise in the reconstruction of sea levels from seismic records. How does changing geoid configuration influence regional sea level variations, and can this effect overprint global eustatic curves? How exact is the timing? Dating of seismic layers is by correlation, and uncertainties in absolute ages can be considerable, using this method.

5.5.3 The Cause of Change. If sea level fluctuated through time, we must search for the cause or causes of such fluctuations.

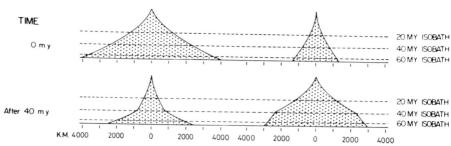

Fig. 5.20. Relationship between volume of Mid-Ocean Ridge and spreading rate. *Upper* cross-sections of ridges spreading at 6 cm/yr *(left)* and at 2 cm/yr *(right); lower* cross-sections of the same ridges 40 million years after a change from 6 cm/yr to 2 cm/yr *(left)* and a change from 2 cm/yr to 6 cm/yr *(right)*. The changes in ridge volume must produce corresponding changes in sea level. [W. C. Pitman, 1979, Am Assoc Pet Geol Mem 29: 453]

The obvious way to change sea level (if not by ice) is to change the average depth of the sea floor: if the floor shallows, the sea level rises; if it deepens, the level falls. We have seen that the depth of the sea floor is tied to its age: to change the depth, we must change the age. To decrease the age (and hence cause a transgression), we must replace old sea floor with young. This can be done by increasing the total mass of new lithosphere formed per year, that is, by increasing sea-floor spreading rates or by increasing the length of the Mid-Ocean Ridge and the trenches, or both (Fig. 5.20).

Global tectonic events which led to a change in the average age of the sea floor, and hence its depth, apparently happened in the past. In the opening of the Atlantic, for example, young sea floor was generated by the new spreading center in the Atlantic, while old floor was subducted elsewhere in the Pacific (or else the entire globe would expand). As the Atlantic grows, the average age of its sea floor increases all the time – at some point it becomes older than the average age in the Indo-Pacific and hence starts increasing the global average. Further growth of the Atlantic then results in a drop of sea level. There are indications, from the magnetic stripes on the sea floor, that global spreading rates in the Late Cretaceous may have been much higher than now. It has been proposed that the high sea level stands of the Late Cretaceous were caused by such a fast spreading rate.

Additionally, large intraplate submarine outpourings of basalts during this period, especially in the southwest Pacific, have to be considered.

The changes which can be produced by replacing old sea floor with young, and vice versa, are large but gradual. How can sea level be changed rapidly?

Mountain-building with shallow ocean crust, or with continental crust, is one way. In this process, shallow crust (which is stacked up within mountain ranges) is removed and replaced by deeper sea floor which will cover itself with a thicker layer of water, drawing down the general sea level. If the Tibetan Plateau represents "doubled" continental crust, for example, the corresponding sea level fall is about 40 m.

The quickest way to change sea level is to fill or to empty an isolated ocean basin. We know (thanks to Leg 13 of the Deep Sea Drilling Project) that the Mediterranean dried up intermittently between 5 and 6 million years ago. The water had to go elsewhere: global sea level was raised by about 10 m whenever the Mediterranean dried up. Conversely, when ocean water rushed in to fill the empty basin, global sea level must have dropped by the same amount. Since salt deposits have been found, or are thought to be present, in other ocean basins also (North Atlantic, South Atlantic), geologically instantaneous transgressions and regressions may have been quite common after the break-up of Pangaea.

Even a "geologically instantaneous" transgression would not necessarily be spectacular to a human observer. A transgression which is geologically very fast (although not produced by the hypothetical isolated basin effect) is proceeding right now in Venice. To this phenomenon we turn in Section 5.6.

5.5.4 Geologic Aspects. Many sedimentary rocks on land were formed in shallow seas, and therefore depend greatly on tectonically driven sea-level fluctuations, whether global or regional in origin. The foregoing sections illustrate the complexity of the processes involved, which geologists face when attempting to interpret the record. The coastal zone, we have seen, is an extremely active environment, where there is intense interaction between the forces of land and sea. Even small morphological features – a hill, a trough – become very important in modifying the wind-driven waves and currents which shape the bottom and control sedimentation. Climatic variations and hinterland relief dictate sedimentary conditions, as shown in the overall contrast between low-latitude carbonate shelves and high-latitude shelves receiving continental debris, or being deeply eroded by ice. Marginal seas, separated from the main ocean, are even more strongly influenced by regional climatic and tectonic circumstances. Generally, faunal diversities decrease in these environments, compared with the open shelf, due to more extreme fluctuations in living conditions.

In our task to learn from the present for the past, we face an important obstacle. The present sea floor is far from being an ideal training ground for aspiring geologists, because of the dramatic cycles of geologically rapid sea level changes during the last half million years, and especially the rise within the last 20 000 years. Equilibrium between the parameters of the system water/depth/climate/sediment rarely obtains as already mentioned in Section 2.7. To demonstrate this basic fact around the world would require a detailed description of a large number of case histories – which is quite beyond the scope of this book. Our philosophy, instead, is to stress principles, but to alert the reader to the complications which make their application difficult.

The postage stamps reflect the international concern about the fate of Venice. They picture the Palace at the Canale Grande and San Marcus Cathedral, threatened by the sea.

5.6 Sea Level and the Fate of Venice

5.6.1 Venice Is Sinking. Ancient Venice, with its San Marcus Cathedral, palaces, and canals, is slowly sinking below sea level.

Can Venice be saved? Let us take a closer look at the formidable scale of problems associated with the task of keeping the sea out. The necessary information has been collected by the *Laboratorio per lo studio della dinamica delle grandi masse* of the *Consiglio Nazionale delle Ricerche.*

Venice is at sea level. The city lies at the rim of the Po delta, in a lagoon protected by a barrier island, the Lido. The Lido is breached by tidal inlets (Fig. 5.21). The signs of sinking of the city are everywhere: docks must be built up; entrances are bricked in; steps are under water; during high water the city is flooded. The extent of sinking was determined by careful geodetic measurements starting from Treviso as a fixed point. The rate increased dramatically during the 1960s. Between 1908 and 1925, Mestre-Marghera sank 0.15 mm per year, on the average. Between 1925 and 1952 it was 0.7 mm, and between 1952 and 1968, 3.8 mm per year. In parts of the region around Venice, the ground sank by 120 mm between 1952 and 1968, that is, 7.5 mm per year!

5.6.2 What Are the Causes for the Subsidence?

1. One reason is the general sinking of the Po Delta. By drilling, it was found that there are more than 3 km of Quaternary sediment below the central delta. If we set the Quaternary approximately equal to 1.8 million years (see Appendix A3), we obtain an average rate of subsidence of almost 2 mm per year. Venice lies at the rim of the delta – hence it sinks a little more slowly, say, 0.5 mm per year on average.
2. The Quaternary sediments underlying the region are partly marine and partly continental, indicating fluctuations in the position of the coastline relative to sea level. Tectonic movements and changes in sediment supply of the Po brought about these fluctuations. At the present time, the shoreline tends to advance inland.

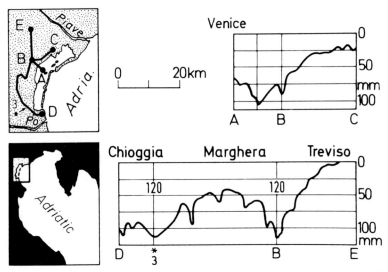

Fig. 5.21. Amount of subsidence of ground in the vicinity of Venice. *Upper left* Location map of Venice and surroundings; *Right* subsidence profiles along the lines *A B C (upper)* and *D B E (lower)*. Difference in elevation (in mm) refers to period between 1952 and 1968. Treviso *(E)* is taken as fixed point for the geodetic measurements; hence its change in elevation (if any) is defined as zero. Note high rates of subsidence (about 7.5 mm/yr) in the industrial region of Marghera (near *B*) and in the Po delta near Chioggia *(D)*. The center of the city of Venice subsided by about 80 mm, i. e., 5 mm/yr. [Data from R. Frasetto, 1972, CNR-Lab Stud Din Masse Tech Rep no 4].

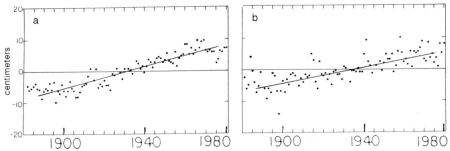

Fig. 5.22. Recent rise of sea level in two regions. **a** Europe and Africa. **b** West coast of North America. *Points* are regional averages of the annual sealevel anomaly. Long-term trends fitted by eye (note difference in slope). [T. P. Barnett, 1984; J Geophys Res 89: 7980; trends added]

3. Compaction of the sediments adds a component to subsidence, which can be locally reinforced by building and by infilling of lagoon areas. This may be the reason for increased sinking around the city's railway station and harbor.
4. The sand layers, of which most of the underlying sediment consists, are groundwater reservoirs *(aquifers)*. When groundwater is removed faster than it is replaced, the ground sags. The sag can reach over considerable distance – a kilometer or more – if the pumping is done from depths of several hundred meters.

Fig. 5.23. The flooding of San Marcus Square by more than 1 m in November 1966.
[Photo courtesy A. Stefanon, Venice]

In the last 20 years the groundwater level was lowered by 5 m in and around Venice, by 20 m in the industrial zone of Marghera. During this period, subsidence accelerated markedly. Perhaps this factor alone can account for the high sinking rates.

5. For the last 100 years the sea level has been rising globally by 1 to 2 mm per years as mentioned in Section 5.3.1 and illustrated in Fig. 5.22.

6. The normal tidal range is about 1 m at Venice. Storms can increase a high tide level considerably. On November 4, 1966, a storm tide reached + 1.9 m, flooding the San Marcus Square (Fig. 5.23). Normally, the lagoon would act as a buffer receiving the incoming water from the Adriatic Sea. However, large parts of the lagoon were filled above sea level to gain land for industrial purposes. The lagoonal area therefore is only about 70 % of what it was some 200 years ago. Shipping channels were deepened partly to more than 12 m. The concern that these actions will increase the influence of storms and storm tides appears justified.

5.6.3 What Can Be Done About It? The results obtained from the studies summarized suggest that the pumping of groundwater must be regulated. Also, the effects of filling the lagoon, deepening the channels, and changing the shapes of the inlets (providing ready access for storm tides) must be studied. Of course, the problem of removing waste in an economic and sanitary fashion must not be forgotten: the tidal action flushes the city's polluted canals. The possibility of regulating tidal flow and blocking storm surges from reaching Venice is being investigated using *sluices* at the Lido. Building and maintaining gates that are up to 300 m long and over 15 m high, and exposed to the sea while anchored in soft sediment, is a great technical challenge.

Not only Venice is threatened by these and similar processes, but many other large coastal cities such as Bangkok in Thailand and Manila in the Philippines.

Nothing can be done about the general regional sinking. We cannot control such geologic factors; we must live with them.

Further Reading

Reineck HE, Singh IB (1975) Depositional sedimentary environments. Springer, Berlin Heidelberg New York

Payton CE (ed) (1977) Seismic stratigraphy – applications to hydrocarbon exploration. AAPG Mem 26, Am Assoc Petrol Geol Tulsa, Okla

Einsele G, Ricken W, Seilacher A (eds) (1991) Cycles and events in stratigraphy. Springer, Berlin Heidelberg New York

Berg OR, Woolverton DG (eds) (1986) Seismic stratigraphy II – an integrated approach to hydrocarbon exploration. AAPG Mem 39, Am Assoc Petrol Geol, Tulsa, Okla

Wilgus CK, Hastings BS, Kendall CG, Posamentier HW, Ross CA, van Wagoner JC (eds) (1988) Sea level changes: an integrated approach. SEPM Spec Publ 42, Soc Econ Paleontol Mineral, Tulsa, Okla

6 Productivity and Benthic Organisms – Distribution, Activity, and Environmental Reconstruction

6.1 The Ocean Habitat

6.1.1 Diversity of Organisms. In the enormous living space provided by the sea, there are *plankton* (drifters), *nekton* (swimmers), and *benthos* (bottom-living organisms). Much of the benthos releases eggs and larvae into the plankton – this is the *meroplankton,* abundant in coastal waters. The larvae are dispersed by currents; they settle when their time has come, and start growing on the appropriate solid substrate. The meroplankton feeds planktonic and nektonic predators. Conversely, the plankton feeds the benthos. Thus there is an intimate ecologic relationship between free-swimming and bottom-living organisms. Ultimately, of course, benthic organisms rely on food produced in surface waters in the sunlit zone (Fig. 6.1). The notable exception is the deep-sea benthic community at the hot vents of the Mid-Ocean Ridge (see Sect. 6.9).

On the whole, fewer species developed in the ocean than on land. The living space of the ocean has fewer nooks and crannies than that on land, and the various parts of the world ocean are interconnected. Thus there is less opportunity for separate evolution from populations to subspecies to species. The greater diversity of land animals is entirely due to the proliferation of insects (> 75 % of one million animal species). Of the marine animal species (180 000) 98 % are benthic; only 2 % are planktonic or nektonic.

These numbers, of course, are subject to revision by continuing investigations. The general trends seem secure, although we do not share the opinion of Pliny the Elder (23–79 A.D.) who stated, ". . . in the whole ocean, as large as it is, there lives nothing that we do not know."

6.1.2 Productivity and Supply of Organic Matter. Sunlight and nutrients determine the growth and distribution of organism, because the food chain has its base in the marine algae. Matter is transferred along this chain from primary production to herbivores, to carnivores, and back to bacteria for recycling. Also, material is transferred downward in the water column, from the productive zone to the sea floor, mainly by *fecal matter* and by *aggregates* containing planktonic remains.

For every 100 gC produced in the sunlit zone, roughly 30 gC reach the sea floor on the shelf and the upper slope, but only 1 gC reaches the deep-sea floor (as indicated by studies with sediment traps). During the long transit downward (order of 2 weeks), most of the sinking material is oxidized and *remineralized*. (See Fig. 6.2.)

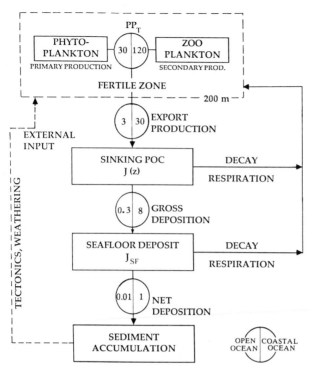

Fig. 6.1. Sketch of transfer of particulate organic carbon in the ocean, from primary production to burial in the sediment. *Numbers* are fluxes in gC/m^2/yr; within each circle the value to the *left* refers to typical open ocean conditions, that to the *right* to the coastal ocean environment. [W. H. Berger, G. Wefer, V. S. Smetacek, 1989, in Productivity of the ocean: present and past. Wiley, Chichester]

Despite the fact that so little organic matter arrives at the deep-sea floor, the abundance of benthic organisms rather faithfully reflects the patterns of primary productivity (shown in Fig. 6.3). Also, the depth effect itself is strongly reflected – the shorter the distance to the productive zone above, the more food arrives on the sea floor, and the more benthic biomass can be supported. Since productivity is highest along the ocean margins, the effects from organic matter production and depth-dependent delivery reinforce each other, so that there is a great contrast between benthic activity (including bacterial growth) on the sea floor of the margins and that of the deep sea.

This contrast is readily seen in the abundance patterns of organic carbon within the surface sediment, in the upper 5 cm, where most of the benthic activity takes place (Fig. 6.4). The low concentrations in deep sea sediments ($C_{org} < 0.25$ %) are typical for this environment, where bacterial growth is slow, and large benthic organisms are quite rare. Underneath the Equator, the concentrations increase somewhat, because *equatorial upwelling* leads to a greater supply of organic matter to the sea floor. The highest concentrations (up to several percent of carbon) are strictly tied to *coastal upwelling* areas along the margins. The contribution of both nutrients and organic

Fig. 6.2 a–d. Fecal pellets retrieved by trapping off California. **a** Tabular pellet (*bar* 200 μm) **b** Ellipsoidal pellet with organic coating (*bar* 200 μm) **c** Surface of ellipsoidal pellet, with coccosphere, diatom fragments and terrigenous particles (*bar* 10 μm) **d** Surface of tabular pellet, with diatom (*Nitzschia* sp.) and coccoliths (*Emiliana huxleyi*) (*bar* 10 μm). [R. B. Dunbar and W. H. Berger, 1981, Geol Soc Amer Bull 92: 212]

carbon from the continents must not be neglected, of course. Off estuaries, algal growth is stimulated both by supply of nutrients from the river, and by estuarine circulation, that is, a type of upwelling. Quite generally, also, organic matter is delivered from the erosion of soil on land, and such matter can make up a substantial portion of the carbon buried in the margins. The ratio of carbon isotopes in such land-derived material is much more in favor of ^{12}C than that in marine-derived matter (–25 to –30 permil in terms of d ^{13}C, versus –20 permil). This isotopic difference (as well as chemical differences, represented by *molecular markers*) allows assessment of the terrestrial contribution.

The large supply of organic carbon along certain margins generates an *oxygen minimum,* due to the high oxygen demand, and this can result in *anaerobic conditions* on the sea floor. Under such conditions, preservation of organic matter is enhanced, as the anaerobic (sulfate-reducing) bacteria are less efficient in destroying organic matter than are those bacteria which use free oxygen for the purpose. The relationship

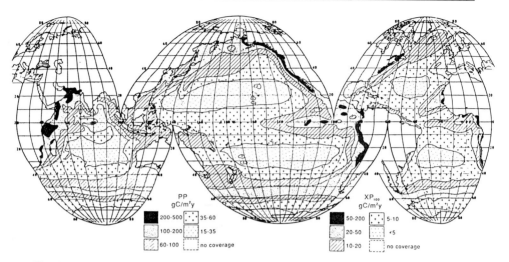

Fig. 6.3. Distribution of primary production (PP) in the world ocean, based on radiocarbon uptake experiments. XP_{100}= Export production, flux at 100 m waterdepth. [W. H. Berger 1992, Z Deutsche Geol Ges 142: 149–178.]

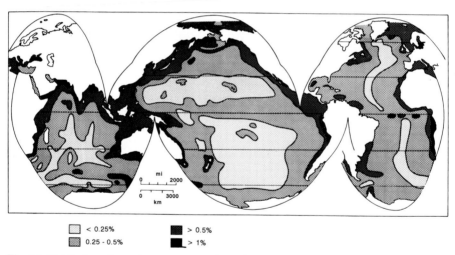

Fig. 6.4. Distribution of organic matter content in marine sediments: percent C_{org} dry weight. [W. H. Berger, J. C. Herguera, in P. G. Falkowski, A. D. Woodhead (eds) 1992, Primary productivity and biochemical cycles in the sea. Plenum Press New York. Data mainly from E. A. Romankevich, 1983.]

between anaerobic sediment and organic carbon content in sediments, after burial, is of special interest in this context (Fig. 6.5). The lack of correlation between sedimentation rate and C_{org} in anoxic conditions, and the good correlation in oxic conditions, is evident when all available data from Neogene marine sediments are considered. It

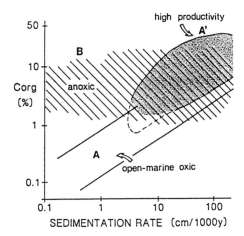

Fig. 6.5. Relationship between organic carbon content and sedimentation rates in anoxic and oxic marine sediments of Neogene age. The main fields are: *A* normal open ocean environments, *A'* high productivity regions, *B* anaerobic conditions. [R. Stein, 1991. Accumulation of organic carbon in marine sediments. Springer, Heidelberg.]

appears that in one case carbon is readily preserved even when sedimentation rates are low. In the other case, rapid burial is necessary to preserve a large portion of the organic carbon supplied to the sea floor.

The correlation between percent C_{org} and sedimentation rate is, at first glance, surprising: the organic matter is being diluted by inorganic matter, yet its concentration rises. The reason is twofold: (1) high sedimentation rates are almost automatically associated with an increased supply of organic matter, because both rates of sedimentation and rates of production are high in continental margin areas, and (2) rapid burial removes the organic matter from the oxygen-rich sediment-water interface, enhancing preservation.

6.1.3 Sunlight and Nutrients. As mentioned, sunlight and nutrients are necessary for marine production in the opean ocean, and this is equally true for the benthic environment, wherever primary production occurs (Fig. 6.3). The sunlit or photic zone is only about 100 m thick. Below that depth little sunlight remains, perhaps 1 % in very clear water. The angle of incidence of the sunlight, the cloud cover, and the amount of suspended matter in the water all influence the depth of penetration of the light. Benthic algae, then, can only occur on the upper half of the shelf area, say, over no more than 2 or 3 % of the sea floor. Planktonic algae *(phytoplankton),* of course, occur everywhere in open surface waters. However, this does not mean that the production by benthic organisms can be neglected. Typically, phytoplankton production is of the order of 50 g carbon per square meter per year (50 g C/m^2yr). For benthic algae the production values can be up to 100 times greater! Much benthic production takes place in salt marshes and in kelp forests, but also in the symbiotic algae of coral reefs.

While the ocean offers plenty of some of the materials necessary for growth, potassium and sulfate, for example, it is decidedly short of others, such as phosphorus, fixed nitrogen, silicate, and trace elements (iron, molybdenum etc.). These nutrients have especially low concentrations in surface waters, because they are constantly being removed by algae. In fact, their availability controls the growth of the algae; thus, they are *limiting nutrients*. Algal material, and also animal matter, sinks below the photic zone and undergoes bacterial decay. Nutrients are released in the process, that is, *remineralization* occurs. The nutrients are not being used at depth, in the absence of light. They become concentrated, therefore.

Thus, the ocean has a deep reservoir of nutrients below the surface waters. The boundary between this reservoir and the nutrient-poor sunlit waters is the top of the thermocline, usually between 50 m and 100 m deep.

When the thermocline is eroded by storms or when the deep water comes to the surface through the pumping effects of eddies or through *upwelling* (Sect. 4.3.3), the nutrients return to the surface water, into the sunlit zone. This fertilization by admixture of deep waters results in increased productivity, both in the algae and in the animals that feed on them. All the good fishing areas are in such regions of vertical mixing.

6.1.4 Salinity. Life in the sea depends on other factors besides sunlight and nutrients. A large number of organisms can tolerate *salinity fluctuations* only if salinity stays between 30 and 40 ‰ (*stenohaline* forms). Examples are radiolarians, reef corals, cephalopods, brachiopods, and echinoderms. In general, their remains indicate marine conditions for the sediment that contains them. However, there are exceptions. A few representatives of the above groups may have relatively wide tolerances: for instance, there is a starfish that lives in the brackish waters of the Baltic Sea. Also, fossils embedded in sedimentary layers do not necessarily show the environment of deposition: they may be redeposited.

With increasing salinity, for example in lagoons of arid regions, fewer and fewer species are able to survive, until only a few representatives remain – an *impoverished* fauna and flora. Some ostracods (bi-valved crustaceans of millimeter size) can tolerate salinities of more than 100 ‰. As these hardy forms have few competitors (or predators), impoverished faunas can be very rich in individuals. Quite generally, of course, species-poor but individual-rich faunas indicate special, *restricted environments*. Restricted does not apply to space here, but to unusual temperature ranges, oxygen deficiencies, and other stress-producing factors.

6.1.5 Temperature. In the open ocean, where salinities are well within the tolerance of all marine organisms, temperature plays a decisive role in controlling distributions (Fig. 6.6).

In polar areas the temperature can fall to -1.5 °C; in marginal seas in the subtropics it can rise to well over $+30$ °C, as in the Red Sea and the Persian Gulf Outside the tropics temperature can vary strongly with the seasons in the upper 100 to 200 m of the water column. Below this depth it varies little, however, being rather low throughout the year.

Most of the ocean is extremely cold: in the deep sea the temperature stays below 4 °C. The cold temperature in itself does not reduce diversity: it has recently been shown that an astounding variety of small benthic organisms occur even at abyssal depths. Large, unpredictable fluctuations of environmental factors appear to be much more restrictive than just extreme cold. Such fluctuations can work damage, especially on eggs, larvae, and the young recruits of a population.

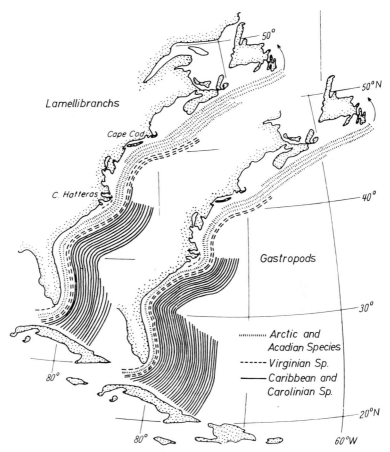

Fig. 6.6. Diversity of benthic shelf organisms as a function of latitude, shelf off the East Coast. An Arctic fauna with few species in the north is being replaced by species-rich temperate and tropical faunas to the south. (Each *line* equals ten species.) Note the sharp break at Cape Hatteras, where warm Southern waters (Gulf Stream) leave the shelf. Arctic waters penetrate down to Cape Cod. Gastropods show a greater change than bivalves (= lamellibranchs or pelecypods). The clams live mostly *within* the sediment, in contrast to the snails. Hence they are more protected from inclement conditions. [A. G. Fischer, 1960, Evolution 14: 64; modified]

Can we reconstruct the temperature distributions from a study of the remains of organisms?

In shallow seas and on the shelf, benthic carbonate-secreting organisms (Fig. 6.7) leave a record of changing temperatures in the oxygen isotope composition of the calcareous shells and skeletons, as first demonstrated by the geochemist S. Epstein and his associates (Chap. 7.3.2). In colder water, relatively more of the heavier isotope ^{18}O is incorporated into $CaCO_3$, changing the $^{18}O/^{16}O$ ratio accordingly (Fig. 6.8).

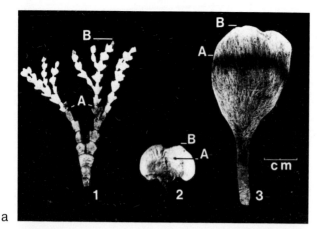

a

b

Fig. 6.7 a, b. Sediment production and benthic photosynthesis. **a** Production of algal carbonate, measured in the field by staining live specimens and observing growth until collection. *A* Termination of staining; *B* collection. *1 Halimeda; 2 Padina; 3 Penicillus;* all common benthic algae in Bermuda. [Photos and experiments by G. Wefer, 1980, Nature (London) 285: 323.] **b** *Acropora* meadow, Florida Keys. The staghorn corals produce abundant carbonate, with the aid of symbiotic algae (dinoflagellates) which live within their bodies. Growth rates are of the order of 1 cm/year [Photo W. H. B.]

Organisms are not necessarily faithful recorders of temperature, of course. In the case at hand, it will be noticed that the coral grows preferentially in the warm season, giving a record that is biased against cold periods.

For open ocean conditions, temperature reconstruction on long time scales has been honed to a fine art by the CLIMAP group (Climate Long-Range Investigation Mapping and Prediction) (Sect. 9.2.3). For shelf seas and enclosed basins the reconstructions are more difficult. Here, environmental factors besides temperature (salinity, muddiness of the water, seasonal bad weather) are much more important and unpredictable than in the open ocean. A change in a faunal assemblage, therefore, may be due to stress (or release of stress) in any one of these factors. One difficulty

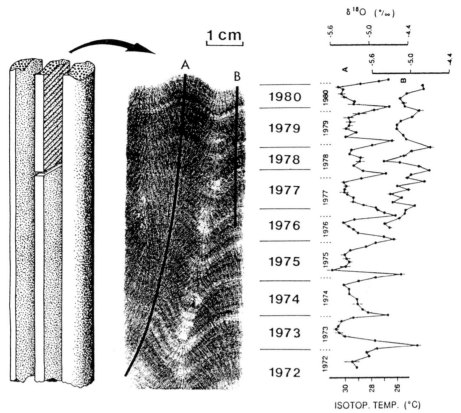

Fig. 6.8. Temperature reconstruction using *Porites lobata,* which forms massive heads of coral off the Philippines. Sampling is along profiles *A* and *B*, in a slab cut from a core. Results of isotopic analysis are to the *right*. Temperature varies roughly between 30 and 26 °C, and this is recorded by the coral. [J. Pätzold in G. Wefer and W. H. Berger, 1991, Marine Geology 100]

which exists for both deep sea and shallow sea, as far as paleotemperatures estimates, is the *selective preservation* of faunal (and floral) assemblages within the sediment. Thus, changes in the assemblages may be due to changing conditions of preservation, rather than to changes in the conditions of survival and growth. To separate the effects of chemistry (that is, early *diagenesis*) from those of biology can be a formidable task.

Additional clues to sea surface temperatures have recently been derived from long-chain (C_{37}–C_{39}) *alkenones* which are produced by coccolithophorids (see Sect. 7.7.5).

6.1.6 Oxygen. The content of *dissolved oxygen* in the water is another environmental factor of great importance, especially when concentrations fall to critically low levels of a fraction of one milliliter of gas per liter of water (normal: 4 to 7 ml/l). In cases where the oxygen concentration becomes low enough, all higher organisms, and even shell-bearing protists, will succumb, and only anaerobic bacteria remain. Such conditions – called *dysaerobic* when moderate, and *anaerobic* when severe – can produce *varved sediments* (Fig. 3.14), since there is no disturbance by burrowing organisms. Much can be learned from varves regarding climatic change, on the scale of decades to millenia, especially with regard to changes in the supply of organic matter and oxygen.

At the present time we have to go to certain special areas to study this phenomenon – fjords in Norway and Alaska, the Black Sea, the Santa Barbara Basin off California. In the geologic past, however, when the poles were not icy cold and did not therefore deliver oxygen-rich water to the deep ocean, conditions of oxygen deficiency were widespread. Much of the petroleum we burn today was formed in oceans with a low oxygen content, in regions where oxygen dropped below critical values and where organic matter did not readily decay, therefore.

Basically, oxygen deficiency arises when oxygen demand is strong, and supply is weak. For example, in the Black Sea the salt water filling the basin (through the Bosporus, Chap. 4.3.5) comes from the Mediterranean and is covered with a layer of freshwater brought in by the Danube, the Dnepr, and other rivers (Fig. 7.12). The light freshwater forms a lid on the heavy deep water, cutting off exchange with the atmosphere. Growth continues in the upper waters, delivering organic matter to the water below. Here it is eaten by animals, which use up oxygen by respiration, and is decayed by bacteria, also using oxygen. Thus, the deepest water becomes entirely anaerobic.

A somewhat analogous process can be observed in organic-rich muds. Free oxygen is present only in the uppermost layer of sediment, millimeter-thick (or centimer-thick when sandy). Burrowing animals must set up air-conditioning by pumping oxygen-rich water through their burrows – otherwise they must suffocate.

The geohistorian attempts to reconstruct the degree of oxygenation at the time and place of deposition of a given sediment layer, from clues such as lamination or nature of burrowing, and from chemical indicators such as sulfides and types of organic matter present. The remains of certain benthic organisms also provide information regarding the level of oxygenation (Fig. 6.9).

6.2 Benthic Life

6.2.1 Types. By far the greatest part of the sea floor is teeming with benthic organisms. Benthos which stays put is called *sessile*. All sponges, corals, brachiopods, and bryozoans are sessile. Benthos which moves about is termed *vagile*. It can move rapidly, like a startled crab, or slowly, like the sluggish sea urchins, starfish, most bivalves, snails, and worms. Both groups have members living *on* the floor or on top of other organisms (e. g., shells or kelp): *the epifauna*. Or they live hidden within rocks and sediment: *the infauna*.

Compared with the 125 000 marine species of epifauna there are only a paltry 30 000 species of infauna. Why? Are there more *niches* in the epifaunal way of life,

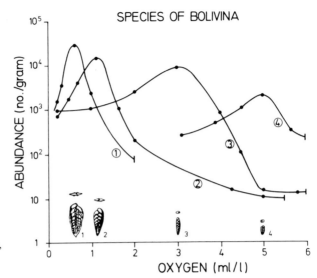

Fig. 6.9. Abundance of *Bolivina* species as a function of dissolved oxygen, California Borderland. *1:B. argentea; 2:B. spissa; 3:B pacifica; 4:B vaughani.* [R. G. Douglas, 1979, SEPM Short Course 6: 21]

that is, more opportunities for different approaches to making a living? We do not know. So far, biologists have not succeeded in determining the number of niches independently from the number of species occupying them.

6.2.2 Food and Substrate. Benthic animals, ultimately, live on food-stuff falling down through the water or coming in from upslope along the sea floor. Both living plankton and dead *detritus* drift in the water: the *seston*. The detritus consists of organic and inorganic particles. One consequence of the dependence of the benthos on this "rain" of seston is a pronounced decrease in benthic biomass with depth: the deeper the sea floor and the farther away from the fertile coastal zone, the smaller the amount of nutritious rain reaching it. Although so very little reaches the abyss (order of 1 % of the production (Fig. 6.4)), hundreds of species – tiny crustaceans and worms – make their living from the scraps coming down. How do these organisms manage to spend less energy in the search of food than the food is worth? They must be extremely energy-efficient.

The *sessile benthos,* of course, does not seek out its food. It waits for the water to bring suspended material from which to take nourishment. The *suspension feeders* (sponges, corals, brachiopods, crinoids, bryozans, and many others) filter the water, either passively, using the natural water flow, or actively, moving water past their straining apparatus. *Commensals* may seek a free meal in addition to shelter: for foraminifera living in sponges, it is convenient that the host provides both protection from enemies and brings in a steady stream of food particles. Sessile benthos is especially abundant on rocky bottom, e. g. in agitated environments, where the water is not too muddy.

The *vagile benthos* on the rocky substrate – starfish, sea urchins, gastropods, ostracods – feeds on epibenthic organisms, for example, by scraping off algae, or preying on sessile animals. It protects itself from storms and predators by hiding in nooks and crannies, growing thick shells, or by clinging to the rock, as do the chitons and patellas using their strong sucker foot. On soft bottom (in less agitated environments), most vagile benthic animals ingest sediment, especially surface detritus (*deposit feeders*). Others hunt for prey.

We see that life in the bottom is greatly influenced by the type of the substrate, which, like the organisms themselves, reflects environmental factors: temperature, salinity, oxygen content, currents, microtopography. Thus, sedimentary processes and life on the sea floor are intimately associated. The remains of organisms within their habitat yield valuable clues to conditions of both growth and sedimentation, therefore. In fact, benthic organisms may *produce* much or all of the substrate, in places.

One aspect of ecology which is of great interest to geology is the *production rate* of hard parts, that is, carbonate (Fig. 6.7). Off Miami, the macrobenthos produces annually about 1000 g carbonate per m^2 in the tidal zone, between 1 and 400 g per m^2 in the deeper water offshore. On shallow areas of the Persian Gulf, a single species of foraminifer, *Heterostegina depressa* (Fig. 6.10), delivers annually 150 g carbonate per m^2. This protist lives in symbiosis with photosynthesizing algae, as do corals. Questions regarding type and rate of mineralization are obviously of importance to paleoecology; here biological and geological research are closely intertwined. Benthic foraminifera whose shells provide a lasting record are ideally suited to define the limits of many of these conditions (Figs. 6.9 and 6.10.).

For an appreciation of the relationships between organisms and environment of deposition, let us have a closer look at benthic organisms and their substrates, hard and soft.

6.3 Organisms and Rocky Substrate

Rocks in the ocean floor outcrop where currents or waves keep the floor free from sediments, or in areas of active erosion. Steep-sided walls in fracture zones and other rugged topography are also likely places for rock outcrops.

A great number of different materials can be called rock: ancient sandstones and limestones of wave-cut platforms or along submarine canyons, basalt scarps or manganese pavements, submerged dead coral and cemented beach sand (beach rock), boulders dropped from icebergs or left in the shelf in moraines, sunken ships, and other man-made objects. All such substrates can be densely covered by benthic organisms under the right conditions, in some cases (drop-stones, ships) forming epibenthic "oases" on an otherwise rather empty muddy sea floor. The most abundant type of rocky substrate undoubtedly is provided by the pillow basalts of the Mid-Ocean Ridge. These basalt outcrops are commonly rather barren, except in local spots of hydrothermal exhalation (see Sect. 6.9).

In shallow areas, rocky bottoms are commonly covered by algae. The algae can be microscopic diatoms (many of these move about on the rock), or up to several-

Fig. 6.10. Examples for different types of foraminifera. *1* Arenaceous form which agglutinates available particles (*Reophax,* x 50); *2, 3* calcareous forms (*Trimosina,* x 250; *Spiroloculina,* x 90). The species of *Trimosina* shown lives at a depth of greater than 20 to 30 m, in the Indian Ocean; *Spiroloculina* prefers shallow water and coarse substrate. *4: Heterostegina depressa* (diameter of test: 0.84 mm) a large tropical form living in shallow water in symbiosis with photosynthesizing algae. The "pseudopods" radiating from the test are used for anchor and also for locomotion. [SEM photos C. Samtleben and I. Seibold. Microphoto R. Röttger]

meters-long soft ribbons waving in the currents, offering hideouts and hunting grounds for fishes. Hence, rocky areas are good fishing grounds (although they can be hard on a fisherman's net).

Encrusting algae tend to cover rock outcrops permanently. Depending on water temperature and depth, *epifaunal* associations develop: sponges, corals, tubeworms, oysters, barnacles, bryozoans, encrusting foraminifera (Fig. 6.11). The *infauna* bores actively into the substrate. This remarkable ability has been acquired by many types of organisms, including sponges, worms, and mollusks. Some sea urchins with unpleasantly (for swimmers) long and brittle spines live in custom-made cavities. At first glance such burrowed rock gives the appearance of being quite solid, since the

Fig. 6.11. Dense epifaunal growth. **1** Edge of coral patch reef, Lizard Island, Great Barrier Reef (14°45'N, 2 m). [Photo E. S.]. **2** Epifauna on a mound of the Ross Sea shelf (Antarctic, 76°59'S, 167°36'E). No algae are present at this depth (110 m) so that the dense growth here is all from animals: large fingery sponges, small bushy bryozoans, finely branched horny corals. A sea lily (echinoderm) is seen in the *upper left*. [Photo J. S. Bullivant, 1967, N Z Dep Sci Ind Res Geol Surv Paleontol Bull 176]

entrances to the burrows are usually quite small. However, in reality, the outer rind of the rock may resemble Swiss cheese in its structure. Storm surf can then destroy this outer layer. Thus the rock wears away, largely through bio-erosion. Limestone is especially susceptible to this type of attack. Debris from boring sponges grinding holes into calcareous rock can make up some 30 % of the sediment in certain lagoons of the Pacific coast.

In shallow rocky areas, organisms commonly are removed periodically by wave action, and their remains collect in depressions or at the foot of the rocks. The *reef-talus* deposits are of this origin. They are coarsely layered thick sequences of sediments surrounding the reef structure, and are familiar from ancient exposed reefs on land (Permian El Capitan Reef, west Texas; Triassic Dolomites in the Alps; Jurassic Malm reefs in middle Europe; Cretaceous Albian reefs in southern Arizona). The coarse-grained material of the reef talus is highly porous and permeable, and can serve as reservoir rock for petroleum – as in the oil fields of the Middle East.

We shall return to reef formations in Chapter 7 when discussing the significance of marine deposits as climate indicators.

6.4 Sandy Substrate

Rocky ground quite commonly has a rich, "flashy" assemblage of diverse organisms, from tropical reefs to Antarctic sponge forests (Fig. 6.11). Substrate of pebbles and boulders is much like rocky bottom provided there is no movement. If waves move the rocks, most sessile benthos cannot survive.

The benthic communities on sandy sea floor are much less flashy. Commonly one sees rather little – the organisms are mostly hidden within the sand. The reason why sessile epifauna is largely absent is the instability of the substrate. Where it is stable, sea grass can take hold and further stabilize the ground. Epibenthic diatoms, foraminifera, and bryozoans grow on such grass.

Vagile benthos, with or without burrows to return to, can be quite abundant on the sandy bottom. Crabs and snails are common sights. In the intertidal range there is a large variety of nonmarine invaders during low tide, mainly birds. These hunt for the hidden infauna. Conversely, during high tide the invasion is from the sea: sting-rays and other fishes digging up worms and mollusks.

The presence of so many predators, and the shifting of the sand which can suddenly expose the infauna, require the ability to burrow very rapidly. Crabs demonstrate this adaptation very obviously, but also many clams can burrow quickly – as clam diggers well know. Burrowing clams have a strong long foot which they extend into the sand below, then inflate by water pressure to anchor it, and pull the rest of the body down through muscle contraction. A smooth outer shell, usually quite sturdy, characterizes such clams. The study of burrowing mechanisms and the resulting tracks in the sediment is a research field by itself (Sect. 6.6).

Burrowing clams are *suspension feeders,* with an inhaling and an exhaling siphon (Fig. 6.12). After death, the shells of such bivalves – hydraulically quite different from the surrounding sand – are sorted out and concentrated into layers of *coquina* (Fig. 6.13). Such coquina deposits are widespread in the geologic record and can be used as marker beds locally.

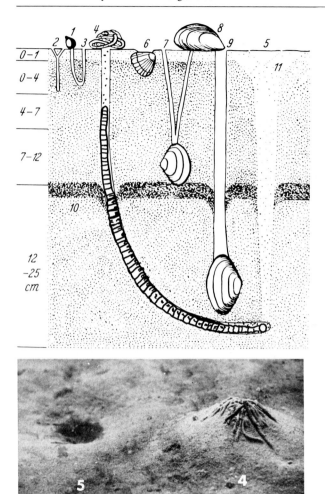

Fig. 6.12. Sediment reworking by benthic organisms in tidal flats of temperate regions. *1* Snail *Littorina; 2* worm burrow *(Pygospio); 3* crustacean burrow *(Corophium), 4* and *5* worm *Arenicola* with fecal mound and funnel (see photo *below*); *6* to *9* bivalves *Cardium, Scrobicularia, Mytilus,* and *Mya; 10* originally horizontal layer, disturbed by burrowing; *11* light brown surface sediment. [H. M. Thamdrup, in R. Brinkmann (ed), Lehrbuch der allgemeinen Geologie, F. Enke, Stuttgart 1964, modified. Photo E. S.]

6.5 Muddy Bottom Substrate

Muddy and clayey substrates present different kinds of challenges for their benthic denizens. Thus the faunal assemblage of this habitat differs greatly from those of the sandy and rocky environments. Muddy substrates are rich in organic and inorganic particles of extremely small size – less than one to a few microns (1 micron = 0.001 mm). This has two consequences. First, the substrate is rather less shifting than the

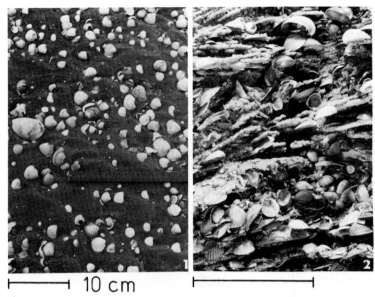

Fig. 6.13. Shell pavement in the upper intertidal, German North Sea. (N of Weser estuary). The shells form low mounds, built by waves and tidal currents. The orientation of the shells on the sea floor indicates current directions [note valves *pointing to right* i. e., downstream in **1** and shingling in box corer profiles in **2**. [Photos E. S. **1** and H. E. Reineck **2**]

sand, due to the high cohesion of the clay particles. It is more difficult to burrow into than sand, but burrows have a better chance of persisting. Second, the high content of organic matter makes it worthwhile for many organisms to pass it through their guts and extract the digestible fraction. Hence, *deposit feeders* are commonly found on and in this type of substrate. Of course, *suspension feeders* would thrive on the organic matter also. However, their filtering apparatuses tend to become clogged with the abundant (and to them useless) clay. Furthermore, by constantly reworking the mud, the deposit feeders make it difficult for sessile benthos to find a stable foothold. Thus, deposit-feeding benthos dominates: about three fourths of the benthos belong to this group.

Deposit feeding can take place on top of the sediment, which may also yield some diatoms, and other algae growing there. Snails typically represent this type of detritus behaviour, for example, *Littorina* in tidal flats (Fig. 6.12). Also, burrowing clams may extend their feeding siphons out to the top, to pipet off the food on the surface. Examples are *Scrobicularia* and *Macoma*. On the tidal flats of northwestern America, the small species *Macoma secta,* with a length of only 6 to 7 cm, can extend its siphon up to 1 m – covering a good-sized territory. Various types of worms and sea cucumbers (holothurians) are deposit feeders sometimes seen in deep-sea photographs. They leave tracks and fecal strings on the surface, and burrows within the sediment (Fig. 6.14). In shallow water, certain deposit feeders can become extremely abundant. The lugworm *(Arenicola),* typical for muddy intertidals, in many areas can

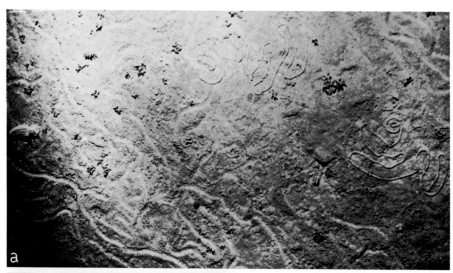

Fig. 6.14 a, b. Tracks and burrows in deep-sea clay. **a** Furrow-like feeding trails of holothurians (deposit feeders). Note range of states of preservation, fresh trail at *lower left*. Also meander-like fecal strings extruded onto sediment surface. Small compact structures probably are *agglutinating foraminifera*. Tracks and trails are not preserved in the record, unless burial is by sudden influx of sediment (such as turbidites). Northern Solomon Trench, water depth 8500 m, field of view in upper photo 1.5 x 3 m. [Photo courtesy R. L. Fisher.] **b** X-ray radiograph, positive, of deep-sea Red Clay. Dense areas are *dark*. Network of burrows (mainly deep-reaching burrows are preserved) illustrates intensity of bioturbation. Typical burrow diameter, 0.5 to 1 cm. *Left* Small manganese nodule. Core taken about 1000 km SE of Hawaii, water depth 5000 m, 49–60 cm below sediment surface. [Photo F. C. Kögler]

attain densities of 200 individuals per square meter (Fig. 6.12). Each worm can ingest several hundred grams of sediment per day. At high densities, the entire sediment, down to about 20 cm depth, may be worked over in a matter of weeks by deposit-feeding organisms.

Muddy bottoms, then, are largely fecal material, many times recycled. This is also true for the deep-sea floor, although here the cycling takes thousands of years rather than months as in the mudflats – a difference of a factor of 10 000. Of course, sedimentation rates also differ considerably, by a factor of up to 1000. The greater number of recycles in the shallow water, before burial, reflects the much higher supply of organic carbon – and hence energy supply – to the coastal sea floor. The low level of benthic activity in deep-sea sediments and the relative greater importance of the infauna is illustrated by the fact that on more than 100 000 photographs from 2000 different deep-sea stations, only about 100 visible animals were counted.

If fine sediments are fecal matter, should one not see more evidence of fecal pellets and fecal strings? Indeed, fecal pellets are extremely abundant, especially toward the surface of muddy sea floor. Here they alter entirely the hydraulic properties of the sediment, give rise to bacterial growth, and to development of fungal mats in places. In the case of calcareous mud, the fecal pellets can harden by cementation and can thus fossilize more readily. They are very abundant in many limestones, as near-spherical and oval grains of diameters between 0.03 and 0.1 mm, a fact that is only realized after careful microscopic study.

The biological reworking of soft sediment on the sea floor is a process of global geochemical significance. Without such *bioturbation,* the sediment would quickly disappear from the marine chemical system. Only a thin upper layer would readily react with the seawater. Through bioturbation, however, a several-centimeter-thick layer of sediment keeps exchanging matter with seawater, the falling organic material remains available for some time before burial, the nutrients in it are remobilized and are given back to the seawater. Thus, the overall fertility of the ocean is closely linked to bioturbation.

A surprisingly large proportion of Phanerozoic sediments on land originate from marine muds. Fossil shales – produced from muds – make up about 50 % of the sedimentary record. Limestones (in large part originating from calcareous muds) provide 20 %, and sandstones 30 %.

6.6 Trails and Burrows

6.6.1 Trace Fossils. In the foregoing, we have looked at sea floor and organisms with an appreciation for the biological viewpoint. However, the geologist's ultimate concern is the *final record:* how is it made? What can it tell us about conditions of the past? An entire branch of geology – *ichnology* – has grown from the study of the tracks, trails, burrows, and other sedimentary disturbances made by organisms. We next turn to some of the problems arising in the study of such trace fossils. We also have to ask just how bioturbation disturbs the orderly recording of events in deep-sea sediments – the sediments from which we hope to extract the most complete and accurate information about ancient climates (as discussed in Chap. 9).

There is a great variety of traces, which eventually become trace fossils: trails, feeding tracks, and fecal strings on the surface, burrows stuffed with fecal mud, burrows used for housing and later filled by washed-in sediment, and others (Fig. 6.15). Various kinds of worms, snails, bivalves, crabs, holothurians, sea urchins, and star fish make such tracks and burrows. Crabs and shrimp make the longest and deepest burrows. In certain cores from off Northwest Africa, we have observed vertical burrows more than 3 m long! Such "lebensspuren" have been and are being, intensively studied by marine geologists in Wilhelmshaven on the North Sea, in Kiel, and in Tübingen. Thus, the German word *Lebensspuren* which translates into *life-traces* has been adopted into English usage.

6.6.2 Lebensspuren are, in a sense, fossilized behavior of benthic organisms. Much like shells, behavior represents adaptation to the environment; hence we can read

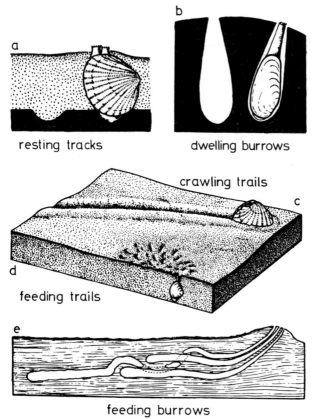

Fig. 6.15 a–e. Types of traces by function, and organisms producing them. **a–d** Bivalves. **e** Worms. [A. Seilacher, 1953, Neues Jahrb. Geol. Palaeontol. Abh. 96: 421 and 98: 87]

environmental information from tracks and burrows. Unlike shells, lebensspuren cannot be transported by currents, but stay in the sediment where they were made. Thus they indicate in situ conditions, whether it be food supply, stability of the sediment, wave action, even depth of water.

The tracks and burrows are useful not only for paleoecologic reconstruction but also for physical geology, and even for tectonics. In the deep sea, for example, vertical burrows do not last if the sediment has a tendency to shear horizontally. Through systematic box coring of deep-sea carbonates in recent years, it was found that vertical burrows are extremely abundant in the west equatorial Pacific, in shallow areas with sandy, stiff sediment. In deep areas, where dissolution of carbonate breaks down the sand, the sediment loses strength. It can flow when shaken by earthquakes. There are no vertical burrows here, only horizontal ones.

On land, the field geologist sometimes deals with badly jumbled marine strata – he may not even be able to tell which side of a layer is up! Burrows, if present, can help decide the issue. Many burrows are U-shaped. If one finds them as an arch, that is, upside down, it proves that the layer has been inverted by tectonic forces. Sometimes there is a question of *lacustrine* (lake-) versus lagoonal or marine sediments. The marine environment has large long burrows in contrast to the less diverse and generally smaller ones of lakes.

6.6.3 Preservation. Not all tracks and burrows are equally well preserved, of course. A delicate surface trail made by a star fish or a mussel has little chance of entering the record. A 30-cm-deep *Arenicola* or *Mya* burrow or especially a 2-m-deep crab burrow has an excellent chance: we should have to remove a thick layer of sediment in order to obliterate it. In general, surface tracks can only be preserved by a catastrophic deposition event such as a flood or a turbidity flow. In the case of continuous accumulation, when sediment builds up gradually and there is a mixed layer from bioturbation, surface tracks are destroyed while only burrows penetrating the mixed layer are preserved in the record (Fig. 6.16).

6.7 Bioturbation

6.7.1 Effects of Mixing. Bioturbation prevents the preservation of thin layers, that is, annual layers, thin turbidite layers, or layers produced by contour currents alike. Thus, today's well-turbated slope sediments are poorly layered. However, this need not always have been the case. During times when the ocean was not well oxygenated, burrowing organisms could not as readily do their work of homogenizing the incoming sporadic supply of sediment. Thus, quite commonly, we see finely layered slope sediments in the geologic record, deposited in warm oceans with low oxygen content. (Warm water dissolves much less oxygen than cold water: only since the Earth has grown ice caps is the deep ocean highly aerated.)

Bioturbation alters and smooths the record. In a finely layered sediment, each layer has a message about its origin and its environment of deposition. But when the layers are mixed, the messages are mixed also, and we obtain a signal which is some

Homogeneous layer

Scolicia

Planolites

Helminthopsis
Chondrites
?Daedalus

Zoophycos

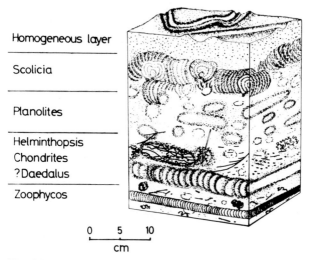

0   5   10

cm

Fig. 6.16. Biogenic sedimentary structures in deep water off NW Africa (about 2000 m). Tiered arrangement of active burrowers reaches 30 cm into the sediment. Bioturbation homogenizes the uppermost 3 cm of sediment completely, destroying surface tracks. Sea urchins (*Scolicia* burrows) and other burrowers produce typical traces (*Planolites* etc.) in different depths within the sediment. The tiered system of burrows is ultimately dependent mainly on organic matter supply, oxygenation, and sedimentation rates. Thus several parameters can potentially be extracted from the ichnological record. [A. Wetzel, 1979, Ph. D. Thesis, Geol Inst Kiel]

sort of an average of the conditions which existed at the time. How large a time interval is being averaged? How does this averaging affect the *dating* of sediments? These questions are being investigated on deep-sea box cores, because of the great importance of the deep-sea record for detailed paleoclimatic reconstruction. (For dating by radioisotopes see Appendix A8).

One way to proceed is to determine the exact concentration of radioactive carbon, as a function of depth in the sediment. The ^{14}C stratigraphy should give us a clue to both the depth of mixing and the effect on age determination (Fig. 6.17).

The ^{14}C enters the sedimentary record within the $CaCO_3$ of the carbonate shells. A certain proportion of the CO_2 in the air (and hence of the HCO_3^- in the surface water) contains ^{14}C, which is produced in the atmosphere through the activity of cosmic rays, from the normal nitrogen-14 atom. The radiocarbon is incorporated into living matter and shells. It decays back to nitrogen. Thus, young shells have more radiocarbon than old ones. The decay rate is such that one-half of the C-14 is gone after 5700 years. This is the *half-life*. Thus, we can predict that a shell will retain one half its radiocarbon after 5700 years, one fourth in 11 400 years, one eighth in 17 000 years, and so on. The limit of measurement is near 35 000 years – one sixty-fourth of the original concentration.

How do we know the initial concentration of radiocarbon? We assume, for simplicity, that it was the same then as it is today in freshly forming shells (although we now know that radiocarbon concentrations were higher during glacial time; see

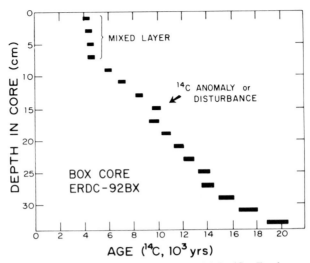

Fig. 6.17. Carbon-14 stratigraphy of a box core from the western equatorial Pacific. *Top layers* (= mixed layer) shows similar ages down to 7 or 8 cm, due to mixing. Overall sedimentation rates is near 1.7 cm/1000 yrs. An anomalous ^{14}C sequence appears at 15–17 cm depth, presumably caused by a redeposition event. [Data in T. H. Peng et al., 1979, Quat Res 11: 141]

Fig. 5.6). Using this assumption, we can calculate the *apparent age distribution* down the core. This is the distribution produced both by the continuous sedimentation of radioactive sediment (i. e., calcareous shells) and the mixing on the sea floor. Here is an interesting trend in the ages: they change very little in the uppermost part of the sediment, the *mixed layer,* and then show a regular progression to higher values downcore.

6.7.2 Mixing Model. Sediment mixing is a complicated process involving the burrowing activities of various types of organisms, disturbing the sediment to various depths (Fig. 6.16). It is not yet possible to describe the physics of mixing in a way which is both correct and useful. However, the pattern shown by the ^{14}C-distribution suggests a very simple (somewhat too simple, for sure) model of how mixing operates on sediment, as follows.

We assume that the sediment consists of two layers only: a mixed layer at the very top and a historical layer below. At the top of the mixed layer the newly arriving material builds up the sea floor at a rate corresponding to the prevailing *sedimentation rate.*

At the bottom of the mixed layer, an equal amount of sediment leaves, per unit time, and enters the historical layer. The mathematical formulation of this model is straightforward. For example, a tracer such as microtectites (from a meteor impact) or volcanic ash, which is added to the sea surface only once, will show a distribution following the decay equation:

$$C_L = C_0 \cdot e^{-L/M}, \tag{6.1}$$

where C_L is the concentration of the tracer at a distance L above the first occurrence, C_0 is the original concentration in the mixed layer, right after introduction of the tracer, and M is the thickness of the mixed layer. Application of this formula to the distribution of volcanic ash, observed in cores from the North Atlantic, suggests that M is in the neighborhood of 6 cm. This estimate agrees well with the mixed-layer thickness suggested by the ^{14}C-stratigraphy shown in Fig. 6.17, for a core from the west-equatorial Pacific. It should be noted, however, that radiocarbon of the sand fraction may yield results that differ from those based on bulk ages.

6.8 Limits of Paleoecologic Reconstruction

We have discussed environmental factors, especially physical ones, and we have looked at the production of lebensspuren and what they may tell us about the environment. Also, we have introduced the problem of bioturbation, the process which alters and smooths the record. What about the organisms themselves, their interactions, their reproduction rates, their larval dispersion? How do larvae know where to settle? How many survive? Who competes with whom? Who eats whom? Which are the symbiotic relationships, which parasitic? How do they control distributions? And especially: how much of all this information enters the geologic record?

These problem are constantly encountered when studying communties, here, *benthic communities*. Already the definition of communities is difficult – the biologists use "organisms that are often found together". Presumably, their web of interaction provides the checks and balances which keep a community *stable,* that is, the same kinds of organisms in roughly the same proportions live together over extended periods of time.

The *record* also has "communities", that is, co-occurring fossils. Not surprisingly, the difficulties in guessing the *interactions* of organisms long dead are virtually insurmountable.

Even seemingly simple questions are hard to answer. For example, how large a territory does a benthic foraminifer feed on? On the shelf of the Arctic there are up to 50 individuals per square meter of the foraminifer *Astrorhiza*. Their shells are about 5 mm in diameter; but, when alive, the net of pseudopodia takes up an area 6 cm in diameter, an area more than 100 times greater than that of the shell. A large part of the sea floor may be "occupied" by these tiny organisms. How would we know this if we only had their shells?

It does not help, of course, that only a selected portion of the community is eventually fossilized. Let us inspect some shallow water communities recognized especially through studies of Danish biologists (Fig. 6.18). In many Arctic and cool-temperate regions, a *Venus* community lies offshore in 10 to 20 m water depth. *Venus,* a bivalve, is a conspicuous member. Others are the star fish *Astropecten,* the sea urchin *Echinocardium,* and the polychaete tube-building worm *Pectinaria koreni.* Species may change, but the genera and the structure of the community are similar everywhere *(parallel communities).* As geologists, we can expect the preservation of *some* of the shells, *some* of the burrows. Nothing recognizable will be left of the

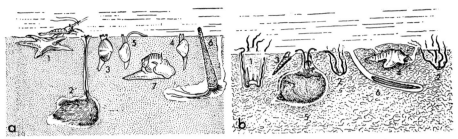

Fig. 6.18 a, b. Typical infauna associations in shallow water in cold to temperate regions. Examples are from the entrance to the Baltic Sea (Kattegart). **a** *Venus* association in sandy substrate at 10–20 m water depth, with the sea star *Astropecten (1)*, the sea urchin *Echinocardium cordatum (2)*, the bivalves *Venus gallina (3)*, *Spisula elliptica (4)*, *Tellina fabula (5)*, the polychaete tube-building worm *Pectinaria koreni (6)*, and the gastropod *Natica (7)*. **b** *Amphiura* association in muddy substrate at about 20 m water depth with the brittle stars *A. chiajei (1)* and *A. filiformis (2)*, the snails *Turitella communis (3)* and *Aporrhais pespelicani (4)*, the sea urchin *Brissopsis lyrifera (5)* and the polychaete *Nephthys (6)*. Coastal regions in other regions of the world have other species, but the basic makeup of the associations persists ("parallel communities"). [G. Thorson, 1972, Erforschung des Meeres, Kindler, München]

Fig. 6.19. Brittle star community. Uppermost continental slope off Senegal, Northwest Africa. [Photo E. S.]

worms or of the shrimps. Nothing also of the predators which invade the area from the sea and from the air.

Muddy sand and deeper waters are settled by the *Syndosmya* community. The name-giving clam is a favored food item of flat-fish – something that certainly would be hard to tell from the fossil record.

Muddy bottoms in depths of about 20 m are conspicuous by their dense cover of brittle stars (*Amphiura* community, Fig. 6.19). Up to 500 individuals lie on a square meter. They dominate the ecology of the sea floor here, but it is doubtful that this information would be preserved in the record.

6.9 "Hot Vent" and "Cold Seep" Communities

Animals – thus the conventional wisdom – ultimately live at the expense of photo-synthesizing organisms which use the sun's energy to produce organic matter from CO_2 and H_2O. There are exceptions, however. Some animals live strictly on bacteria which derive their sustenance from *chemosynthesis* rather than *photosynthesis*. In some cases, such bacteria are symbionts.

The most spectacular examples of this type of association have recently been discovered at deep-sea vents, where hot seawater leaves the basaltic sea floor, after reacting with it deep inside. During these reactions, oxygen is lost, and sulfate is reduced (Sect. 10.4.4). Where the water exits, e. f., in *"black smokers"* (Fig. 6.20) there is an opportunity to derive energy from the oxidation of the sulfide contained. Thus, sulfide-oxiding bacteria can thrive here. Certain types of these bacteria live in great abundance as symbionts within vestimentiferan tube worms (e. g., *Riftia pachyptila*), and giant clams (e. g., *Calyptogena magnifica*), among others, and thus *"hot vent"* chemosynthesis gives rise to a thriving community of strange (and not so strange) creatures (cover and Fig. 6.20). The tube worms live entirely on the bacteria which they harbor – they have no means of extracting food from outside, or digesting it.

The story of the discovery of vent communities is complex, and begins in the 1970s with a search for undersea hot springs, whose presence was deduced from temperature anomalies and chemical measurements. The first glimpse of unusual communities came from a deep-sea photograph (taken by P. Lonsdale) showing remains of large clams in unexpected concentrations, near a spreading center. The living vent community, in all its exuberance, was first seen in 1977 by the geologist J. Corliss and associates, diving on the Galapagos spreading center at 2.5 km depth, in Woods Hole's deep submersible *ALVIN*. (Corliss, J. B., and others (1979) Submarine thermal springs on the Galapagos Rift. Science 203: 1073–1083.)

A search for similar hot springs elsewhere, and associated vent communities, was highly successful, and we now have descriptions of a variety of such faunas from different regions, in both the Pacific and the Atlantic. Some 160 species new to science have been described, mainly mollusks, annelids, and arthropods. Some appear to be closely related to certain Mesozoic shallow water forms.

Fig. 6.20. "Black smokers", made of precipitated metal sulfides at an East Pacific Rise Vent site. Pillow lavas are covered by mussels and tube worms. [E. Seibold, Das Gedächtnis des Meeres, 1991, Piper, München. After P. Giese, 1983]

Many exciting questions were raised as a result of these discoveries. How do these communities survive, given the fact that hot vents exist for only a short time, on the order of decades? How do these organisms cope with highly toxic fluids containing H_2S and trace metals in high concentrations? What are the implications for the evolution of life, especially in the distant Precambrian with reducing sea-water conditions? And for those who wonder who might be out there in space – is the sun necessary for the origin of life?

In the meantime, additional exploration by *ALVIN* and other submersibles, e. g. *CYANA,* showed that these unusual communities are not restricted to hot vents on spreading centers, but occur in back arc basins and other places with hydrothermal activity as well. More importantly, related but somewhat different types of communities were found on the continental margin, sitting on *"cold seeps"* off Florida and, together with carbonate *"chimneys"* off Oregon (see Fig. 2.8b). Here bacteria oxidize not just H_2S but a variety of other reduced compounds such as methane (CH_4), ammonia (NH_4)$^+$, and light hydrocarbons. Again, these bacteria form the basis of a chemosynthetic food chain.

The vent communities are geologically old: in Oman and on Cyprus in Cretaceous ophiolites, and even in Carboniferous rocks as in New Foundland, fossil examples of tube worm communities were found, in part in association with massive sulfide ores. These fossils now appear in an entirely new light.

Further Reading

Hedgpeth JW (ed) (1957) Treatise on marine ecology and paleoecology, vol 1. Geol Soc Am Mem 67
Schäfer W (1972) Ecology and palaeoecology of marine environments. Oliver & Boyd, Edinburgh
Haq BU, Boersma A (eds) (1978) Introduction to marine micropaleontology. Elsevier, New York
Suess E, Thiede J (eds) (1983) Coastal upwelling – its sediment record. Part A: Responses of the sedimentary regime to present coastal upwelling. Plenum Press, New York
Jannasch H, Mottl MJ (1985) Geomicrobiology of deep-sea hydrothermal vents. Science 229: 717–725
Rona PA (1986) Mineral deposits from sea-floor hot springs. Sci Am 251(1) 66–74
Berger WH, Smetacek VS, Wefer G (eds) (1989) Productivity of the ocean: present and past. Dahlem Konferenzen. Wiley, Chichester

7 Imprint of Climatic Zonation on Marine Sediments

7.1 Overall Zonation and Main Factors

7.1.1 The Zones. The chief factor in producing climatic zonation is the amount of energy received from the sun – it is high in the tropics, low at the poles (Fig. 7.1). A coarsely latitudinal zonation of the oceans generally employs the categories *tropical, subtropical, temperate,* and *polar,* whereby the poleward part of temperate and the more temperate part of polar could be distinguished as *subpolar* (Fig. 7.2).

The *tropical zone* has an excess of heat, which it exports. Seasonal fluctuations are minimal; average temperatures are near 25 °C, with sustained open ocean maxima close to 30 °C. Near the equator, daily rainfall, cloud cover, and weak winds, lead to excess precipitation over evaporation. At the equator proper (± 2° latitude) fertility is high because of equatorial upwelling. Elsewhere it is low, except near continents.

The *subtropical zone* is the broad region between the tropics proper and the temperate areas. It is the desert belt, both on land and in the sea. Cloud cover is low, evaporation rates are high, and salinities, therefore, attain values well above average. This zone is invaded by the neighboring climate regimes depending on seasons. Annual temperature ranges can be very high. Coastal areas show seasonal upwelling, depending on wind strength.

The *temperate zone* is strongly influenced by seasonal change. Rainfall generally exceeds evaporation, and salinities are correspondingly reduced. Being the transition between the warm and the cold regions of the planet, the temperate zone has strong temperature gradients, hence strong winds. These force mixing of surface waters with (nutrient-rich) waters in the upper thermocline, along the west-wind drift. The temperate zones are fertile regions, therefore. In the poleward parts of the temperate zones, mixing is further enhanced by seasonal break-down of the thermocline.

The *polar areas,* finally, take up the smallest portion of the globe, but they are of prime importance as makers of climate. Their ice rim fixes the endpoint of the overall temperature gradient, which ultimately controls winds, currents, and evaporation-precipitation patterns (Figs. 4.14 and 7.2).

The climatic zones are not exactly parallel to latitudes: note how the boundaries are shifted by the currents of the subtropical gyres, especially the Gulf Stream and its west-wind extension.

7.1.2 Temperature and Fertility are the most important climatic factors in the ocean, as far as the production and distribution of biogenous sediments. In the

a

b

Fig. 7.1 a, b. Climatic extremes in the present ocean. **a** Fringing reef and lagoons, Huahine, Society Islands, tropical South Pacific. [Courtesy D. L. Eicher] **b** Tabular iceberg, Antarctic Ocean. [Courtesy T. Foster]

tropics and subtropics nutrients are limiting to growth. Here the surface waters become depleted of nutrients because the density gradient (light warm water on top of cold heavy water) hinders upward diffusion of nutrient-rich deep waters. Generally, therefore, production is low. In high polar areas, light is the limiting factor: productivity also is low. In the coastal zones of the tropics and in temperate latitudes,

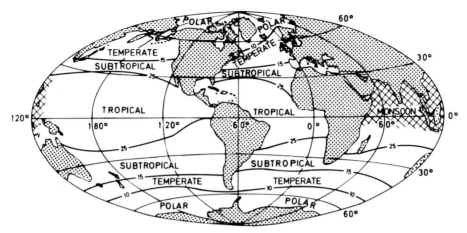

Fig. 7.2. Climatic zonation of the open oceans. Zone boundaries tend to follow latitudes and line up with climatic belts on land. Temperature, seasonality, and water budget (evaporation-precipitation balance) are the most important descriptors. Temperate and polar can be separated by another zone: subpolar. Approximate temperatures of surface waters in °C shown at the boundaries.

the supply of light and of nutrients both is adequate; here production is high (see Sect. 6.1).

Temperature imprints itself on sediment patterns in several ways, either directly (through association with ice-generated debris or warm-water coral, for example) or indirectly (through correlations with the productivity of the ocean). Very warm waters lack nutrients, very cold ones lack sunlight, as mentioned.

More subtle is the relationship between *temperature profile* and production (and hence output of biogenous sediment). To grow, marine algae need to stay close to the surface of the ocean, in the sunlit zone. To replenish nutrients lost from this zone, by the settling of fecal material and organic aggregates, we need deep mixing. However, such mixing also removes a large portion of the algae from the sunlit zone, taking them to greater depth, where they cannot grow. Clearly, a *succession* of nutrient delivery and stable stratification in time or space provides the most favorable conditions for growth, and hence for sediment production.

Blooms of coccolithophorids or diatoms must be contemplated with these mechanisms in mind. Such blooms can be extensive ("whitings" seen from satellites in the case of coccolithoporids). They may, in fact, deliver much of the biogenous sediment found on the sea floor – coccoliths in the low latitudes, diatoms in high latitudes and coastal regions. Recent flux measurements with sediment traps support this concept of *episodic flux*.

Temperature anomalies, specifically cold anomalies within the subtropics and along the equator, denote regions of upwelling, which are of special interest as producers of organic-rich sediments (and as fishing areas) (Sect. 4.3.3). *Equatorial upwelling,* which is due to divergence of westward-moving currents where the Coriolis Force changes sign, leaves a strong imprint in the shape of increased sedimentation rate of calcareous, siliceous, and phosphatic materials. Also, certain metals are concentrated below the equator, by the upwelling mechanism. We shall return to these subjects when discussing deep-sea sedimentation (Chap. 8) and ferromanganese nodules (Sect. 10.4).

Nutrient supply, as already mentioned, is increased through vertical mixing and upwelling. Upwelling is highly seasonal, being induced by the wind field (Fig. 4.19), which is quite variable, of course. Thus, the major upwelling regions – off Peru, California, northwestern Africa, Namibia, and also in the Arabian Sea – show high but greatly varying productivity.

Seasonal upwelling is largely tied to monsoonal activity. On the continents, seasonal contrasts are exaggerated, compared with the oceans (continental climate). Around the large and high land masses of East Asia, this seasonal contrast affects the entire northern Indian Ocean (monsoonal climate, see Fig. 7.2, hachured area). During winter, cold air masses flow off the continent (northeast monsoon in India) bringing drought and upwelling. During summer, the circulation is reversed: the southeast monsoon brings rain from the ocean. In the eastern Pacific, upwelling virtually ceases every 3 or 4 years for a year or so. *El Niño conditions* are said to prevail during such events. "El Niño" means "the child" and refers to the Christ child, in reference to the observations, by Peruvian fishermen, that conditions of poor fishing start around Christmas. The El Niño phenomenon is not well understood, although it is of great importance even on a global scale. The history of such oscillations can be studied in sediments with annual layers (varves) off California (Fig. 7.17), off Peru, and in the Gulf of California. In the Santa Barbara Basin off Southern California, El Niño layers are characterized by unusually large proportions of warm-water diatoms, and reduced supply of organic matter. Because of this reduced supply, oxygenation increases, and in some cases bioturbation sets in, destroying a number of varves. Thus, the stratigraphy of bioturbation in the Santa Barbara Basin contains valuable clues to the long-term workings of the Pacific wind field, and indeed the global heat budget.

7.2 Biogeographic Climate Indicators

7.2.1 Paleotemperature from Transfer Methods.
Biogenous sediments are excellent indicators of climate: many organisms have narrow ranges of temperature and fertility, and the presence of their remains immediately yields excellent clues for climate reconstruction. Also, the chemistry of the shells can give indications of temperature, salinity, and growth rate. The most widespread biogenous deposit on the present sea floor is calcareous ooze, which consists of plankton shells. Of the shell-making plankton, the foraminifera were found to be the climate indicators par excellence. They are reasonably diverse, widespread, and easily identified under a low-power microscope (that is, after some practice). Planktonic foraminifera live mainly in the uppermost 200 m of the water column. There are about 20 abundant species, and as many rare ones. Each climatic zone has one or more characteristic species and a typical abundance pattern of species within it. Some temperature-sensitive species

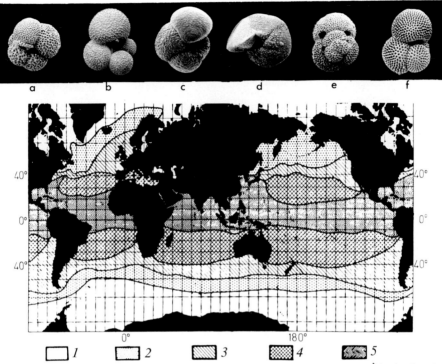

Fig. 7.3. Major distributional zonation of planktonic foraminifera. *1* arctic and antarctic; *2* subarctic and subantarctic; *3* transition zone; *4* subtropical; *5* tropical. Dominant foraminifera: *a: Neogloboquadrina pachyderma, zone 1* left-coiling, *zone 2* right-coiling; *b: Globigerina bulloides, Zone 2; c: Globorotalia inflata, Zone 3; d: Globorotalia truncatulinoides, Zone 4; e: Globigerinoides ruber, Zones 4* and *5; f: Globigerinoides sacculifer, Zone 5.* [Map from A. W. H. Bé, D. S. Tolderlund, 1971, in B. M. Funnell, W. R. Riedel, ed, 1971. The micropalaeontology of oceans. Univ Press, Cambridge. SEM photos courtesy U. Pflaumann, Kiel]

are shown in Fig. 7.3, with the areas they characterize. The general agreement of the faunal zonation with the climate zonation of Fig. 7.2 is obvious – the role of latitudinal bands and of the deflection by surface currents again is apparent.

7.2.2 Transfer Equations. The agreement between faunal and climatic zonation suggests that it should be possible to reconstruct climate by *counting the relative abundance* of the species in the sediment assemblages. This is indeed the method most commonly applied. The counts are used to calculate the original surface water temperatures by means of a *transfer equation,* a term coined by the paleoecologist John Imbrie. To illustrate how this works let us consider a very simple transfer equation, namely the weighted average of the optimum temperature:

$$T_{est} = \Sigma \ (p_i \cdot t_i)/\Sigma p_i \tag{7.1}$$

Here, T_{est} is a temperature estimate based on an assemblage of foraminifera (or other shells), p_i is the proportion of the i-th species, and t_i its temperature optimum. The optimum, in this context, is the temperature at which we find the highest proportion of the species in the calibration set.

The *calibration set* consists of sea-floor surface samples and corresponding surface water temperatures. For example, we may find that *on the sea floor* species no. 3 has its highest proportion underneath surface waters with an annual average temperature of 20 °C. We call this its optimum, and designate it as t_3. For the temperature estimate of a given sample, the product $(p_3 \cdot t_3)$ is one of the terms of the above summation. The larger the p_3, the closer the sum will be to 20 °C. The presence of species with a higher optimum temperature will pull the result up above 20 °C, and vice versa. Under favorable conditions, this simple equation will estimate the correct temperature to within better than 3 °C. The technique can be improved by considering the *temperature range* of each species, and by regression of the estimated temperature on the expected temperature, in the calibration set. The method then gives excellent results. More powerful transfer techniques also exist, based on the same principle of calibration. Factor-regression analysis is one of these techniques. It was used by the CLIMAP group to map conditions in the ice age ocean.

7.2.3 Application of Transfer Methods: Climatic Transgression.

Great progress has been made in determining the temperatures of the sea surface for the latest Pleistocene, especially in the North Atlantic. The information is retrieved from numerous sediment cores distributed throughout the region. An example of such reconstruction, by the CLIMAP group, is given in Fig. 7.4.

The distribution of surface water temperatures – and hence of surface currents and water masses – differs greatly for the last ice age maximum (~ 20 000 years ago) from the present situation.

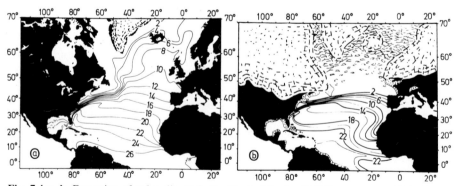

Fig. 7.4 a, b. Expansion of polar climate in the North Atlantic, 17 000 years ago. **a** Today's sea surface temperature in winter. **b** Transfer map, winter surface temperature 17 000 years ago (in ocean), and distribution of ice sheets and pack ice. [Data from CLIMAP in E. Seibold, 1975, Naturwissenschaften 62: 321; see also Science 191: 1131, 1976; Geol Soc Am Mem 145: 464 pp, 1976]

Today, the polar front is just south of Greenland (Fig. 7.4, 2 °C isotherm). During the glacial maximum, the front ran from New York to the Iberian Peninsula. Norway, and even England, were largely cut off from the warming influence of the Gulf Stream, with dramatic climatologic consequences.

The motion of the polar front in the North Atlantic, as reflected in deep-sea cores, is a prime example for climatic transgression. Such a transgression can be used to determine the *rate* of change. How fast does a glacial period change into an interglacial one? How fast can the glaciers build up at the end of a warm period? These are questions of considerable interest, because mankind is involved in *climate modification,* by industrial CO_2 emission and by deforestation, and because the present unusually warm period (since 10 000 years ago) has been about as long as comparable ones in the past, throughout the Late Pleistocene. We shall return to these questions in Chapter 9 when discussing the geologic record.

7.2.4 Limitations of the Transfer Method. The method outlined (or others like it) allows an estimate to be made of any environmental variable which shows a correlation with foraminiferal abundances (or abundances of other plankton: coccoliths, radiolarians, diatoms). However, the significance of such estimates is in doubt where the environmental variable is not in control of growth. A prime example is salinity. There is little evidence that salinity has any influence on plankton distribution within the normal range of seawater salinity.

There are some other problems with the transfer method. First, exact calibration is difficult. Sediment arriving on the sea floor becomes thoroughly mixed with older sediment. Thus, the calibration set of sea-floor samples contains information spanning thousands of years. Yet, the calibration set of present sea surface conditions is based on a very few years at most. Are these last few years – the ones whose data enter the present-day temperature atlas, for example – representative for the last two or three thousand years? Perhaps.

Second, there is the problem of *differential dissolution,* hence *selective preservation* of plankton shells on the sea floor. Most of the calcareous ooze on the deep-sea floor is exposed to at least some dissolution. Thus, the assemblages from which we wish to draw paleoclimatic information are being altered on the sea floor. The more delicate shells are being removed, while the thick-walled forms are being concentrated (Fig. 7.5). This is also true for siliceous shells, although dissolution patterns are different.

For older deposits the question arises: to what degree has evolution changed the optima and ranges of the species? The question is ever-present in paleoecologic research, and is always difficult to deal with.

7.3 Diversity and Shell Chemistry as Climatic Indicators

7.3.1 Diversity Gradients. One way to map changing climatic patterns is to focus not on the species themselves but on the statistical patterns they make: their *diversity* and their *dominance.* Diversity can be measured in various ways; the most realistic

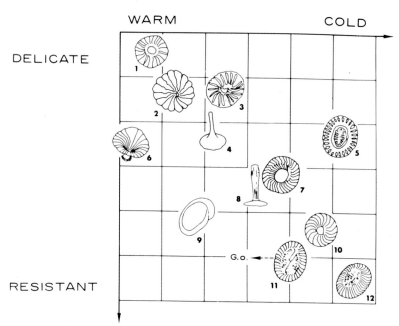

Fig. 7.5. Effect of selective dissolution on the temperature aspect of a mixed coccolith assemblage. Warm-water forms are removed and cold-water species are enriched in the assemblage as dissolution proceeds. *1: Cyclolithella annula; 2: Cyclococcolithina fragilis; 3: Umbellosphaera tenuis; 4: Discosphaera tubifera; 5: Emiliania huxleyi; 6: Umbellosphaera irregularis; 7: Umbilicosphaera mirabilis; 8: Rhabdosphaera stylifera; 9: Helicopontosphaera kamptneri; 10: Cyclococcolithina leptopora; 11: Gephyrocapsa sp.* (includes *G. oceanica* and *G. caribbeanica;* the latter is shown); *12: Coccolithus pelagicus.* [W. H. Berger, 1973, Deep-Sea Res. 20: 917]

definition is the number of species in a standardized sample. For any sufficiently large group of shallow-water organisms, both benthic and planktonic, the number of species is high in the tropics, low toward the poles. There is, then, a diversity *gradient,* which runs roughly parallel to the global temperature gradient. In coastal areas, however, anomalously low salinity values or high influx of mud can adversely affect the marine habitat which can lower diversity independently of temperature.

Quite commonly, abundances of certain species are very high in areas of low diversity, because the physical rigor of the environment excludes competition, or enemies, or both. The great dominance of such species indicates environmental stress. Isolated lagoons, brackish marginal seas, and Arctic shelves illustrate the principle of low diversity and high individual abundance under conditions of physical stress. So do polluted lakes, for example.

7.3.2 Oxygen Isotopes. Another way to track climatic change in paleoclimatic reconstruction is to find relationships between shell chemistry and environment. The most widely applied method in this regard is the *oxygen isotope technique* of the famous chemist H. C. Urey. Urey showed, in 1947, that carbonates precipitated from the same aqueous solution should have different ratios of oxygen-18 to oxygen-16, depending on the temperature at which the precipitation proceeds.

The usefulness of this concept to paleoclimatic research was established soon after, by S. Epstein, R. Buchsbaum, H. A. Lowenstam, and H. C. Urey. They compared the isotopic composition of mollusk shells which grew at various temperatures, in their natural environment. Results showed that there was an orderly relationship between the isotopic composition and the temperature of growth. A fit to their data points yielded an equation which is widely used for paleo-temperature determination:

$$t = 16.5 - 4.3 \, (\delta_s - \delta_w) + 0.14 \, (\delta_s - \delta_w)^2. \tag{7.2}$$

The terms are as follows: t is the temperature, δ_s is the oxygen isotope composition of the shell sample, δ_w is that of the water the shell grew in. The δ notation describes the deviation of the ratio of oxygen-18 to oxygen-16 from that of a standard (for example mean ocean water), as a fraction of that of the standard:

$$\delta^{18}O = \frac{^{18}O/^{16}O \, (\text{sample}) - \, ^{18}O/^{16}O \, (\text{standard})}{^{18}O/^{16}O(\text{standard})} \cdot 1000 \, . \tag{7.3}$$

For convenience, it is expressed in *"per mil"*; that is why the above fraction is multiplied by 1000.

Whenever an observed relationship between temperature and isotopic composition follows Eq. (7.2), the shell is said to have grown in *isotopic equilibrium* with seawater. While mollusks (and planktonic foraminifera) generally precipitate shells in oxygen isotope equilibrium by this criterion, many organisms do not. Furthermore, the δ_w is variable geographically and through time, especially in coastal regions. It is closely related to the salinity of the ocean, because evaporation and precipitation affect both $^{18}O/^{16}O$ ratio and salinity patterns. Also, it depends on *ice volume*. Unless these effects can be excluded, the paleotemperature cannot be deduced from the oxygen isotopes, other than within rather broad limits (see Fig. 5.15 and Sect. 9.3.4).

7.3.3 Carbon Isotopes. Besides oxygen isotopes, there are two stable *carbon isotopes* in calcareous shells: carbon-12 and carbon-13. These also have a temperature dependence, but the main effect on their ratio is the $^{13}C/^{12}C$ ratio in the dissolved inorganic carbon of the seawater and the metabolic activity of the shell-secreting organisms. Fast-growing organisms, and those harboring symbiotic unicellular algae (zooxanthellae), precipitate shells with anomalous values of carbon-13. However, many shells yield information about the $^{13}C/^{12}C$ ration in the seawater they grew in.

The $\delta^{13}C$ signal in planktonic and benthic foraminifera – an index for the ratio between the isotopes ^{13}C and ^{12}C in their shells – has recently emerged as a major tool in reconstructing the marine carbon cycle. In surface waters, the habitat of planktonic foraminifers, carbon is extracted by photosynthesis ("fixed") into organic matter. This process runs slightly faster for ^{12}C than for ^{13}C, and hence ^{12}C is incorporated preferentially. Basically, because of this preferential fixation, ^{13}C is ultimately enriched in the surface water, and this enrichment is reflected in the shells precipitated by the planktonic forams. Organic matter – enriched in ^{12}C – tends to sink and is oxidized at depth, releasing fixed carbon back to the water as CO_2. Thus, deeper water is richer in ^{12}C than surface water, and benthic shells acquire correspondingly lower $\delta\ ^{13}C$ values than planktonic ones. At the same time, the deeper water also has more CO_2 by the same process. In principle, therefore, we can say that the difference in $\delta\ ^{13}C$ values of co-occurring benthic and planktonic foraminifers is a measure of the efficiency of extracting carbon out of surface waters. This fundamental connection between the *carbon isotope compositions of benthic and planktonic shells* on the one hand, and the *pumping of carbon* from the surface to depth on the other, was first pointed out by the geochemist and oceanographer W. S. Broecker. It has important implications for the reconstruction of the CO_2 pressure in surface waters (and hence the atmosphere) from the $\delta\ ^{13}C$ values in fossil foraminifera on the sea floor.

7.3.4 Other Chemical Markers. Other useful markers include both metals and certain organic compounds, which are incorporated differently into shells or into organic matter, depending on productivity or temperature. For example, *cadmium* has been used for the reconstruction of nutrient patterns; its concentration parallels that of phosphate, and it is incorporated into foraminiferal shells in accordance with its abundance in seawater.

Magnesium in calcareous shells tends to increase with temperature. This effect appears to be more pronounced in the lower organisms, such as calcareous algae and foraminifera, than in the higher ones, for example the barnacles. The relationship has not been exploited for paleoclimate work to any significant degree.

Mineralogy follows temperature to some degree. There is a tendency for relatively more *aragonite* to be precipitated by benthic organisms in warm waters than in cold waters, where *calcite* is distinctly favored. Aragonite is the less stable form of calcium carbonate and changes to calcite during diagenesis. Thus, aragonite content of ancient calcareous sediments is largely a result of post-depositional alteration, rather than of conditions of formation.

7.4 Coral Reefs, Markers of Tropical Climate

7.4.1 Global Distribution. The classic geologic method of mapping climatic belts independently of individual species is to take typical shelf assemblages, such as *coral reef associations.* Even if the species within the associations have changed their preferences, as a whole, the reef association is believed to indicate tropical shallow waters. Today, the tropical ocean is the core region for coral reefs (Fig. 7.6). Reefs extend into the subtropics on the western sides of the ocean basins, but are greatly restricted on the eastern sides. They cover about 4 % of the total shelf area at the present time. The warm western boundary currents are responsible for carrying the reef limits to their latitudinal extremes, while eastern boundary currents bring cold water from high latitudes and are responsible for the restricted distribution of reefs on the eastern side of ocean basins. The reefs of the west Atlantic island Bermuda reach to about 30° N thanks to the Gulf Stream. On the eastern margin, in the sphere of

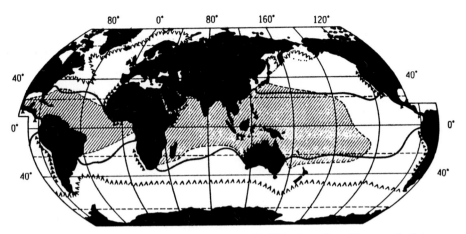

Fig. 7.6. Distribution of coral reefs *(hachured areas)*. 20 °C isotherm for coldest month of the year shown as *heavy line*. Note the general correspondence. Areas of coastal upwelling *(triangles)* inhibit coral growth. Hence the eastern margins are unfavorable for reef development. Present maximum equatorward extent of iceberg drift also is shown. [E. Seibold in R. Brinkmann [ed]. 1964, Lehrbuch der allgemeinen Geologie. F. Enke, Stuttgart; coral reef distribution after J. W. Wells, 1957, Geol Soc Am Mem 67: 609]

influence of the Canary Current and of upwelling, reef organisms begin to grow only south of Dakar (15° N). However, on the whole, the equator does denote the mid-line of the distribution, an observation which is useful in paleogeographic reconstructions.

7.4.2 Reef Building. *"Coral reefs"* are built up by many different kinds of organisms, and corals do not necessarily contribute the bulk of the material. Up to 10 000 g of calcium carbonate per square meter can be produced annually, with a rate of upbuilding of 10 mm per year. Calcareous ooze, for comparison, builds up a thousand times more slowly, with accumulation rates near 10 $g/m^2/yr$. Part of this high production of calcium carbonate is based on photosynthesis by unicellular symbiotic dinoflagellates living within the corals. During daytime, they use CO_2 and thereby promote precipitation of calcium carbonate. Skeletons are deposited up to five times faster during daytime than during the night. Seasonal temperature variations are indicated in annual layers (see Fig. 6.8).

About 50 % of the reef is solid matter and 50 % is empty space (that is, pore space filled with water). Where the sea floor sinks, reefs can build enormous shelf bodies (on continental margins), or mountain peaks (on sunken volcanoes).

How is the reef built? To begin with, we have to distinguish between the *framework* (the supporting structure) and the *fill,* much as in building construction. The corals proper, and, just as importantly, the calcareous algae, are involved in producing the framework. Massive or (in protected places) branching stony corals can grow up by more than 2.5 cm/yr, as individual colonies. In the shallowest water, encrusting algae occur which grow over dead portions of the reef and cement it to such a degree that it can resist even the strongest surf. These encrusting algae, too, can reach a high rate of upward growth over considerable areas. The fill between the framework consists of reef debris and the hard parts of various other reef inhabitants. The fill can provide up to nine tenths of the reef mass.

The reef dwellers are legion, for the reef offers an extraordinary variety of life styles (Fig. 7.7). Solid substrate favors epibionts such as bryozoans and sessile foraminifera. Cavities offer shelter for fish and crustaceans. The rich and colorful mollusk fauna is familiar from underwater photography and (rather unfortunately) from souvenir shops.

The tropical coral reef is the shallow-water habitat with the greatest number of species, that is, with the greatest diversity. More than 3000 species live in Indo-Pacific reefs. The skeletal remains produced in rich variety are broken up in various ways, by boring algae, sponges, and sea urchins. Much is crushed by fishes and crustaceans. In many skeletons, calcite crystals are held together by organic material. When it decays, such skeletons disintegrate to lime mud (*micrite*). Finally, waves and tidal currents grind up the relatively soft calcareous matter. The debris fills the interstices within the reef structure, collects in the lagoons behind the reefs, and washes up on the beach. Also, debris trickles down the sides of the reef into greater depths, building the reef flanks. Dissolution of aragonite, recrystallization and cementation of the debris start very soon after deposition. These processes (important in the investigation of fossil reefs as oil reservoir rocks) are influenced by microbial activity as well as interaction with rainwater. Already a slight drop in sea level would expose large areas to rain water. Thus, phreatic processes (karstification, cementation) were extremely important throughout the Pleistocene.

There are other concentrations of skeletal carbonates sometimes called "reefs" even in arctic waters (red algae, mollusks in shallow waters or corals of the genus *Lophelia* in deeper waters). These are quite distinct from tropical reefs, however.

7.4.3 Atolls. The various processes of reef building have their classic representation in atoll development. Palm-studded rings of coral and shell debris, enclosing a lagoon and sheltering it from the pounding surf, are quite common in the tropical South Pacific. Charles Darwin first explained their origin (Fig. 7.8). His theory was built on the idea of a sequence going from *fringing reef* to *barrier reef* to *atoll*. The fringing reef grows immediately next to the land. The barrier reef is separated from the land by a shallow lagoon. The atoll, finally, is an isolated quasi-circular ring of reefs, with an emergent portion of reef debris, the atoll island. In the center of the ring is a lagoon. Dimensions vary; a large atoll can be up to 40 km in diameter. Since the algae, both as carbonate producers and as symbionts of corals, need light, the coral reefs can grow only in very shallow water. An atoll, therefore, cannot have grown up from deep waters. This is the reasoning that led Darwin to propose a combination of slowly sinking sea floor and upbuilding of reef.

Darwin's theory has since been proven correct by drilling into several atolls. In the drillhole on Enewetak Atoll (Marshall Islands of the tropical Pacific) the basaltic basement, of Eocene age, was hit at about 1400 m depth. Detailed investigation of the cores recovered from the Enewetak drill hole indicated that the reef had on occasion emerged from the sea: land snails, pollen, and spores from terrestrial plants are

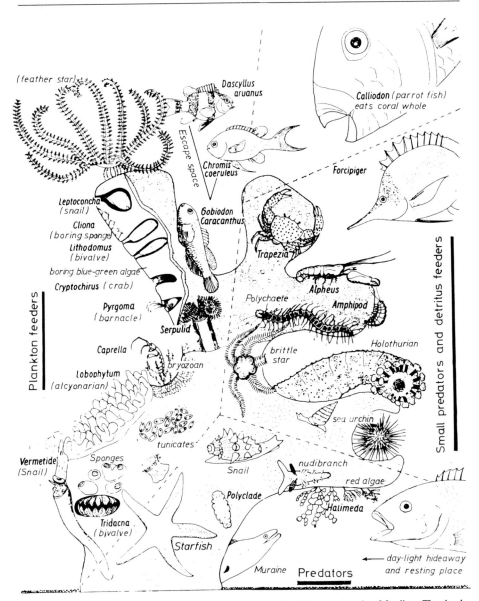

Fig. 7.7. Stone coral colony with associated macrofauna, classified as to mode of feeding. The dead basal part of the coral forms a hard substrate on which epibenthos settles. Note the boring organisms (e. g., *cliona*) which help destroy coral structures. [S. Gerlach, 1959, Verh Dtsch Zool Ges 356]

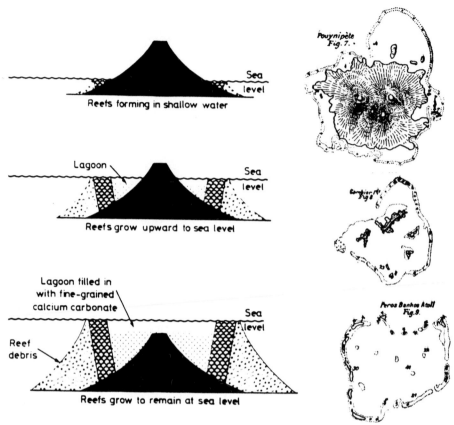

Fig. 7.8. Darwin's hypothesis of atoll formation. Redrawn, with maps by Darwin (right) to show the resemblance in form between barrier coral reefs surrounding mountainous islands, and atolls or lagoon islands. [C. Darwin, 1842. On the structure and distribution of coral reefs, Ward Lock & Co, London]

present. Also, there are signs of freshwater influence in the chemistry of the limestones, and there are hiatuses – gaps in the record – indicating that erosion took place at times during the Tertiary. A similar situation is indicated on Bikini Atoll, 200 miles away. In both cases, the hiatuses are at depths of around 200 to 300 m below today's reef surface, although on Enewetak there are also deeper ones.

Are *flat-topped seamounts* ("guyots") in the western Pacific sunken atolls? At least some probably are – shallow-water reef rocks have been dredged from their upper flanks, the tops being covered by pelagic sediment. Alternatively, they are simply eroded volcanoes, planed off at sea level, by waves that readily eat into volcanic ashes and debris. Perhaps the reef organisms were unable to keep up with the rate of sinking, as the dead volcano rode downward with the subsiding crust. Was it too cold? Or were the waters too poor in nutrients? There are many indications that reef building was greatly reduced at times, especially in the middle Cretaceous, and that many guyots originated at that time. The reasons remain obscure, so far. In any case, whether a volcano ends up as an atoll or a seamount may largely depend on the climatic conditions which allowed or prevented the growth of coral reefs.

7.4.4 Great Barrier Reef. The 2000-km-long *Great Barrier Reef,* lying 30 to 250 km off eastern Australia, is the most impressive reef structure in the world today, and demonstrates how biological sedimentation increases the size of continents. The reef grows on subsiding crust, and is most massive in the north (around 10° S), reaching about 1500 m thickness. In the south (at 24° S), it is only 150 m thick. The difference may be related to the migration of the Australian continent. Plate tectonics requires that Australia moved some 12° to the north during the last 24 million years (the "Neogene"). Coral growth could begin as soon as the northern margin crossed the southern limit of growth, about 20 million years ago. The more southern portions crossed this limit later, offering less opportunity for growth. The importance of plate motion for the architecture of the Great Barrier Reef is diminished, however, by the finding (from recent drilling) that the main reef is less than 1 million years old.

7.5 Geologic Climate Indicators

7.5.1 Chemical Indicators. We discussed some of the aspects of biological indicators of climate. Coral reefs, of course, can equally well be considered geologic indicators: they are calcareous sediment bodies whose patterns can be mapped. In fact, simple compilations of reef occurrences through time have been made from time to time to indicate the position of the equatorial belt and, ultimately, the cross-latitudinal motion of the continents. Biogenous carbonate deposits in general yield similar clues (Fig. 7.9). Although the method is not very reliable, it does provide some checks on geophysical results.

Salt and dolomite deposits also have been used to delineate climatic belts. They are characteristic for the subtropics and result from excess evaporation as discussed (Sects. 3.7.3 and 7.6). In tropical areas, laterite and the clay mineral kaolinite are typical weathering products. They find their way into deep-sea deposits, where they can be used to trace climatic changes on land (see Sect. 8.4.3).

7.5.2 Physical Geologic Indicators are especially obvious in high latitudes, due to the effects of ice (Fig. 7.9).

On shelves entirely covered by ice, as in the Antarctic, erosion predominates. Bare rock, polished and scratched, dominates; morainal debris fills depressions. Icebergs calving from glaciers take enclosed debris with them, and release it as dropstones in melting (Fig. 7.10). During the glacial periods of the Pleistocene, the area of the sea floor affected by iceberg sedimentation was considerably expanded (Sect. 3.2.2). Also, extensive shelf areas were exposed at the time due to the sealevel drop, and glacial ice could deposit *end moraines* in their typical lobe-shaped morphology, as on the East Coast north of New York and on the area of today's North Sea and Baltic. These moraines, then, are glacial relicts which are being worked over by shelf currents at the present time (Sect. 5.4).

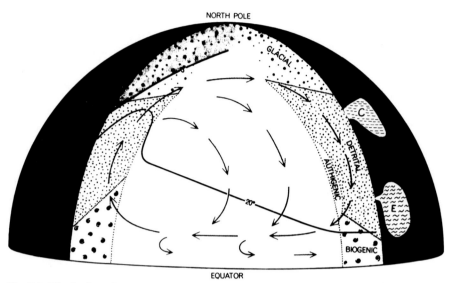

Fig. 7.9. Distribution of shelf deposits in relation to climatic zonation, in the northern half of an idealized ocean basin. Continents *black*. Current shown as *arrows*. *Detrital* refers to terrigenous river influx, *authigenic* to phosphatic deposits associated with organic-rich sediments, *biogenic* to carbonates (mollusks, foraminifera, corals, and algae). Shell carbonate is produced outside the tropics as well but tends to be masked by terrigenous supply. Restricted shelf seas collect carbon-rich *(C)* or evaporitic sediment *(E)* depending on climate zone. [K. O. Emery, 1969, Sci Am 221 [3] 106, modified]

Many high-latitude shelves (off Siberia, and off northern Alaska) are covered seasonally by sea ice. The rivers entering these shelf seas are commonly ice-covered, except for a few summer months. When the ice breaks up and the flood sets in, the freshwater may stay on top of the sea ice, off Alaska up to 10 km offshore. It penetrates downward through cracks in the ice and can dig 10 to 20-m-wide *potholes* on the sea floor below the places of entry.

The sea ice, 1 to 4 m thick, hinders the transport of sediment by dampening waves and longshore currents. The absence of sufficient longshore transport can perhaps explain the scarcity of lagoons in the Arctic. The sea ice also shields the entire Arctic sea floor from pelagic sediment supply: sedimentation rates in the central Arctic ocean are extraordinarily low (~ 1 mm/1000 years).

The ice can pile up, compressed by ocean currents, as so vividly described by Fridtjof Nansen (1861–1930), the great Arctic explorer. On the shelf, the ice can then scrape the ground and produce deep, long ravines. These shelf areas are evidently distinctly unfavorable for the laying of pipelines or the building of platforms. Furrows produced by drift ice during the last glacial have been discovered by side-looking sonar (Sect. 4.3.2) on the Scottish and Norwegian shelves out to the shelf rim.

Arctic environments are very demanding as regards benthic life. Paucity of species is characteristic. One lagoon in Alaska southeast of Point Barrow, about 8 km wide

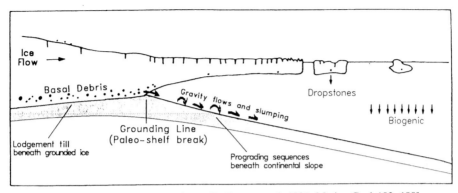

Fig. 7.10. Ice-sheet depositional model. [A. K. Cooper et al. 1991, Marine Geol 102, 180]

and 4 m deep, is frozen over for 9 months in the year. On freezing, salt is largely excluded from the ice; thus the salinity of the remaining water is increased to more than 65 ‰. When the ice and the snow cover melts, in early summer, the water is diluted to a salinity of 2 ‰. When the sea ice melts, in mid-summer, normal marine waters enter the lagoon and salinity increases to 30 ‰. In October, the lagoon freezes over again, the cycle starts anew. Clearly, only very few types of shell-producing organisms (or burrowers) can be expected under such extreme conditions.

The supply from rivers in high latitudes consists mainly of silt, with clay and sand being distinctly subordinate. This is an interesting sedimentologic phenomenon, presumably caused by mechanical weathering through the freeze-thaw cycle. The process is important in producing *loess,* the silty sediment blown out from glaciated areas and piled up in their perimeter.

7.6 Climatic Clues from Restricted Seas

7.6.1 Salinity Distributions and Exchange Patterns. In contrasting tropical reef and ice-carved shelf, we have compared the warm and the cold extremes in the global ocean. There is another important contrast, that between regions of excess evaporation and of excess precipitation. These conditions are known on land as *arid* and *humid.* The ocean being water, the words "arid" and "humid" might seem to make little sense in describing the corresponding climatic belts in the ocean. However, we retain the terms as convenient descriptors of the balance between evaporation and precipitation.

Climate-produced salinity differences in the open ocean are present; they outline the major patterns of evaporation excess (central gyres) and precipitation excess (equator, temperate to high latitudes) (see Fig. 7.11). Evaporation on land is estimated to reach 71 000 km^3/yr, at sea much more, some 524 000 km^3/yr.

The salinity differences in the open sea are too small to leave much of a direct record via inorganic or biogenous sedimentation. However, the differences are ampli-

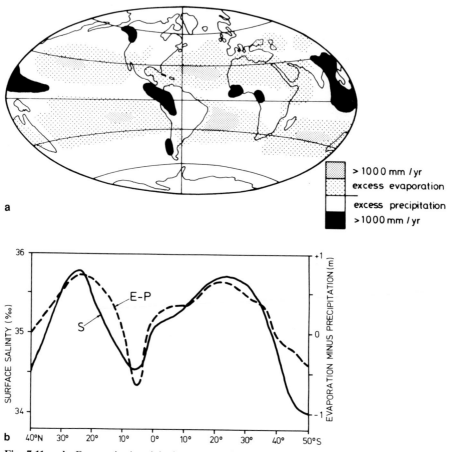

Fig. 7.11 a, b. Evaporation/precipitation patterns in the world ocean, and corresponding salinity distribution. **a** Map showing difference between precipitation and evaporation (E. Seibold, 1970, Geol Rundsch 60: 73, based on G. Dietrich, 1957). **b** Surface water salinities of the ocean, averaged by latitude, and compared with evaporation-minus-precipitation values. [H. U. Sverdrup et al., 1942, The oceans. Prentice-Hall, Englewood Cliffs p. 124]

fied in the restricted seas adjacent to the ocean basins. We shall illustrate this amplification and its effects on sedimentation in some detail, believing that these concepts are able to guide the interpretation of a large class of ancient marine sediments from both shallow and deep basins (see also Sect. 4.3.5).

The seas in the arid belt, with excess evaporation, have a common typical exchange pattern with the open ocean: shallow-in, deep-out (Fig. 7.12). This circulation pattern is called *anti-estuarine*. Prime examples are the Mediterranean, the Persian Gulf, and the Red Sea. Here, the loss of water by evaporation greatly exceeds the influx from rain and rivers. Thus, the sea level drops and water enters through a

Atlantic – Gibraltar – Mediterr.   Atlantic-Sill-Norwegian and Greenland Fjords
Indic – Bab el Mandeb – Red Sea   Mediterr. – Bosporus – Black Sea
Indic – Hormus – Pers. Arab. Gulf   North Sea – Belts – Baltic

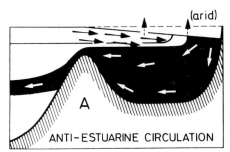
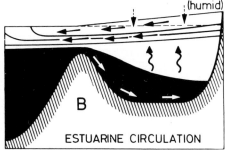

Fig. 7.12. Anti-estuarine and estuarine circulation in basins with excess evaporation and with excess precipitation, respectively. The arid basin (**A**) is characterized by downwelling, hence low fertility and high oxygen content. The estuarine basin (**B**) is characterized by upwelling and salinity stratification, hence high fertility and low oxygen content. The *geographic names above the graph* give three examples each for anti-estuarine and estuarine circulation. [G. Dietrich, K. Kalle, 1957, Allgemeine Meereskunde. Boerntraeger, Berlin, modified]

passage from the adjacent open ocean. The incoming water, of course, is derived from the surface water of the ocean, because it is at the surface that the downhill gradient exists. The excess evaporation within the arid sea increases the salinity of the inflowing water, which increases its density and makes it sink. Thus, the arid basin fills up with heavy, saline, surface-derived water. Being heavier than the open ocean water at the same elevation outside, the deep saline water pushes out and spills over the sill, at depth (Fig. 7.12). In this fashion the circulation of shallow-in, deep-out, is generated.

Where rain and river influx exceeds evaporation, sea level rises, and the water in the humid sea is freshened out and becomes *lighter* than the open ocean water.

Thus, the heavier ocean water pushes in at depth and displaces the less saline water from below. The incoming water is *subsurface* water. The outflow proceeds on the surface, where the gradient makes the water run down into the ocean. Examples for this type of circulation (deep-in, shallow-out) are the Black Sea, the Baltic, and the fjords of Alaska and Norway. This circulation pattern is called *estuarine*, or *fjord-like* (see also Sect. 4.3.5).

Whether a marginal basin has an "estuarine" or "anti-estuarine" circulation has profound consequences for its fertility and sedimentation. A comparison between the Baltic as the humid model and the Persian Gulf as the arid model will highlight the differences.

7.6.2 The Baltic As Humid Model of a Marginal Sea. First, the salinity distributions. In the Persian Gulf, *surface salinity* increases in going from the entrance near Hormus into the inner Gulf. The high salinity water *flows outward along the bottom*. In the Baltic, salinity *decreases* at the surface, from the entrance toward the inner bight. The heavy marine water *comes in at the bottom,* spilling from one depression to the next. In the Baltic, low salinities are prevalent, generally less than 18 ‰, due to dilution from rain and rivers. In the Persian Gulf, salinities reach 40 ‰.

Both excessively low and excessively high salinities are unfavorable for normal marine organisms (Table 7.1). In the Baltic, the fully marine fauna of the entrance is gradually reduced in going deeper into the basin. Correspondingly, the dominance of certain brackish-tolerant species greatly increases. The marine mollusks become smaller and their shells thinner.

Another consequence of the humid condition is that the lagoons at the rim of the Baltic can easily freshen out, depending on precipitation, influx, and degree of isolation. In such enclosed embayments, reeds and other plants grow to build up peat, eventually to be fossilized to coaly layers. The "Wealden" of England and NW Germany, a sedimentary sequence at the Jurassic-Cretaceous boundary, is a deposit of such an environment. The analogous boundary conditions in the Persian Gulf, of course, are hypersaline lagoons and evaporite pans.

Finally, *stratification of the water column* is much more stable in the Baltic than in the Persian Gulf, due to the high density contrast between brackish surface water and saline deep water. In summer, when the surface waters in the Baltic are heated to 15 °C or so, density difference and stability are further increased. Thus, vertical mixing virtually ceases; but without such mixing, no new oxygen can be supplied to the deep waters from the surface waters which are in contact with the atmosphere.

We can map the stratification using benthic foraminifera distributions (Fig. 7.13). Where the boundary between the more saline bottom waters and the more brackish surface waters touches the basin slope, a sharp boundary between upper and lower benthic fauna develops. A change in carbonate content and other sediment properties also occurs here – clues which can be useful for the interpretation of ancient marine sediments.

Productivity in the Baltic sea is high because nutrient-rich *subsurface waters* enter from the North Sea. The nutrients are trapped within the Baltic through organic sedimentation (algae, fecal matter) and are partially recycled from the sea floor. They reach surface waters through mixing by winter storms and by diffusion. The high supply of nutrients within the basin, however, leads to high productivity which in turn delivers much organic matter to the sea floor. Oxygen is rapidly used up by decay of this matter. Especially during periods of stable stratification, the bottom water de-

Table 7.1. Decrease of species from open ocean into Baltic. (After Remane 1958)

Categories	North Sea	Baltic entrance	Bay of Kiel	Middle Baltic
Marine Bivalves	189	42	32	5
Marine Snails	351	68	49	9
Cephalopoda	32	5	4	–

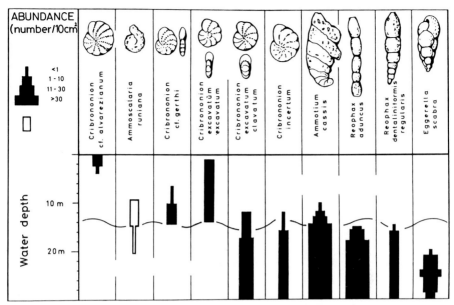

Fig. 7.13. Bathymetric distribution of the abundant benthic foraminifera in the western Baltic. *Width of black columns* indicates number of living specimens found per 10 cm^2 seabottom *(Ammoscalaria: only empty shells were found in 1963/64)*. The *wavy line* at about 14 m depth marks the boundary between outflowing surface water and incoming marine water. [G. F. Lutze, 1965, Meyniana 15: 75]

velops a serious oxygen deficiency, with values of less than 10 % saturation. Under certain condition *all* the oxygen can be used up, and H_2S (the foul-smelling gas) develops from bacterial sulfate reduction.

7.6.3 Condition of Stagnation. As the oxygen content drops, the bottom fauna changes. At a concentration of less than 1 ml O_2 per liter of water, shelled organisms disappear. Only a few species of metazoans remain: mainly worms (annelids, nematodes) and crustaceans. They stir the uppermost sediment, feeding on deposits, leaving burrows but essentially no hard parts. Below 0.1 ml O_2 per liter, the environment becomes hostile to all metazoans; only some protists survive, and the various anaerobic bacteria.

As mentioned, both high productivity and stable stratification are responsible for the oxygen deficiency in Baltic bottom waters. If nutrient supply is then involved in producing oxygen deficiency, could man's activities increase "stagnation" of the Baltic? Certainly. Deforestation and agriculture greatly increase soil erosion and hence the input of soil extracts, that is, nutrients. Also, modern agricultural methods in surrounding regions rely on fertilizer, much of which enters the Baltic Sea with the runoff. Sewage and industrial waste is yet another source. Thus, stagnation is indeed favored by man's activities. However, sediment cores from the central Baltic show that stagnation occurred frequently even before human factors could play a role. This is reflected in the portions of the record which consist of finely laminated sediments, with layers of millimeter dimensions (Sect. 3.9, Fig. 3.13). These layers indicate the absence of deposit feeders and other burrowers; hence they show the absence of oxygen.

The increase in stagnation in the central Baltic during the last 50 years, then, *could* be largely due to human activities, but it could also be a natural phenomenon which will eventually reverse itself. This example illustrates a very general problem in *environmental research and engineering,* namely that it is generally extremely difficult to keep apart the natural background fluctuations and the human influences.

The chemistry of anaerobic sediments is complex. Only a few aspects can be mentioned here. The CO_2 concentration in the bottom-near water with very low O_2 content is high: CO_2 is produced as the O_2 is used up. CO_2 and water produce carbonic acid, and the pH drops (to less than 7, compared with open ocean values near 8). Consequently, calcareous shells are dissolved on the sea floor. High CO_2 content of interstitial waters also means that hydrogenous carbonates can form. Manganese is readily mobilized under conditions of oxygen deficiency, and migrates upward in the sediment as Mn^{2+}. It is kept on the sea floor both by oxidation and precipitation as oxide (MnO_2) and as *manganese carbonate* ($MnCO_3$), which forms under conditions of anaerobism and high CO_2 concentrations. Iron carbonate may also form, although iron is less mobile, being readily precipitated both as sulfide and as oxide. When free oxygen is gone, bacteria produce sulfide by the reduction of sulfate ion

$$SO_4^= + 2\ CH_2O \rightarrow H_2S + 2\ HCO_3^-. \tag{7.4}$$

Iron sulfide forms under these conditions; hence pyrite crystals (FeS_2) are common in black anaerobic sediments. If sulfate reduction occurs at the very surface of the sediment, calcareous shells are preserved, because the destruction of sulfate greatly increases alkalinity.

7.6.4 The Persian/Arabian Gulf As Arid Model of a Marginal Sea. None of the conditions typical for the Baltic Sea develop in the Persian Gulf, because oxygen is brought to the sea floor by sinking saline waters, and because fertility is generally low. The *incoming ocean surface waters* are low in nutrients, as are warm surface waters everywhere. Hence the Persian Gulf is nutrient-starved. One place where anaerobic conditions can develop is in hypersaline lagoons, within the sediment. Sulfate reduction then occurs and organic-rich layers with pyrite can develop. However, such layers should be readily distinguishable in the geologic record from the stagnant basin layers in humid conditions.

The sediments of the Gulf proper are very different from those of the Baltic: organic contents (0.6 to 1 %) are about five times lower, and carbonate content (> 50 %) is more than ten times higher. Benthic organisms exist at all depths and at all times; any lamination is quickly destroyed by bioturbation. Mobilization of heavy metals stops within the sediment because of the high oxygen supply in the mixed layer.

In the Persian Gulf, the incoming surface water brings plankton from the Indian Ocean; some of it survives as the salinity increases within the Gulf; the remainder dies off. Planktonic foraminifera gradually diminish toward the inner Gulf, as seen in the ratio planktonic foraminifera/total foraminifera (Fig. 7.14). The deep-water outflow may help to transport pollutants to the Indian Ocean. Its geologic effects can be seen in shallow-water sediment particles on the continental slope below the Oman shelf break.

7.6.5 Application to Large Ocean Basins. The contrast between estuarine (humid) and anti-estuarine (arid) circulation is apparent even on the scale of very large basins (Sect. 4.3.5). The Mediterranean is the obvious example. Right now it is anti-estuarine, or arid, with sediments high in carbonate, low in organic matter and heavy metals. During the early Holocene, black layers (sapropel) developed in the eastern Mediterranean. The most reasonable explanation is a freshening of surface waters by runoff, and a reversal of the circulation (leading to increased productivity and diminished oxygenation) – a suggestion first made by B. Kullenberg, the Swedish oceanographer who developed the piston corer (Fig. 9.1). His hypothesis has recently received much support from oxygen isotope studies indicating lowered salinities in surface waters during sapropel periods.

On an even larger scale, the North Atlantic Basin is an anti-estuarine or arid basin, with sinking surface water, well-oxygenated bottom water, high carbonate and low organic matter content in sediments. Also, sediments have low opal content, as in the Mediterranean and in the Persian Gulf: diatoms and radiolarians are readily dissolved. In contrast, the entire North Pacific can be seen as an estuarine-type basin,

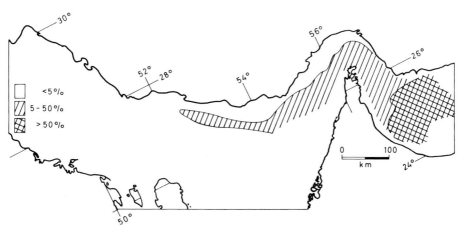

Fig. 7.14. Distribution of planktonic foraminifera abundance in the sand fraction of surface sediments of the Persian Gulf. The number shown is the percentage of planktonic foraminifera, of all foraminifera. The pattern indicates influx of plankton with surface waters entering from the open ocean *(right)* into the Gulf. [Data M. Sarnthein, 1971, Meteor Forschungsergebn Reihe C 5: 1]

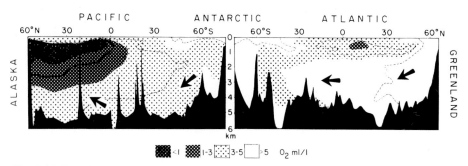

Fig. 7.15. Estuarine *(left)* and anti-estuarine *(right)* aspects of the deep circulation in Pacific and Atlantic ocean basins. The age of the deep waters is reflected in the distribution of the dissolved oxygen concentration. [W. H. Berger, 1970, Geol Soc Am Bull 81: 1385]

because of the low salinities reached in its northern belt (Fig. 7.15). Its sediments, in keeping with the Baltic model, are low in carbonate, high in diatoms and organic matter (at the slopes especially), and much of its deep ocean floor is covered by heavy metal concretions, the manganese nodules (Sect. 10.4).

Together, the North Atlantic and the North Pacific mark the endpoints of a global system of deep circulation that is driven by export of water vapor from the North Atlantic to the North Pacific. This export is sensitive to climatic change. It was reduced during glacial time, and so was the deep circulation which depends on the surface salinity contrast produced by the vapor transport. The replacement of the deep cold water flowing out of the North Atlantic by warmer near-surface water implies a large gain of heat for the North Atlantic. When the anti-estuarine circulation of the North Atlantic is turned down, in glacial times, this heat flux is greatly diminished, favoring the expansion of sea ice (Fig. 7.4 a, b).

7.7 Global Change: the Problem of Detection

7.7.1 An Unusual Decade. A global experiment is now in progress, whereby green-house gases (CO_2, CH_4, and others) are being added to the atmosphere in increasing amounts. Almost all atmospheric physicists agree that this should lead to warming. Indeed, we observe a trend of general increase of global temperatures since 1970 (Fig. 7.16) and a global sea level rise since 1880 (Fig. 5.22). However, the question is how much of this is due to human activity. In fact, based on the greenhouse gases now present in the atmosphere, a warming should already be in evidence, according to some calculations. The unusually high temperatures of the 1980s were widely heralded (especially in the media) as the first results of the increase of greenhouse gases (Fig. 7.16).

Cimatologists have been much more cautious, and for good reason: there is a large natural variability in climate from one decade to the next. An unusually warm

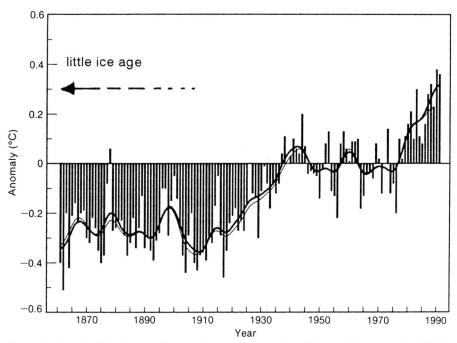

Fig. 7.16. Combined land, air, and sea surface temperatures, for 1861 to 1991, averaged globally, and shown as anomalies relative to the average of 1951–1980. Note the warming since the end of the 19th century, which followed the "little ice age". Temperature data chiefly based on compilations by P. D. Jones and co-workers. [C. K. Folland et al., 1992, in J. T. Houghton, B. A. Callandar, S. K. Varney, eds, Climate change 1992 – Supplement. Cambridge Univ Press, Cambridge, U. K.]

decade cannot be assigned a "cause" unequivocally – it might have been a chance occurrence. How likely is an "unusually" warm decade, under natural conditions?

To answer this question, we have to study climate history on a rather finer scale than is usually done by geologists. *Varved sediments* (Sect. 3.9) and growth of coral (Sect. 7.5) can deliver background on this matter, from the oceanic record. Let us have a brief look at what kind of information we might obtain.

7.7.2 Santa Barbara Basin. Varved sediments are found where a strong oxygen minimum intersects the continental slope in an upwelling region, or where there is an oxygen deficiency in a marginal basin. Such a basin lies off *Santa Barbara,* California (Sect. 3.9). It is about 600 m deep, and its sill intercepts the general oxygen minimum off California, at 460 m. Dissolved oxygen is usually too low for large benthic organisms; therefore the seasonal flux to the sea floor is preserved in finely laminated layers (varves).

The laminated sediments here have a record of productivity, on a year-by-year scale. This record can be read from the content of plankton remains, such as diatoms

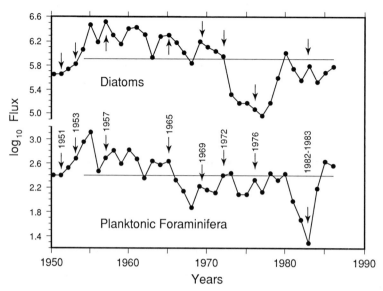

Fig. 7.17. Productivity record in Santa Barbara Basin. The flux values were obtained by estimating the total number of individual fossils within all taxonomic groups preserved in a given varve sample and divided by the depositional surface area of the sample. *Horizontal lines* indicate average fluxes for 1954–1986. *Arrows* mark the occurrence of ENSO events. [C. B. Lange, W. H. Berger, S. K. Burke, 1990, Climatic Change 16: 319]

and planktonic foraminifers (Fig. 7.17). One notes that productivity varies greatly, and that it has been quite low since the early 1970s, compared with the previous two decades. The El Niño event in 1983 resulted in an extremely low supply of planktonic foraminifers, while the El Niño in 1976 was centered on a period of low diatom flux.

7.7.3 El Niño. "El Niño" originally refers to a warm current off southern Ecuador and Peru, which flows southward (see Chap. 7.1.2). During periods when the warm current is especially strong (every 3 or 4 years) large-scale climatic anomalies occur all over the tropical Pacific. The largest El Niño event on record is that of 1982–1983, which was felt all over the globe. It started as an invasion of the eastern tropical Pacific by warm waters from the west, which essentially shut down equatorial upwelling in the east, as well as upwelling off Ecuador and Peru. It was felt even in Tahiti as an unusual warming of the sea surface, resulting in whitening of coral, that is, in stress to its dinoflagellate symbionts. Similar *bleachings* occurred around the Galapagos Islands, and even in Indonesia, the Arabian Gulf, and the Caribbean. Extensive flooding occurred along the west coast of South America. Forest fires extensively damaged the Indonesian rainforest, because of a severe drought: the region of high rainfall had moved east. The list goes on . . . .

Was the 1982–1983 El Niño (or better: ENSO Event, for "El Niño-Southern Oscillation", referring to large-scale air pressure changes) a result of the "global experiment"? This cannot be answered. Perhaps (just perhaps) this event was so unusually severe because (a) it would have been strong in any case (b) it was reinforced by effects from a volcanic eruption (El Chichón, in Mexico) (c) it occurred within the warmest decade on record. Which brings us back to the original question about the unusualness of the 1980s. More will be learned as we study longer records.

7.7.4 The Coral Record. Wherever there are corals in the tropical oceans, there is a record of climatic change within the skeletons of these small colonial organisms (Fig. 6.8). These records are now being actively collected, from Bermuda, the Galapagos, Fanning Island, the Great Barrier Reef, and other places. There are two pieces of information that are especially useful: rate of growth and oxygen isotope composition. There may be other types of useful indices as well, related to biomarkers and trace elements deposited within the skeleton – research on these has just recently started.

Time resolution is in some cases being pushed toward daily fluctuations. The layers representing daily growth are typically less than 20 μm thick. The measurement of daily growth rates should greatly expand the use of corals as indicators of environmental change.

7.7.5 Global Change: the Role of Marine Geology. "Global change" has become the catch phrase expressing the notion that human industrial activities are impacting the planetary environment in major ways, with important consequences for the future of global climate. The health and wealth of nations depend largely on climate, as becomes obvious especially during long spells of drought. Thus, any long-lasting significant changes in weather patterns are of prime concern. At present, *greenhouse gases* (carbon dioxide and methane, among others) are added to the atmosphere in substantial amounts that increase from one year to the next. Extrapolating into the future, computer modeling of the radiation balance of the Earth suggests that a doubling of carbon dioxide over the natural background (expected within the next 50 years or so) should increase global temperatures by some 3 °C with an uncertainty of about 1.5 °C. High latitudes are expected to be much more affected than low latitudes.

There is some evidence that the expected global warming has, in fact, begun: the last decade was the warmest on record since global records have been kept (about the last 100 to 150 years) (see Fig. 7.16). Clearly, an important question is whether this recent increase in global temperatures could be just a chance occurrence, unrelated to human activities. To check this possibility we need *"climate proxies"* that allow us to determine past fluctuations in weather patterns. Also, once we are able to establish the amplitude of natural fluctuations (before the industrial revolution), we should be able to improve the computer models; to be trustworthy, these should generate climate predictions that are tagged with the appropriate uncertainties.

Climate proxies can be derived from a great variety of high-resolution records, such as tree rings, ice cores, and cave deposits. In marine geology, coral growth bands (Sect. 7.4) and varved sediments are of special importance in this context (Sect. 3.9).

A number of temperature proxies are available for extracting climate fluctuations from varved sediments, including microfossil assemblages (warm-water and cold-water indicators; Sect. 6.1.5), oygen isotopes (Sect. 7.3.2), and *alkenone* content of organic matter. The *alkenone* method (see S. C. Brassell et al., 1986, Nature 320: 129) relies on the observation that double-bond abundance within certain long-chain lipid molecules (unsaturated methyl ketones, C_{37}) decreases with temperature. The alkenones are contributed to the sediment by the remains of coccolithophores, in coastal regions predominantly by *Emiliania huxleyi*.

When reading the record of corals or varves to assess climate variability, we come up against a major difficulty. As we go back in time within such high-resolution records, we soon enter the *"Little Ice Age"*, which is a climatic period known to be highly "untypical" when compared with the last several thousand years. The Little Ice Age lasted from roughly 1450 to 1850 A.D. It was characterized by advances of mountain glaciers in most parts of the world, and by occasional spells of unusually cold winters in North America, Europe, and Asia. In Europe, we commonly see paintings from the 18th century which show ice-covered rivers and lakes that have not been frozen for a life time! Thus, a portion of the recent warming we have experienced (Fig. 7.16) may simply be due to a return to more "normal" conditions, such as existed in the early Middle Ages before the Little Ice Age. How, then, are we going to tell which part of the recent warming is "natural" and which (if any) is due to human influence? Clearly, this problem can only be attacked by looking at even longer records, to extract the long-period variability of climate.

It is usually assumed that the Sun's radiation is constant – in fact, the average irradiation of a surface facing the Sun at the mean Earth-Sun distance is referred to as the *"solar constant"* (ca. 1.95 cal/cm^2/min). Unfortunately, the convenient assumption contained in this label is unfounded: the Sun's output is variable. On the scale of centuries this is documented, for example, for changes in abundance of sun spots. Using spectral analysis, we can see the effects of *sun-spot cycles* (22 yr and 11 yr) in many tree-ring data, and even in some marine varves (although less distinctly). So, perhaps the Little Ice Age is part of a long-term solar cycle. Alternatively (or in addition), it may owe its excess of severe winters to unusually intense volcanic activity between the 15th and 19th centuries.

The methods of marine geology – or paleoceanography – can add to the store of knowledge which we need to be able to say, eventually, "yes, we see the global warming". The records recovered from corals and varves establish not only the amplitudes and frequencies of large-scale climatic variations (such as the *"El Niño"* phenomenon in the Pacific). They also tell us about the likelihood of abrupt climatic change. Such changes, which may be defined as unusually fast transitions from one set of climatic conditions to another, occur on various time scales, with various amplitudes. In Section 9.2.4, a dramatic change taking but decades at the beginning of the Holocene will be discussed. The last such change took place in the 1830s, perhaps as a result of unusual volcanic activity. This change is seen both in tree rings and in marine varves (in Santa Barbara Basin, off California). There is a possibility that such climate changes are not completely reversible: a new condition may arise and stabilize itself, although the factor which produced the change has ceased to operate.

Further Reading

Urey HC (1947) The thermodynamic properties of isotopic substances. J Chem Soc: 562–581

Epstein S, Buchsbaum R, Lowenstam HA, Urey HC (1953) Revised carbonate-water isotopic temperature scale. Bull Geol Soc Amer 64: 1315–1325

Ekmann S (1953) Zoogeography of the sea. Sidgwick & Jackson, London

Purser BH (ed) (1973) The Persian Gulf – Holocene carbonate sedimentation and diagenesis in a shallow epicontinental sea. Springer, Berlin Heidelberg New York

Berger A, Schneider S, Duplessy JC (eds) (1989) Climate and geo-sciences. Kluwer Academic, Dordrecht

Bleil U, Thiede J (eds) (1990) Geological history of the polar oceans: Arctic versus Antarctic. Kluwer Academic, Dordrecht

Summerhayes CP, Prell WL, Emeis KC (eds) (1992) Upwelling systems: evolution since the early Miocene. Geol Soc, London

8 Deep-Sea Sediments – Patterns, Processes, and Stratigraphic Methods

8.1 Background

As mentioned in the introduction, deep-sea deposits were first explored in a comprehensive fashion during the British *Challenger* Expedition (1873–1876). Many thousands of samples were subsequently studied by John Murray (1841–1914), naturalist on the *Challenger*. He and his co-worker A. F. Renard published a weighty report on the results, which laid the foundation for all later work in this field of research. The first great step beyond Murray's work were the results of the German *Meteor* Expedition, almost half a century later (1927–1929). A new branch of oceanography started with the recovery of long cores by the Swedish *Albatross* Expedition (1947–1949), that is, Pleistocene oceanography. It revolutionized our understanding of the great Ice Ages. Another great step came in 1968 with *Glomar Challenger* and the Deep Sea Drilling Project, which provided the samples for Tertiary and Cretaceous ocean history.

We shall treat ocean history separately (Chap. 9). Here we summarize the present-day patterns, as they appear in surface samples and in box cores (Fig. 8.1), and discuss some of the important processes of sedimentation, as well as aspects of dating.

8.2 Inventory and Overview

8.2.1 Sediment Types and Patterns. Let us first make an inventory of the types of sediment that are present (Table 8.1). The main types are already familiar from the last chapter. *Pelagic clays* are extremely fine-grained lithogenous and volcanogenic deposits. *Oozes* consist of biogenous materials: shells of planktonic foraminifers, radiolarians, coccolithophores, and diatoms. *Hemipelagic deposits* are the same as clays and oozes except with large admixtures of shelf-derived sediment and of continental material. The list in Table 8.1 is not exhaustive, but includes the bulk of types found on the sea floor. Categories may vary between authors.

The terms *"ooze"* and *"clay"* were introduced by Murray to describe pelagic deposits (*"Globigerina* ooze" and *"red clay"*). Modern pelagic clays of the deep sea are usually reddish brown (rather than red).

The broad outlines of deep-sea sediment patterns are rather simple (Fig. 8.2). The major facies boundary in deep-sea deposits is the calcite compensation depth, that is,

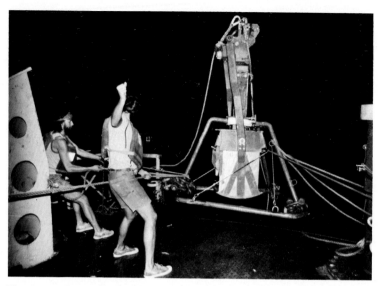

Fig. 8.1. Recovery of deep-sea sediment by box corer. The coring device is a steel box with sharp edges which is pressed into the sea floor by heavy weights on top. The box is closed by a shovel which rotates around two bolts just above the box. The shovel is pushed downward and sideward when the box is pulled out, by pulling on the arm opposite the shovel. The frame with the three legs steadies the corer before it penetrates the sediment. The many lines visible in the photo are used to prevent the heavy device from swinging on deck. This type of box corer was first used by H. E. Reineck. [Photo T. Walsh, S.I.O.]

the boundary between calcareous and noncalcareous sediments. Essentially, the calcareous facies characterizes the oceanic rises and elevated platforms, while the "red clay" facies is typical for the deep basins. Thus, the overall pattern is depth-controlled. Superimposed on this pattern are the siliceous deposits, which accumulate below areas of high fertility, that is, the oceanic margins, the equatorial belt and the polar front regions. The oozes and clays are made of particles which fall down through the water column as a kind of "rain". However, along the margins the deep sea is being invaded by relatively coarse terrigenous materials (mainly silts, but also sands), in places to a considerable distance from the shelf, far into abyssal plains ("m" in Fig. 8.2, also "glacial debris").

8.2.2 Biogenous Sediments Dominate. The bulk of the deep sea deposits consists of biogenous sediments, notably plankton shells (Table 8.1; Fig. 8.3). About one half of the deep-sea floor is covered by *oozes,* that is, sediments formed from plankton remains: coccoliths (ca. 5 to 30 μm), foraminifers (ca. 50 to 500 μm), diatoms (ca. 5 to 50 μm), radiolarians (ca. 40 to 150 μm). The remains of coccolithophores (which are part of the "nannoplankton") are also referred to as "nannofossils".

   The organisms producing the shells drift passively with ocean currents. Some migrate up or down, thus catching different horizontal currents at different depths.

Table 8.1. Classification of deep sea sediments. [W. H. Berger 1974 in C. A. Burk and C. L. Drake. The geology of continental margins. Springer Heidelberg Berlin New York]

I. (Eu-) pelagic deposits (oozes and clays)
    < 25 % of fraction >5 μm is of terrigenic, volcanogenic, and/or neritic origin
    Median grain size 5 < μm (excepting authigenic minerals and pelagic organisms)
    A. Pelagic clays. $CaCO_3$ and siliceous fossils < 30 %
       1. $CaCO_3$ 1–10 %. (Slightly) calcareous clay
       2. $CaCO_3$ 10–30 %. Very calcareous (or marl) clay
       3. Siliceous fossils 1–10 %. (Slightly) siliceous clay
       4. Siliceous fossils 10–30 %. Very siliceous clay
    B. Oozes. $CaCO_3$ or siliceous fossils > 30 %
       1. $CaCO_3$ > 30 %. < 2/3 $CaCO_3$: marl ooze. > 2/3 $CaCO_3$: chalk ooze
       2. $CaCO_3$ < 30 %. > 30 % siliceous fossils: diatom or radiolarian ooze
II. Hemipelagic deposits (muds)
    > 25 % of fraction > 5 μm is of terrigenic, volcanogenic, and/or neritic origin
    Median grain size > 5 μm (except in authigenic minerals and pelagic organisms)
    A. Calcareous muds. $CaCO_3$ > 30 %
       1. < $2/3$ $CaCO_3$: marl mud. > 2/3 $CaCO_3$: chalk mud
       2. Skeletal $CaCO_3$ > 30 %: foram ~, nanno ~, coquina ~
    B. Terrigenous muds. $CaCO_3$ < 30 %. Quartz, feldspar, mica dominant
       Prefixes: quartzose, arkosic, micaceous
    C. Volcanogenic muds. $CaCO_3$ < 30 %. Ash, palagonite, etc., dominant
III. Special pelagic and/or hemipelagic deposits
       1. Carbonate-sapropelite cycles (Cretaceous)
       2. Black (carbonaceous) clay and mud: sapropelites (e. g., Black Sea)
       3. Silicified claystones and mudstones: chert (pre-Neogene)
       4. Limestone (pre-Neogene)

Except for some of the radiolarians, virtually all plankton organisms making sediment live in surface waters. Coccolithophores and diatoms need light for photosynthesis. This is also true for many planktonic foraminifers, because of symbiotic algae living within their bodies. Besides, food supply is highest in surface waters.

Shells of most perished organisms never reach the sea floor, and most of those that do reach it are destroyed by dissolution. This is true both for calcareous and siliceous material, but especially for the latter. Very few of the ubiquitous open ocean diatoms are seen in sediments; they are mostly too delicate to be preserved.

In summary, the most important factors controlling the composition of biogenous deep sea sediments are *productivity* and *preservation* (Fig. 8.4). Productivity, of course, controls the supply of plankton remains, while depth of deposition controls the preservation of carbonate, which constitutes the bulk of biogenous sediments. Dissolution at depth is enhanced by the high pressures and low temperatures there.

8.2.3 Sedimentation Rates. How fast do deep-sea sediments accumulate? The first widely used rate estimates were those by W. Schott (1935). He identified the thickness of the post-glacial sediment in the central Atlantic by noting the presence of *G. menardii* (Fig. 8.3), a tropical foraminifer that is absent during glacials, in the Atlantic. Thickness of uppermost sediment with *G. menardii* divided by the age of the

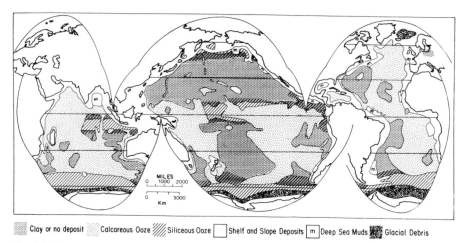

Clay or no deposit   Calcareous Ooze   Siliceous Ooze   Shelf and Slope Deposits   m Deep Sea Muds   Glacial Debris

Fig. 8.2. Sediment cover of the deep-sea floor. The chief sediment types or facies are pelagic clay and calcareous ooze. [W. H. Berger, 1974, in C. A. Burk, C. L. Drake (eds) The geology of continental margins. Springer, Heidelberg]

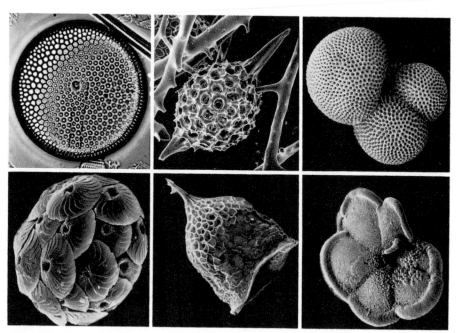

Fig. 8.3. Shell-bearing planktonic organisms. *Clockwise from upper left* Siliceous diatom (x 600), centric warm-water form; siliceous radiolarian (x 180); calcareous warm-water foraminifer *Globigerinoides sacculifer* (x 55); tropical subsurface foraminifer *Globorotalia menardii* (x 28); organic-walled tintinnid (x 480); calcareous coccolithophore (x 2100) with interlocking platelets ("coccoliths"). [Diatom microphoto by H.-J. Schrader; all others: SEM photos by C. Samtleben and U.Pflaumann]

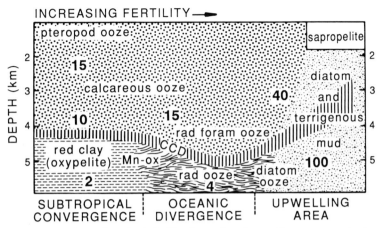

Fig. 8.4. Distribution of major facies in a depth-fertility frame, based on sediment patterns in the eastern central Pacific. *Numbers* are typical sedimentation rates in mm/1000 yr (which is the same as m/million yr). [Source as for Fig. 8.2]

Holocene (the time since deglaciation, established on land) yields sedimentation rates. His method is still useful today for shipboard determinations in the central Atlantic. Published rates show maximum values for terrigenous muds off river mouths (up to several meters per thousand years), and very low ones for "red clay" (a few millimeters per thousand years, typically ranging from 1 for the central South Pacific to 3 in the central Atlantic and in the northern North Pacific). Calcareous ooze has intermediate rates, typically between 10 and 30 mm/1000 yr, depending on productivity (Fig. 8.4). The various rates, ultimately, are based on dating from measurement of radioactive isotopes within the sediment (see Appendix A8) and using correlation with biostratigraphic markers or paleoclimatic and paleomagnetic signals (see below).

As a rule of thumb, rates are high close to important sources (continents, volcanoes), and below regions of high productivity. Clay rates also depend strongly on dust supply and wind directions.

8.2.4 Thickness of Deep-Sea Sediments. An important step in the investigation of the sea floor was the assessment, in the 1950s, of the total sediment thicknesses by seismic (acoustic) refraction and reflection methods. Early results of such surveys sent shock waves of surprise through the geologic community. The oceans were supposed to be a permanent, stable receptacle of continental and volcanic debris. M. Ewing at Lamont Geological Observatory, and R. Raitt at Scripps, and their collaborators, found that the sedimentary columns for typical basins in the Atlantic are only about 500 m thick, and in the Pacific, a mere 300 m (see Fig. 10.5). It became clear that the old view of the deep ocean floor as a sedimentary memory reaching back for perhaps billions of years had to be revised. In 1959, H. H. Hess could still entertain the idea that a hole from the sea floor to the Earth's mantle might sample primordial sediments on the way down. Shortly after, however (in 1960), he adopted the half-forgotten concept of mantle convection and constructed a model providing for a new sea floor "every 300 to 400 million years", which, he suggested, "accounts for the relatively thin veneer of sediments on the ocean floor" (see Chap. 1).

Before we turn to a closer examination of the various types of deposits, we briefly touch on the subject of how the sediment arrives at the sea floor, in the *pelagic rain.*

8.3 The Pelagic Rain

8.3.1 Importance of Fecal Transport. In the pelagic realm, the bulk of the sediment arrives as a *rain* of particles. (*"Marine snow"* is a technical term for sinking aggregates, so we cannot use it in the general sense for sinking particles, although it would be quite appropriate). The nature of this rain has been studied in recent years by *sediment traps,* which in essence consist of funnels attached to moored buoys, anchored in various places in the open ocean. The funnels are positioned at various depths. They are open at the top (with a hole 30 cm to 1 m in diameter), and they have a collecting cup (or several cups) at the lower end. From such experiments it was found that much of the pelagic rain – perhaps most of it – consists of fecal pellets of various sizes and shapes, and in various states of disintegration. The pellets derive from copepods, salps, krill (in the Antarctic), and other grazing organisms (Fig. 6.2).

The accelerated sinking made possible by *"fecal transport"*, and also by aggregate formation, allows even the smallest of particles (wind-blown dust, coccoliths) to arrive at the sea floor within a week or two. If left to settle on their own, they would take years! In the case of fine calcareous, siliceous, and organic particles, all of which are subject to remineralization in the water column, they would not arrive at all on the deep-sea floor.

8.3.2 Flux of Organic Matter. The proportion of the primary production that leaves the photic zone (the *export production;* see Fig. 6.3) experiences substantial losses during settling, despite fecal transport. Presumambly, this is due to break-up of pellets and aggregates from scavenging and decay. The amount of organic material caught in traps at various depths in the open ocean roughly decreases by the same factor as the depth increases; that is, the amount caught at 1000 m is about one fifth of that trapped at 200 m, given the same overlying productivity in the photic zone (Fig. 8.5).

The high export factor in the coastal ocean (Fig. 6.3) and the short distance to the sea floor are two factors that enhance the burial of organic matter around continental margins (Fig. 6.4). In comparison, rates in the deep sea are extremely low. Other factors causing this contrast are seasonality and "fast stripping" (see below), as well as high rates of burial of organic matter.

8.3.3 Seasonality of Flux. The larger particles (sand-size foraminifera and radiolarians) reach the deep-sea floor within 1 to 2 weeks, much as do the pellets. This rapid settling of the pelagic rain means that seasonal fluctuations of productivity in surface waters produce seasonal food supply, even in the abyssal environment. Thus, benthic organisms on the deep-sea floor are subject to feast and famine, just like the plankton. In high latitudes, seasonal variations in the particle rain is especially pronounced (Fig. 8.6).

CARBON FLUX/PRODUCTION

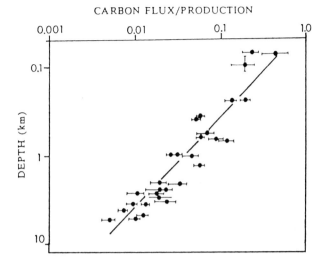

Fig. 8.5. Fluxes of particulate organic carbon found in sediment traps, as fraction of the productivity of overlying waters. [E. Suess, 1980, Nature 288: 260]

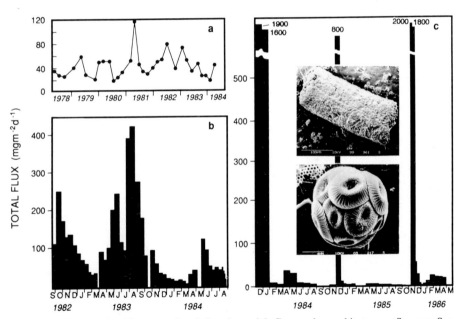

Fig. 8.6 a–c. Seasonal and interannual variations in particle flux as observed in traps. **a** Sargasso Sea (W. G. Deuser); **b** Gulf of Alaska (S. Honjo); **c** Bransfield Strait, Antarctica (G. Wefer et al.). [W. H. Berger, G. Wefer, 1990, Palaeogeography, -climatology -ecology Global Planet Change Sect 89: 245]. Photo shows krill fecal string and close-up of particle (coccosphere) on surface of string. Photo courtesy G. Wefer

The fact that fecal transport makes up a high proportion of the pelagic rain implies that biological patterns in surface waters dominate not only the sources of sediments in the deep sea, but also the patterns of transport. There is one interesting corollary (among many): the more mineral matter that is available for inclusion in fecal pellets (e. g., eolian dust), the more efficiently nutrients are removed from surface waters, by particles. This mechanisms of *"fast stripping"*, by enhanced settling, presumably is responsible for the fact that little nutrient-rich (or muddy) water reaches the open sea, from the coastal zone.

In the following we shall take a closer look at the major types of deep-sea deposits resulting from the pelagic rain: "red clay" (in fact, reddish brown), calcareous ooze, siliceous ooze. Manganese nodules will be discussed in the section on resources (Chap. 10).

8.4 "Red Clay" and "Clay Minerals"

8.4.1 Origin of "Red Clay": the Questions.
Of all deposits, "red clay" is uniquely restricted to the deep-sea environment. The bulk of the components are extremely fine-grained, and the coarse silt and sand fractions consist of particles originating in the ocean: hydrogenous minerals, volcanogenic debris, ferromanganese concretions, and traces of biogenous particles such as fish teeth, arenaceous forams, and in some cases, spicules and radiolarians.

What is the source of "red clay"? This question is really two questions in one.

The first question addresses the ultimate source: to what extent is the clayey fraction in pelagic clays derived from in situ decomposition of volcanic material, and what is the contribution from continents and other sources?

The second question concerns transportation: what is the relative importance of transport by wind versus transport by rivers and ocean currents in bringing continent-derived clay particles to their site of deposition?

8.4.2 Composition of "Red Clay".
To answer the questions posed (how much of the clay is of oceanic, how much of continental origin, and how did it get there?), we need to know the composition of the fine-grained constituents, their distribution on the sea floor and their accumulation rates. Analyses by X-ray diffraction began in the 1930s (R. Revelle in the Pacific, C. W. Correns in the Atlantic) and have been greatly improved and systematically applied to deep-sea deposits since. The following minerals are important in the clay and fine silt fractions of "red clay" (see Appendix A4):

1. Clay minerals: smectite, illite, chlorite, kaolinite, and mixed-layer derivatives;
2. Lithogenous minerals; feldspar, pyroxene, quartz;
3. Hydrogenous (or authigenic) minerals: zeolite and ferromanganese oxides and hydroxides.

The hydrogenous minerals provide much of the X-ray amorphous material present in "red clay" Iron-oxide minerals with Fe^{3+}, typical for oxygen-rich environments, are responsible for the red to brown colors.

Concerning the lithogenous and hydrogenous minerals, they can be identified also in the coarser fractions and compared to their parent rocks (e. g., basic or acidic volcanics, granitic rocks), so that their origin can be readily deduced. The distribution of quartz in the North Pacific suggests eolian transport from desert belts which agrees with evidence on size distribution and chemical composition of the quartz. The Asian highlands are the nearest source of dust.

The clay minerals warrant our special attention, since they make up the bulk of the finest size class of nonbiogenous deep-sea sediments (ca. two thirds of the clay size fraction; median diameters around 0.001 mm, that is, 1 µm). The clay size fraction in turn provides some 90 % of "pure" "red clay". The main groups may be briefly introduced as follows (see Fig. 8.7).

In the mineral group *smectite* (or *montmorillonite*), layers occur in a sequence such that an aluminous octrahedral layer is sandwiched between tetrahedral layers. Abundances of Mg^{2+} (also Fe) and Al^{3+} in the octahedral layer, and of Al^{3+} and Si^{4+} in the tetrahedral layer are such that there is a small net negative charge. This charge is balanced by exchangeable cations between the "sandwiches". The cations are hydrated, thus introducing variable amounts of water into the interlayer positions. This is the reason for the prime criterion for the identification of smectite, that is, its ability to expand. Smectite originates from low-temperature chemical alteration of volcanic rocks, on land and on the sea floor, or from hydrothermal processes on the sea floor. Fine ash particles alter to clay minerals of this group, when exposed to seawater for a long time.

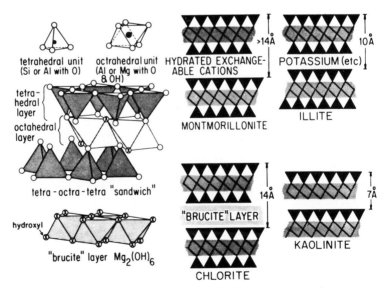

Fig. 8.7. Structure of clay minerals. The thickness of a complete layer is measured in Angstrom units ($\AA = 10^{-8}$ cm), using X-ray diffraction techniques. Note the basic similarity of the clay minerals. [Mainly after R. E. Grim, 1968, Clay mineralogy. McGraw-Hill New York]

Illite is a general term for clay components belonging to the mica group and their derivatives; for present purposes it may be considered a fine-grained degraded muscovite. The muscovite structure again shows an octahedral layer, and the ratio of Si to Al in the tetrahedral layer is exactly 3 to 1. The net negative charge of the "sandwich" is balanced by nonhydrated, firmly bound K^+ ions fitting in the holes left by the hexagonal arrangement of the corners of the silica tetrahedrons.

Chlorite also consists of "sandwiches", but now bound by an additional octahedral layer, the so-called brucite layer. Net charges of tetrahedral layers and of octahedral layers balance each other, and therefore there are no inter-layer ions. Chlorite is a common constituent of low-grade metamorphic rocks which are widely exposed on glacially eroded shield areas and provide much of the material for glacial deposits.

Kaolinite consists of alternating tetrahedral and octahedral layers. It is a product of intense chemical weathering, and represents an insoluble alumino-silicate residue remaining after cations are stripped from feldspars and other minerals by extensive leaching.

8.4.3 Distribution of Clay Minerals. The clay minerals that are most abundant in deep-sea clay are smectite (montmorillonite) and illite (Fig. 8.8). Their distributions suggests that montmorillonite has important sources in oceanic volcanism, at least in the Pacific, while illite is largely derived from the continents. The remaining two important clay minerals, kaolinite and chlorite, also are land-derived; kaolinite from chemical weathering in the tropics and chlorite from deep erosion and physical weathering in high latitudes. Off the west coasts of North Africa and Australia,

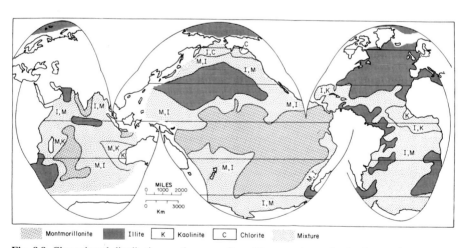

Montmorillonite   Illite   K Kaolinite   C Chlorite   Mixture

Fig. 8.8. Clay mineral distribution on the ocean floor. The map shows the dominant mineral in the fraction less than 2 μm. *Mixture* indicates that no one clay mineral exceeds 50 % of the total. [W. H. Berger, in C. A. Burk, C. L. Drake, eds, 1974, The geology of continental margins. Springer, Heidelberg. Main source: J. J. Griffin et al., 1968, Deep Sea Res 15: 433]

kaolinite actually becomes dominant. Chlorite generally is abundant off continents in high latitudes and is dominant in the Alaska Bight.

From the evidence reviewed, we may conclude that the case for continental sources of most clay minerals is very strong, and it remains only to ask what proportion of the smectite is continent-derived. For the Atlantic, P. E. Biscaye (1965) argued that montmorillonite crystallinity patterns run parallel with detrital patterns, and that this indicates a predominance of continental montmorillonite supply. In the Pacific, oceanic sources and "the ring of fire" supply the volcanic materials which decay to montmorillonite.

8.5 Calcareous Ooze

8.5.1. Depth Distribution. The ocean recieves calcium from rivers and from the hydrothermal alteration of basalt on the young sea floor. The input is such that the amount of calcium in the oceans can be delivered within somewhat less than one million years. Balancing this input, the ocean precipitates calcium carbonate. The precipitation takes place near the surface, within shell- and skeleton-building organisms (coccolithophores, foraminifers, mollusks, corals, algae). Some of these hard parts (in the main planktonic) find their way to the sea floor, where they are preserved on the more elevated parts and dissolved on the deeper ones, because the undersaturation of seawater increases with pressure and with decreasing temperature (Fig. 8.9).

There are considerable differences in the carbonate distribution patterns of the Pacific and the Atlantic Ocean, with the Atlantic having higher carbonate percentages at all depths (Fig. 8.9b). Ultimately, this difference is due to the effects of deep ocean

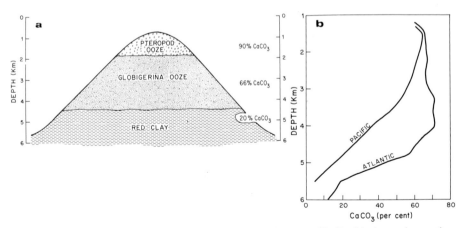

Fig. 8.9a, b. Depth distribution of calcareous deep sea sediments. **a** Idealized bathymetric zonation of deep-sea deposits, produced by increasing dissolution of carbonate with depth. [According to J. Murray, J. Hjort, 1912, The depths of the ocean. Macmillan, New York] Pteropods are pelagic snails with aragonitic shells. **b** Generalized depth profiles for carbonate content in deep-sea sediments. [R. R. Revelle, 1944, Carnegie Inst Wash Publ 556]

circulation, which fills the deep Atlantic with calcite-saturated waters (NADW), and leaves much of the Pacific undersaturated (Sect. 7.6.5; Fig. 7.15).

8.5.2 Dissolution Patterns in the Deep Sea. Most of what is known about global dissolution patterns on the sea floor is represented by mapping the calcite compensation depth (CCD) (Fig. 8.10). The *CCD* or *carbonate line* is analogous to the snow-line on land, which tends to follow a certain elevation contour in a given mountain range, at a given latitude. In concept, the CCD is the particular depth level at any one place in the ocean where the rate of supply of calcium carbonate to the sea floor is balanced by the rate of dissolution, so that there is not net accumulation of carbonate (Bramlette, 1961). In practice, the CCD is mapped as the level at which percent carbonate values drop toward zero. This method can lead to difficulties in areas with exposures of pre-Recent carbonates (e. g., along the Pacific Equator, at great depth).

The CCD topography is quite different for the three great ocean basins, even though depths remain largely between 4 and 5 km. In the Pacific, the surface described resembles a dinner plate with upturned rims and a groove along the Equator. Average depth is near 4.5 km. In the Atlantic, the CCD surface resembles an inclined plane, with the low end in the north. There is a marked E-W asymmetriy in the south. The greatest depth is in the Nord Atlantic (> 5.5 km), where the deep water is young and supersaturated with calcite down to about 4.5 km. The shallowest CCD levels are in the northern North Pacific, where deep waters are old and rich in excess CO_2 (that is, CO_2 added from respiration and decay at depth). Here waters are undersaturated (or close to that state, in the upper portion) for much of the water column below 1 km depth.

8.5.3 Peterson's Level and the Lysocline. A major step forward in the understanding of carbonate dissolution on the deep-sea floor came from field experiments

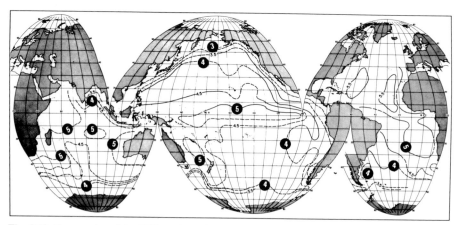

Fig. 8.10. Topography of the CCD surface, that is, the depth in kilometers below which little or no carbonate accumulates. [W. H. Berger, E. L. Winterer, 1974, Spec. Publ. Int. Assoc. Sedimentol. 1:11]

by M. N. A. Peterson, which showed a drastic increase of dissolution rates below 3500 m in the central Pacific (Fig. 8.11a). From these observations (and related ones, using the calcareous tests of foraminifers), it appears that the level of the CCD is strongly controlled by a level of dissolution increase, somewhat shallower in the water column, and possibly coincident with the transition from saturation to increasing undersaturation with calcite.

There is another CCD-like level which can be mapped to describe dissolution patterns on the sea floor: the "*lysocline*". The concept of the lysocline was introduced to denote a contour-following boundary zone between well-preserved and poorly preserved foraminiferal assemblages (Fig. 8.11b). In our snowline analogy of the CCD, the lysocline corresponds to the boundary between fresh high-altitude snow and the wet or refrozen snow on the lower slopes. The preservational lysocline on the sea floor correlates with Peterson's level, as far as this can be checked. In the Atlantic and in the South Pacific, the lysocline marks the top of the Antarctic Bottom Water.

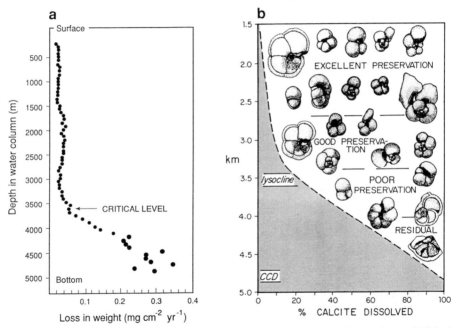

Fig. 8.11 a, b. Dissolution of carbonate as a function of depth. **a** Peterson's experiment. Polished calcite spheres were exposed in a line kept taut by a large buoy submerged below the surface. The line was anchored by a heavy weight. After 4 months, the spheres were recovered and weighed. The diagram shows the weight loss. [M. N. A. Peterson, 166, Science, 154:1542]. **b** Differential dissolution and the lysocline. At the shallower depths on the sea floor, usually above 3000 m or so, pelagic foraminifers are well preserved. Below a critical level, preservation rapidly deteriorates with depth. The boundary between well-preserced and poorly preserved foraminifers on the sea floor is the lysocline. It is closely associated with the critical level of Peterson, and also with the saturation level, in areas of low productivity. [W. H. Berger, 1985, Episodes 8:163]

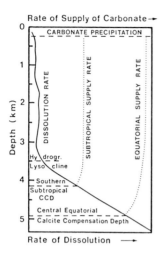

Fig. 8.12. Conceptual model for the origin of the CCD and its relationship to the lysocline. Increased carbonate supply at the equator depresses the CCD, as seen in Fig. 8.10. [Berger et al., 1976, J Geophys Res 81: 2617]

It appears very probable, therefore, that over large areas the lysocline denotes a level of increase in the aggressiveness of the bottom water toward calcareous shells.

Using the concepts of a critical level of dissolution rate increase ("*hydrographic lysocline*") and associated mappable depth level of preservation, a simple model of carbonate deposition and of the depression of the equatorial CCD readily emerges (Fig. 8.12). Increased carbonate supply at the equator in the Pacific translates into the observed depression of the CCD, a major feature of Pacific sedimentation that allows one to follow the motion of the sea floor across the equator for the last 40 million years.

The fact that a level of equal preservation (the lysocline) roughly follows a level of equal calcite saturation in the water column opens the possibility for reconstructing *changes in saturation state* of the deep ocean, through time. This information is crucial in reconstructing the partial pressure of CO_2 in the ocean, which, in turn, is useful information with regard to reconstructing atmospheric CO_2 concentrations for the distant past. For constant total ocean CO_2, a lowering of the average lysocline position by 500 m translates into a lowering of atmospheric CO_2 of about 10 % (that is, 30 ppm for the late Holocene). During glacial time, a lowering of roughly this magnitude did indeed occur (see Sect. 9.3.1 and 9.3.2).

8.5.4 Dissolution Patterns Near Continents. Returning to the CCD map (Fig. 8.10), we note that high fertility along the Pacific equator leads to a depression of the CCD by some 500 m. Paradoxically, high productivity *raises* the CCD in the margin areas around continents. The striking difference in content of organic matter between coastal and deep-sea sediments (Fig. 6.4) offers a clue to this apparent contradiction. In the fertile areas of the ocean margins, the high supply of organic matter leads to highly increased benthic activity as well as to the development of much CO_2 in interstitial waters, which produces carbonic acid. Thus, carbonate ion is destroyed, and calcite shells are attacked even at depths of a few hundred meters on continental slopes.

In contrast, in the equatorial areas of the central Pacific, increased fertility leads to an increased supply of calcareous shells which goes well beyond the increased supply of organic matter in this region. Carbonate readily transits to the sea floor, while organic carbon tends to be filtered out on the long way down (see Fig. 6.1). Consequently, the ratio of calcitic shell to organic carbon is relatively high in the pelagic realm, which is favorable for the preservation of calcite. Nevertheless, some of the carbonate also dissolves above the lysocline, due to organic matter supply. Preservation is never perfect.

8.5.5 Why Is There a CCD? The ultimate reason why abyssal waters dissolve calcite is that organisms supply calcium carbonate to the sea floor in excess of the amount that can be sedimented over the long run. This amount is fixed by the influx from the continents and from hydrothermal sources. The shell supply to the ocean floor that exceeds the overall influx ultimately depletes the ocean of calcium carbonate, which results in bottom waters that are sufficiently undersaturated to redissolve the excess supply of calcium carbonate to the sea floor. Thus, a dynamic steady state is maintained. From this simple "book keeping" concept, it can be readily inferred that, through geologic time, an overall increase in productivity leads to an overall increase in dissolution, and vice versa.

Of course, we must be careful not to extrapolate too far back when using the present ocean as a model for the past. Mass production of coccoliths (or better, *nannofossils*) began sometime in the early Cretaceous, of planktonic foraminifers in the late Cretaceous. Also, if the ocean was less well mixed in the Mesozoic than today (say, because of less vigorous deep water production), this would have greatly affected the general nature of the CCD at the time.

8.5.6 A Global Experiment. At the present time, mankind is engaged in a global experiment involving carbonate dissolution, as well as climatic change. We – the industrial nations mainly – are burning off enormous amounts of coal and oil at an increasing rate. Large-scale deforestation is proceeding in the tropics and elsewhere, for the sake of agricultural development and for wood products and fuel. The resulting carbon dioxide enters the atmosphere. So far, an amount equivalent to almost 50 % of the CO_2 originally present in the atmosphere has been added within the past century. Almost one half of this amount has entered the ocean, the rest has stayed in the atmosphere (Fig. 8.13a). Eventually, over the next few centuries, some *ten times* the original CO_2 in the atmosphere could be added, if the readily available coal and oil is burned.

How will the ocean react to the continuing (and perhaps increasing) input of CO_2?

In the long run, the ocean floor will neutralize most of the industrial CO_2 through the dissolution of carbonate (Fig. 8.13b):

$$CO_2 + H_2O + CaCO_3 \rightarrow Ca^{2+} + 2HCO^-_3 . \tag{8.1}$$

Thus, the industrial CO_2-pulse will produce a hiatus on the sea floor. A thickness of about 1 m of carbonate sediment will have to be dissolved from the carbonate-bearing sea floor, assuming that available coal and oil deposits are burned up. Initially, how-

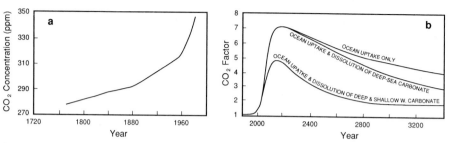

Fig. 8.13 a, b. CO_2 increase in the atmosphere and projected carbonate dissolution on the sea floor. **a** Atmospheric CO_2 increase as seen in ice cores from Siple Station, Antarctica, and in the data of C. D. Keeling, from Mauna Loa in Hawaii. [U. Siegenthaler, H. Oeschger, 1987, Tellus 39: 140.] **b** Calculated factor by which future atmospheric CO_2 will exceed preindustrial CO_2 if present trends of CO_2 input continue till oil and coal are used up. Note that carbonate dissolution on the sea floor does not prevent a strong initial rise of CO_2. [R. B. Bacastow, C. D. Keeling, 1979, U S Dpt Energy Conf 770385: 72]

ever, dissolution will proceed mainly in shallow areas, especially in high latitudes, where the water needs little additional CO_2 to become undersaturated. Subsequently, it will affect low latitudes, presumably interfering with reef growth there. It will take several hundred years for the deep areas of the sea floor to start "feeling" the effects of the increased CO_2 supply. It takes this long to replace existing deep waters with new deep waters that bring the message of new atmospheric conditions to the deep-sea floor.

Can we predict how high the CO_2 level will rise, and what the effects will be on climate? A great number of scientists – meterologists, oceanographers, geochemists, and geologists – are working on this question. It is now generally agreed that a doubling of the CO_2 content would increase the global average temperature by at least 2 °C (4 ° Fahrenheit). Some estimates go as high as double this amount. A general temperature rise of about 0.5 °C, from a century ago, has been observed. The significance of this rise is not entirely clear, however (see Fig. 7.16).

A doubling of CO_2 might come within the next 50 years or so, depending on how the trends of use of carbon fuels develop. Increased release of methane also must be considered in this context (see Epilog). What might be the effect of forcing up the global temperature by several degrees? Nobody knows the answer. It is possible that the climate would simply be warmer and moister over large areas in mid-latitudes, and that the change would be quite beneficial to agriculture in many countries. However, it is also possible that a marked warming will produce undesirable side-effects: drought in the northern grain belts, for example, or a great increase in hurricane activity. A large-scale melting of Antarctic ice is conceivable, which could produce a sea level rise ten times faster than now. Another possibility to be considered is that the (geologically) rapid CO_2 input introduces a strong disequilibrium into the ocean-atmosphere system, which provokes *short-term climatic oscillations,* something that is definitely inimical to economic, and hence political, stability.

The only safe prediction, from the point of view of geology, is that the CO_2 input will decrease again, perhaps within two to three centuries, either from exhaustion of resources, or from economic disruption caused by climatic change, or other environmental problems (even neglecting the possibility of war). At that point, atmospheric CO_2 will start decreasing as the alkalinity of the ocean rises from the dissolution of carbonate.

The CO_2 problem reminds us of the extent to which we are now exploiting the natural cycles of our planet for human benefit – yet a large part of a growing world population remains without adequate food and shelter.

8.6 Siliceous Ooze

8.6.1 Composition and Distribution. In considering the siliceous deposits of the deep sea, many of the geochemical questions reappear which were raised earlier in connection with "red clay" and calcareous ooze. What are the contributions from continental weathering, submarine alteration, volcanic and hydrothermal emanations? What mechanisms control the concentrations of dissolved matter in seawater? That is, what controls the state of saturation?

Are biogenous particles the sole sink for such dissolved matter, or is there uptake by "upgrading" of clays? What is the rate at which redissolution on the sea floor supplies dissolved matter to the overlying waters?

First, let us take a brief look at the distribution, production, and dissolution patterns of the siliceous deposits.

We have already encountered the constituents of such deposits: remains of diatoms, silicoflagellates, radiolarians, and sponge spicules, all of which are made of opal, a hydrated form of amorphous silicon dioxide. Diatom oozes are typical for high latitudes, diatom muds for pericontinental regions, and radiolarian oozes for equatorial areas (Figs. 8.2 and 8.4). Both diatom and radiolarian ooze, of course, are mixtures of various kinds of sediments, with one or the other siliceous form being dominant (Fig. 8.14). The siliceous deposits occur in areas of high fertility; that is, in regions with relatively high phosphate values in surface waters (Fig. 8.15). This overall correspondence between fertility patterns and silica-rich deposits can be considerably modified by redeposition processes within individual regions. The silica frustules are light and easily transported, and the activity of benthic animals, which tends to resuspend fine sediment, is especially pronounced in fertile areas. Thus, aided by bottom currents and gravity, siliceous frustules tend to accumulate in local and regional depressions.

8.6.2 Controlling Factors. In analogy to other kinds of deposits, the concentration of siliceous fossils in the sediment is a function of (1) the rate of production of siliceous organisms in the overlying waters. (2) the degree of dilution by terrigenous, volcanic, and calcareous particles, and (3) the extent of dissolution of the siliceous skeletons, most of which apparently occurs shortly after deposition.

The first variable, production of siliceous shells, attains its maximum in coastal regions (Sect. 4.3.3; also Figs. 8.2 and 8.15). This leads to the formation of a *silica ring* around each ocean basin. *Silica belts* are provided by the latitudinally arranged

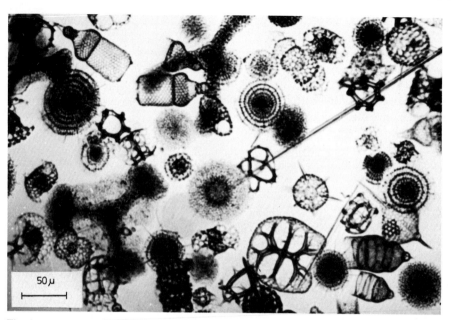

Fig. 8.14. Assemblage of modern radiolarians from sediments recovered in the equatorial Pacific. [Microphoto W. H. B.]

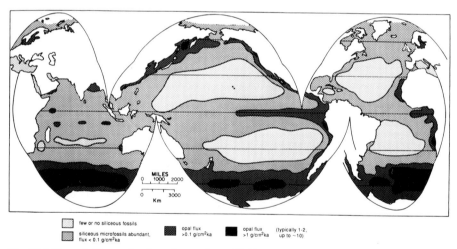

Fig. 8.15. Flux of siliceous fossils to the sea floor. [W. H. Berger, J. C. Herguera, in P. G. Falkowski, A. D. Woodhead, eds, 1992, Primary productivity and biogeochemical cycles in the sea. Plenum Press, New York]

oceanic divergences which are the result of atmospheric circulation. The regions of divergence have nutrient-rich surface waters, hence there is sufficient silica available to make robust siliceous shells. Also, such areas are rich in grazing zooplankton, which pack the siliceous frustules into fecal pellets, thus accelerating delivery to the sea floor (Fig. 6.2).

To obtain an estimate of the amount of silica precipitated in the upper waters, one might multiply the measured amount of organic production with the ratio of solid silica to organic matter found in suspension. This yields only a rough estimate, of course.

A typical fixation rate of about 200 g SiO_2/m^2 is suggested, with a range from less than 100 g (central gyres) to more than 500 g (Antarctic). Of this fixation, only 1 $g/m^2/yr$, that is, 0.5 %, can be incorporated into sediments if the river input is the only source of silica. Twice that (that is, 1 % of fixation) can be sedimented if we assume an equal contribution of silica from seawater-basalt reactions, especially at the hydrothermally active ridge crests. The global map of silica flux (Fig. 8.15) suggests that 2 $g/m^2/yr$ (= 0.2 g/cm^2ka) is indeed a reasonable average value.

The second factor, the degree of solution of the siliceous material, reflects the ratio between accumulation rates of nonsiliceous and siliceous particles. As concerns dilution of silica by carbonate, one might expect that a high supply of calcareous shells would be accompanied by an equally high supply of siliceous shells, because both siliceous and calcareous plankton depend on productivity of upper waters. This is not generally the case, however. Indeed, there is a distinct negative correlation between silica and calcite distributional patterns. This has been ascribed to opposing chemical requirements for preservation. We have seen already that increasing productivity, from some point on, leads to decreasing preservation of calcite, and to increasing accumulation of silica. A similarly opposing trend is indicated for depth relationships, with silica corrosion being greatest in upper waters (due to elevated temperature), that of carbonate being greatest at depth.

The third factor controlling the abundance of siliceous fossils is the extent of dissolution. The preservation of siliceous shells is rather closely correlated with their abundance in sediments. A positive correlation between abundance and preservation may be ascribed to an increased supply of easily dissolved diatom hash in fertile areas, which will "buffer" interstitial waters for the more robust skeletons, and to an otherwise favorable chemical environment in organic-rich sediments with slightly acidic interstitial waters. In general, silico-flagellates and diatoms tend to dissolve well before radiolarians and sponge spicules, and the following dissolution sequence can be established (from least to most resistant): (1) silicoflagellates, (2) diatoms, (3) delicate radiolarians, (4) robust radiolarians, (5) sponge spicules. Because of the overriding importance of Antarctic opal deposition, the preservation of opal in the rest of the ocean must to a large degree depend on how much the Antarctic Ocean is able to extract for deposition on its own sea floor.

8.6.3 Geochemical Implications. The dissolution of opaline skeletons within the surficial sediment layer of the sea floor delivers silica to the deep waters. This flux from the sediment to the water is evident from concentration gradients in interstitial waters and differences in concentration between these waters and the overlying sea-water. By far the greatest reflux is from the uppermost part of the sediment column, where recently arrived shells and frustules are being dissolved. However, the process of dissolution and redelivery to seawater apparently continues downward into the section, and includes sediments as old as 10 million years (Fig. 8.16).

It is possible and even probable that some of the silica released to the interstitial waters reacts within sediments to form new minerals. However, near the sediment surface most of the silica re-enters the ocean water. Thus, "old" bottom water, which has been in contact with the sea floor for a long time, is silicate-rich. The reverse is true for "young" bottom water which has arrived from the surface only recently. Therefore, concentrations of dissolved silica are high in the deep North Pacific ("old" water), and are low in the deep North Atlantic ("young" water).

From this overall distribution of dissolved silica in deep ocean waters, we can draw an obvious conclusion. The reason that silica concentrations in deep water are relatively low cannot be the uptake, if any, of dissolved silica by clay minerals. If it were, the "old" waters should be the more depleted in silica. The reason for the low concentrations must be that the deep water remembers its depleted condition at the surface (from the silica extraction by diatoms) and that it did not have time to saturate itself with respect to the actively dissolving opaline shells.

8.6.4 Deep-Sea Cherts. The discovery of chert within deep-sea sediments has both fascinated geologists and frustated them in efforts to drill and recover complete sections. The formation of deep-sea chert (siliceous sediments cemented by crypto-crystalline and microcrystalline quartz) appears to proceed from mobilization and reprecipitation of opal, generating a disordered cristobalite (= fibrous quartz) which eventually alters toward a quartzitic rock with mostly quartz-replaced and quartz-filled fossils as diagenesis progresses. Recrystallization may proceed at various rates and somewhat divergent patterns, depending on the original sediment present.

If siliceous fossils and/or silica-rich volcanic glass are to be available as a source for later production of chert, the following conditions appear necessary: (1) a sufficiently high supply of opal and low supply of dilutant; (2) silica-rich bottom water; (3) a reasonably high burial rate; and (4) chemical conditions favorable for the preservation of siliceous shells.

During diagenesis, the opaline skeletons dissolve and volcanic material, if any, releases silica during devitrification, and the silica-rich interstitial solutions migrate along bedding planes or fractures and vertically to areas of precipitation in nearby permeable lenses or layers. Why precipitation should be favored in some places and mobilization in others in the same sediment is poorly understood in detail.

In the western equatorial Pacific and in many other areas of the global ocean, massive chert beds first appear in upper Eocene sediments, when drilling down into the sea floor (Fig. 8.16; "Ontong Java Series"). At that level, sound waves are strongly reflected, due to a sudden change in *impedance* (product of density and sound velocity). Interstitial waters tend to lose dissolved silicate to precipitation at this horizon (Fig. 8.16, right). Note that there are other strong reflectors as well: the "Drake Series" (marking the beginning of the Neogene) presumably derives from the

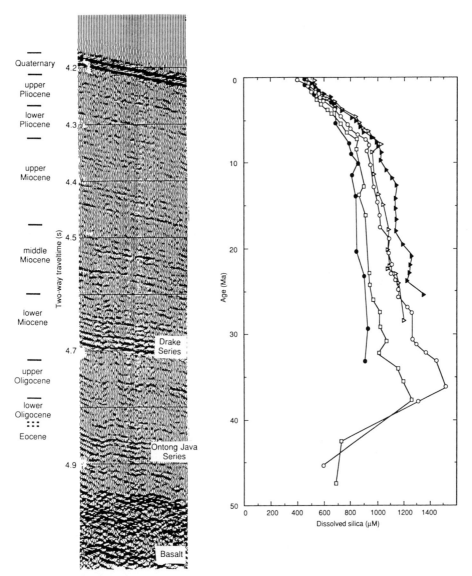

Fig. 8.16. Acoustic reflectors in pelagic sediments on Ontong Java Plateau (Site 805) *(left)* and dissolved silicate in associated drilling sites (Site 805: *open triangles*) *(right)*. Site 805 is just north of the equator. Water depth is 3188 m. The sediment down to the Ontong Java Series reflectors is calcareous ooze and chalk. Total sediment thickness is about 1 km. Chert beds first appear near the top of the Ontong Java Series and are middle to late Eocene in age. Dissolved silicate drops at that horizon, indicating precipitation. [W. H. Berger, L. W. Kroenke, L. A. Mayer, et al., 1991, Proc Ocean Drilling Progr Init Rpts 130: 497; silicate measurements by M. Delaney]

effects of episodic dissolution pulses within the post-Eocene carbonate sequence. The basement reflectors are basaltic layers. The profiles of dissolved silicate indicate reflux of silica to the sea floor from sediments as old as 10 million years. This loss of silica is unfavorable for subsequent formation of chert.

The question of why deep-sea chert deposits are concentrated in some geologic periods and not in others cannot be answered at this time. Presumably, changes in the global silica supply to the ocean (weathering processes, volcanism, ridge-crest hydrothermal activity) changed the amount of silica deposited, and changes in ocean productivity changed the distribution patterns (see Sect. 8.6.2).

Certain ancient cherts, exposed as radiolarites in ophiolites and mélanges of active margins (ribbon cherts), may be turbidites. Such cherts, it may be surmised, could form because opal-rich sediments along active margins were redeposited by turbidity currents into marginal basins, without much dilution from terrigenous materials (due to a high sealevel stand).

8.7 Stratigraphy and Dating

8.7.1 General Considerations. All absolute dating of deep-sea sediments ultimately relies on the measurement of radioactive isotopes and their daughter elements (Appendix A8). However, almost all routine dating is done by these three stratigraphic methods: determination of fossil content, determination of magnetic reversal sequence, and determination of sequence of chemical properties such as stable isotopes of oxygen, carbon, and strontium. To these tools *(biostratigraphy, magnetostratigraphy, chemostratigraphy)* may be added the counting of astronomical cycles observed (or suspected) in paleontological, sedimentological, or physical properties of a pelagic section *(cyclostratigraphy* or *Fourier stratigraphy)*. In all cases except cycle-counting, dating is by *correlation* to a section considered to be well dated.

The best correlation, naturally, is achieved when several methods used simultaneously agree on the placement of a given section, with respect to the reference section.

8.7.2 Aspects of Biostratigraphy. The workhorse of stratigraphic correlation in pelagic sediments is (micro-)paleontology, that is, the identification of fossils, notably nannofossils, foraminifers, radiolarians, and diatoms. The stratigraphic distribution of these fossils was worked out mainly in the last 40 years or so. The work was begun by pioneers such as the American geologists M. N. Bramlette (1896–1977), A. R. Loeblich and H. Tappan, the Swiss paleontologist H. Bolli, and the Australian-American paleontologist W. Riedel, among others, who realized the great potential of microfossils for dating pelagic sequences. Nannofossils and radiolarians are useful throughout the age range of deep-sea sediments (Fig. 8.17). Planktonic foraminifers diversify sometime in the middle Cretaceous, and diatoms in the late Cretaceous.

The principle of stratigraphic correlation goes back to the beginnings of scientific geology and paleontology; it is the observation that sedimentary sections in different parts of the world show a similar sequence of different associations of fossils. Quite

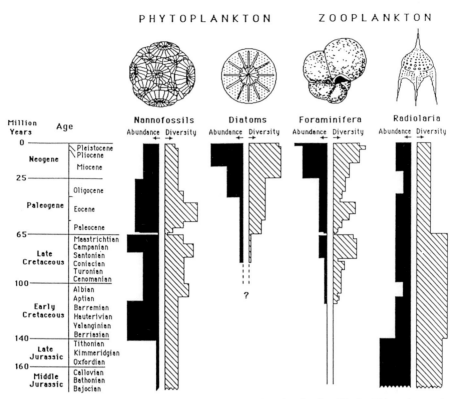

Fig. 8.17. Abundance and diversity patterns of pelagic microfossils. [H. R. Thierstein et al., in G. B. Munsch, ed, 1988. Report of the Second Conference on Scientific Ocean Drilling Cosod II, European Science Foundation, Strasbourg]

generally, there are more different kinds of fossils in tropical regions than in extratropical ones. Thus, much finer subdivisions of sequences are possible in low latitudes than in high ones. Also, *diversity of fossils* changes through time, due to changes in the rates of origination and extinction.

The *stratigraphic resolution* achieved by identifying microfossils depends on circumstances – clearly, where diversity is low, resolution is poor. Under favorable conditions (that is, when the critical species are present) the *zones* of pelagic microfossils in the Neogene allow resolution between one and a few million years in each group. In the latest Neogene, nannofossil zonation is exceptionally detailed, and here the resolution is about 0.5 million years. Additional refinement is possible by using well-dated *first-appearance datums* (FAD) and *last-appearance datums* (LAD), as well as *evolutionary lineages* in the various groups. Besides pelagic forms, benthic species are of interest in biostratigraphy. In the Tertiary, major changes in benthic faunas occurred in the middle Eocene and at the end of the Oligocene.

8.7.3 Magnetostratigraphy. Reversals of the Earth's magnetic field (already encountered when discussing seafloor spreading and age-mapping of the sea floor; Sect. 1.8) impart to deep-sea sediments a sequence of normal (N) and reversed (R) magnetization, which is preserved under favorable conditions and can be measured. The strength of the signal depends on the concentration of iron-rich minerals, mostly. Results – a sequence of N and R tags, one for each sample – can be matched to the established record, as first shown by C. G. A. Harrison and B. Funnell in 1964. For example, on going downward into the sediment from the sea floor, the first clear switch from N-samples to R-samples would be expected to correlate to the *Brunhes-Matuyama boundary*. This reversal event is dated at 790 000 years, based on radioactive decay of ^{40}K and on counting astronomical cycles within the $\delta^{18}O$ record (cyclostratigraphy, Sect. 8.7.5).

When faced with "floating" sections, a match cannot be made uniquely, unless other information is available (fixpoints from biostratigraphy, for example). The more confirmation from other methods, the greater is the confidence that a particular match

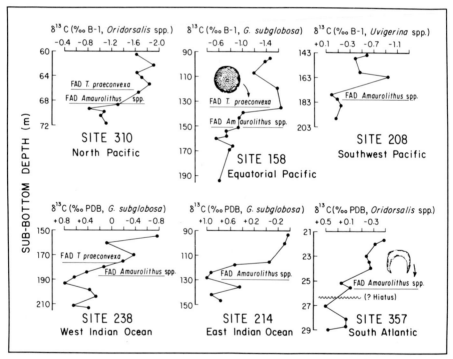

Fig. 8.18. Correlation by chemo- and biostratigraphy. Global ocean shift in $\delta^{13}C$ values as seen in benthic foraminifers, about 6 million years ago, identified by association with biostratigraphic markers. *FAD* First appearance; *T. praeconvexa*: a centric diatom; *Amaurolithus* spp.: a horseshoe-shaped nannofossil. Note consistent relationship of FAD *Amaurolithus* to shift in $\delta^{13}C$. [B. U. Haq et al., 1980, Geology 8: 427; redrawn]

is correct. One problem is that it is quite difficult to detect missing sections in a magnetic reversal sequence. However, in conjunction with biostratigraphy, magneto-stratigraphy has proved to be a powerful tool for global correlation, since reversals are essentially instantaneous and world-wide (which is not generally true for bio-stratigraphic events).

8.7.4 Chemostratigraphy. Many chemical changes in the properties of ocean water are world-wide and are recorded in the composition of microfossil shells. Prime examples are the stable isotopes of oxygen, carbon, and strontium. The records of oxygen and carbon isotope ratios in calcareous fossils show cycles, steps, and trends, which can be used to greatly refine stratigraphic resolution once a general framework has been established. In the example given (Fig. 8.18), the *first appearance datum* (FAD) of two phytoplankton species confirms the global synchroneity of a shift in deep water $\delta^{13}C$ toward more negative values, within the latest Miocene. This shift (which occurred about 6 million years ago) can be used for global correlation, once its identity in a given core is determined. We shall discuss the significance of these types of isotope changes in the context of paleoceanography (Chap. 9).

Changes in *strontium isotopes* show long-term trends in the Cenozoic, and can be used therefore to assign an age range to a post-Cretaceous sample even in the absence of other information (Fig. 8.19). A value of, say, 0.7088 in the ratio of ^{87}Sr to ^{86}Sr, measured in a sample of nannofossils or foraminifera, would uniquely place the sample near the end of the middle Miocene.

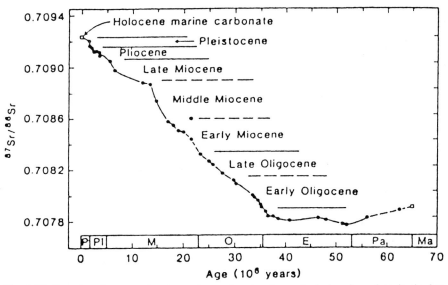

Fig. 8.19. Strontium isotope ratios in marine carbonates. Increasingly heavier values in the late Cenozoic indicate the increasing importance of deep erosion of continental crust relative to stron-tium supply from volcanogenic sources. [F. M. Richter et al., 1992, Earth Planet Sci Lett, 109: 11; modified]

Why do strontium isotopes change in the fashion observed? Presumably, the trend stems from a decrease of basaltic sources of strontium (ratios near 0.702 to 0.703) relative to continental sources (ratios from 0.705 to > 0.712). The reason, it may be assumed, is an overall emergence of continents due to mountain-building since the late Eocene, as well as a decrease in oceanic volcanism.

8.7.5 Cyclostratigraphy. The discovery that pelagic sediment properties show regular fluctuations with periods close to astronomical ones (as shown by spectral analysis: J. Hays, J. Imbrie, N. Shackleton, 1976; N. Pisias, 1976) opened the way for a new dating method: counting of astronomical cycles within a record, extracted by Fourier analysis. The method yields floating chronologies (unless tied into other fixpoints). It has yielded valuable results, especially in Quaternary sediments. We shall encounter the various cycles in Chapter 9, when discussing Quaternary sediments.

Further Reading

Schott W (1935) Die Foraminiferen in dem aequatorialen Teil des Atlantischen Ozeans. Wiss Ergeb Deutsch Atlant Exped Forschungsschiff Meteor, 1925–1927. 3 (3), 43–134

Hill MN (ed) (1963) The sea – Ideas and observations on progress in the study of the seas, vol 3, Interscience, New York

Harrison CGH, Funnell BM (1964) Relationship of palaeomagnetic reversals and micropalaeontology in two Late Cenozoic cores from the Pacific Ocean. Nature 204: 566

Lisitzin AP (1972) Sedimentation in the world ocean. SEPM Spec Publ 17, Soc Econ Paleontol Mineral Tulsa, Okla

Hsü KJ, Jenkyns H (eds) (1974) Pelagic sediments on land and under the sea. Spec Publ Int Assoc Sedimentol 1

Bolli HM, Saunders JB, Perch-Nielsen K (eds) (1985) Plankton stratigraphy. Cambridge Univ Press, New York

Chester R (1990) Marine geochemistry. Unwin Hyman, London

9 Paleoceanography – the Deep-Sea Record

9.1 Background

Paleoceanography, the study of ocean history, emerged in the 1930s and 1940s, when cores became available that provided data from which history could be reconstructed. Initial efforts by W. Schott (1935), using short cores taken by the German research vessel *Meteor,* have been mentioned (Sect. 8.2.3). In essence, the Swedish Deep Sea Expedition (1947–1948) played the same role in launching the new science of historical oceanography that the *Challenger* Expedition had played 70 years earlier, for physical and biological oceanography. The research vessel *Albatross* set out from Gothenburg in 1947, to begin the circumnavigation of the world's tropical environment, under the leadership of Hans Petterson. The expedition used a new device, the *piston corer,* developed by B. Kullenberg in Copenhagen. This technique typically recovered cores of a length of 7 m or so, with the oldest sediment being from 0.3 to 1 million years in age. Kullenberg's device, with modifications, is still used today (Fig. 9.1).

The scientists reporting on these cores became the founders of paleoceanograpy. They defined the fundamental questions about the Pleistocene history of the ocean: regarding changes in productivity (G. Arrhenius), in plankton distributions and surface currents (F. B Phleger, F. L. Parker), in surface water temperature and ice volume (C. Emiliani), and in deep circulation patterns (E. Olausson). Since then, these questions, and related ones, have been pursued vigorously.

Beginning in 1968, when the drilling vessel *Glomar Challenger* (Fig. 0.6) left port in Galveston (Texas) to initiate scientific drilling in the deep ocean, many new dimensions have emerged in paleoceanography: the length of time available for detailed study increased from about 1 million years to the past 100 million years! Since 1985, the *Joides Resolution* (Fig. 0.7) has been employed for deep ocean drilling. In unconsolidated ooze and mud, nearly undisturbed sequences can be recovered by this ship, using hydraulic piston coring. Recovery of several kilometers of core material, from a single 2-month leg, is not unusual. The study of these materials is time-consuming and labor-intensive. It requires the coordination and cooperation of marine geologists from many institutions, in the USA and abroad. The harvest of these efforts has not yet been wholly reaped, and new findings are constantly being reported.

The following summary illustrates some of the highlights of the various discoveries that resulted from coring and drilling. It it as much a survey of questions as

Fig. 9.1. Recovering the Pleistocene record by piston coring. The corer is a wide-diameter model; note the white sediment in the core nose. Note also that the core barrel was bent during the operation (on the *upper end*). Other equipment on deck: deep-sea camera frame (with protective grid), hydrophone for seismic profiling (wrapped on spool), box corer (aft). [SIO *Eurydice* Expedition, 1975]

of answers. We shall first acquaint ourselves with the Ice Age Ocean, whose study defined the scope of paleoceanography, and we will then turn to the task of applying the principles derived from such study to older sediments made available by drilling.

9.2 The Ice Age Ocean

9.2.1 Why an Ice Age? To a geologist, the dominant fact of our time is that we live in an ice age which has been going on for at least two or three million years. In earlier chapters, this reality, which greatly influences the way we interpret the present patterns of erosion and deposition, and the landforms we have inherited, has been referred to several times.

The idea that vast sheets of thick ice once covered northern Europe and northern North America became generally acceptable some time during the middle of the last century. Once this step had been made, it was not long before *several* such glaciations were discovered, with warm intervals separating them. We are now living in such an *interglacial*. How many advances and retreats of continental ice had occurred? What is the time scale of these cycles? How do the cycles relate to astronomical factors, that is, the rotation of the Earth and its path around the Sun? These questions were difficult to even attempt to answer from the land record: each succeeding glaciation erased many of the traces of the previous one. The study of long cores from the deep-sea floor opened the door to a new understanding. Such cores (Fig. 9.1), contain, in places, an uninterrupted record of the alternations of *glacials* and *interglacials*, within the "ice age".

Why is there an ice age at all? We do not know for sure. Clearly, it is colder during an ice age than before or after, so that the Earth's heat budget is involved. This budget is largely controlled by *albedo* (the amount of incoming light reflected into space) and by the *greenhouse effect* (the trapping of infrared trying to leave Earth, by greenhouse gases). Thus (unless we invoke a dimming of the sun), it seems reasonable to assume that albedo was increased and greenhouse gases were decreased, during ice ages. We shall see, in the following, that this assumption is correct. Nevertheless, we still do not know the exact mechanisms responsible. Rather than approaching these very difficult questions of climatology, let us ask some quite simple questions, such as:

1. What did the ocean look like during maximum glaciation?
2. What does the record show about the transition from the last glacial to the present?
3. What is the nature of the ice age cycles, that is, their frequency and amplitude, and their extent in time?

9.2.2 Conditions in a Cold Ocean. How did the ice age ocean differ from the present one? The question is by no means settled, but a consensus has been reached on several aspects.

First, it is generally agreed that *surface currents were stronger.* It is obvious why this should be so: surface currents are driven by winds, and winds depend on horizontal temperature gradients. With the ice rim and polar front much closer to the Equator, the temperature difference between ice (0 °C or less) and the tropics (ca. 25 °C) was compressed into a much shorter distance than now. Hence the temperature gradient was greater, winds were stronger, and so were ocean currents.

Equatorial upwelling was intensified as a consequence, as well as *coastal upwelling*. Thus, at the same time that fertility decreased in high latitudes, due to ice cover, it increased in mid-latitudes (because of intensified mixing) and in the subtropics (due to upwelling).

Second, it is accepted that the *ocean surface was cooler,* on the whole, than today. With a substantial part of northern continents and seas under ice, the Earth reflected the Sun's radiation more readily (had a higher *albedo*) than today, hence it absorbed less of the radiation and its atmosphere was cooler. A cold atmosphere holds less water than a warm one, and large areas on land therefore were drier than today. Water vapor is the most important greenhouse gas, so that the trapping of infrared radiation within the atmosphere was diminished. Dry areas (such as grasslands and deserts) reflect more sunlight than wet ones (such as forests). Also, a more fertile ocean would be slightly more reflective than a clear dark blue one with less algal growth. All these factors favored reflection of the Sun's radiation back into space, and hence favored cooling (Table 9.1).

Table 9.1. Albedo[a] values of ocean and land surfaces

Land	(%)	Ocean	(%)
Desert regions	20 to 30	Low and mid latitudes	4 to 10, max. 19
Grasslands, most forests	15	High latitudes	6 to 55
Tropical forests, wetlands	7 to 10	(high values: sea ice)	
Snow cover (old to new)	35 to 82		

	(%)
Average planetary albedo as seen from space	29
Average surface albedo	14

[a] albedo: proportion (in percent) of incident light reflected upward, and not available for heating.

Third, it is known that some ocean regions cooled more than others, and that the regional degree of change depends closely on the *movement of boundaries of climatic zones*. If, for example, an area is near the boundary of subtropics and temperate belt, it will then belong alternately to one or the other zone. Thus, the changes here are substantial. Conversely, changes can be much less in the center of tropical or sub-tropical climatic regions.

Last but not least, the ocean surface was substantially lower during maximum glaciation, by 120 to 130 m (Sect. 5.4), so that shelves were exposed and the ocean surface was somewhat smaller. This also increased the albedo of Earth, providing for additional cooling.

9.2.3 The 18 K Map. These various concepts were put to the test by a coherent quantitative reconstruction of the ice age ocean, in the 1970s. A. McIntyre, T. C. Moore, and colleagues, using the "transfer" techniques of J. Imbrie (Sect. 7.2.2), produced a map of sea surface temperatures for a typical (northern) summer month, during maximum glaciation. The map is also called *18 K map,* for "18 kilo years ago", a date based on radiocarbon dating (see Fig. 9.2). The construction of the map proceeded from deep sea cores, from which both surface and subsurface samples were taken. Surface samples were used for calibration (Sect. 7.2.1). The abundance patterns were correlated to surface water temperatures in nothern summer, to provide the basis for the reconstruction.

From the information shown in the 18 K map, it can be concluded that the ice age ocean 18 000 years ago was characterized by: (1) increased thermal gradients along polar fronts, especially in the North Atlantic and Antarctic; (2) equatorward displacement of polar frontal systems; (3) general cooling of most surface waters, by about 2.3 °C, on average; (4) increased upwelling along the equator in Pacific and Atlantic; (5) increased coastal upwelling and strengthening of eastern boundary currents; and (6) nearly stable positions and temperatures of the central gyres in the major ocean basins.

Fig. 9.2. Approximate distribution of sea surface temperatures for northern hemisphere summer, during the last ice age maximum (about 18 000 years ago; "18 K map"). Note the extent and thickness of ice. [CLIMAP Project Members, 1976, Science 191: 1131, simplified]

9.2.4 Pulsed Deglaciation. How and why did the last glacial change into the present interglacial? How long did this take? The first question, presumably, must be answered by invoking changes in the seasonal distribution of the Sun's radiation (the Milankovitch Mechanism, to be discussed in Sect. 9.3.5). The second question is the more readily answered: it took between 7000 and 8000 years, approximately, to reduce the ice masses of the last glacial to something like those we have today. In the process, the sea level rose some 120 m (Sects. 5.3.1 and 5.4.2). Recent detailed work on corals in Barbados confirmed earlier indications that deglaciation occurred in *pulses* or *steps* (Fig. 9.3; Fig. 5.6b). Two major pulses are seen: Step 1, from 13.5 to 12.5 ka (also known as Termination Ia), and Step 2, from 11 to 9.5 kyrs (or Ib). These ages are based on dating by thorium (see Appendix A8). In the radiocarbon scale (heavy lines in Fig. 9.3), the ages of the two steps are centered on 12 and 9.5 ka, which agrees exactly with two major warming steps known from northern Europe. Comparison with the summer insolation curve at high northern latitudes (dashed line, Fig. 9.3) suggests that unusually warm summers are responsible for initiating melting.

The retreat of the polar front in the North Atlantic, from its position between Long Island and Portugal in the glacial (Fig. 7.4) to its present one off Greenland, also took place in this discontinuous manner. A substantial re-advance of this front occurred during the so-called *Younger Dryas,* between 11 000 and 10 000 radiocarbon years ago. (According to ice core measurements, some 12 800 to 11 700 calendar years.) The origin of this cold period, which returned much of Europe to glacial-type conditions for several hundreds of years, poses a major problem in paleoclimatology and paleoceanography. The effects of the event are seen around the world; its record was discovered in western Tibet as well as in the Sulu Sea, south of the Philippines, for example. The origin of the Younger Dryas cold spell is entirely unknown (large-scale

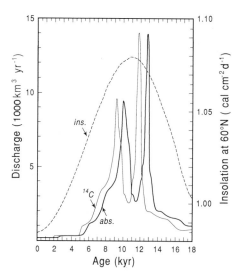

Fig. 9.3. Rate and timing of glacial melt-water discharge compared with subarctic summer insolation. Discharge calculated from a depth-versus-age curve for Barbados corals *(A. palmata). Heavy line* time scale based on radiocarbon dating; *thin line* time scale based on thorium dating by E. Bard and colleagues, Nature 1990, 345: 405; *dashed line* summer insolation at 60° N, based on calculations by A. Berger. [R. G. Fairbanks, 1989, Nature 342: 638, simplified]

ice surges are one possible mechanism), but it is very likely that the ocean's circulation was involved in a drastic change in the heat budget in some fashion. Somehow, the influx of tropical heat from the south, which today warms the northern North Atlantic, was greatly diminished. A cascade of feedback processes then took over (beginning, for example, with a great expansion of sea ice), plunging the world back into the ice age, from which it had just escaped. This miniature ice age only lasted about one millenium and ended abruptly, perhaps within decades.

In the northern Atlantic, a series of distinct layers of the last glacial period, rich in ice-rafted debris and poor in foraminifera, suggests the occurrence of other abrupt climatic events, perhaps triggered by short pulses of massive discharges of icebergs.

Our inability to explain the mysterious Younger Dryas cold spell points out some fundamental limitations to paleoceanography (and paleoclimatic) research. The climate system is extremely complex (see Epilog). Many of the important elements of the system are not well constrained. The ocean's functions (such as redistribution of heat by currents) cannot be isolated and determined as long as the functions of the other elements of the system remain uncertain. We know that the ocean plays a dominant role in the heat budget, not only because of currents, but also because of its control on albedo (e. g., via sea ice cover) and water vapor (the dominant greenhouse gas), and also because it controls short-term changes in atmospheric CO_2 (via its carbon cycle and the biological pump, discussed in Sect. 9.4.2).

9.3 The Pleistocene Cycles

9.3.1 The Evidence. The marine geologists who studied the cores recovered by the *Albatross* Expedition soon recognized that the Pleistocene record shows a long series of alternating climatic states. This finding has attained great importance in the earth sciences, especially in the study of climate dynamics. The cycles express themselves as fluctuations in faunal and floral composition, in the abundance of carbonate, and in the ratio of oxygen-18 to oxygen-16 in foraminiferal shells, as well as other properties (see Fig. 9.4).

9.3.2 The Carbonate Cycles of the equatorial Pacific (first described by G. Arrhenius) are commonly interpreted as *dissolution cycles,* with high dissolution in interglacials (low $CaCO_3$ values) and low dissolution in glacials (high $CaCO_3$ values). Productivity variations (which were once thought to be solely responsible for the carbonate fluctuations) apparently play a secondary role in producing the cycles. What is the evidence for this interpretation? The argument is based on the vertical fluctuation of levels of equal preservation, measured as the degree of shell damage, above the lysocline. A vertical range of 0.5 to 1 km (as observed) is sufficient to produce very pronounced carbonate cycles, in depths below 3500 m.

The causes for the dissolution cycles are not clear. Fluctuation of the sea level must be important. During low sea level stands, shelves – the preferred sites for carbonate deposition – are exposed. Thus, they cannot receive carbonate. The balance piles up on the deep-sea floor from deposition of plankton shells. Conversely, during

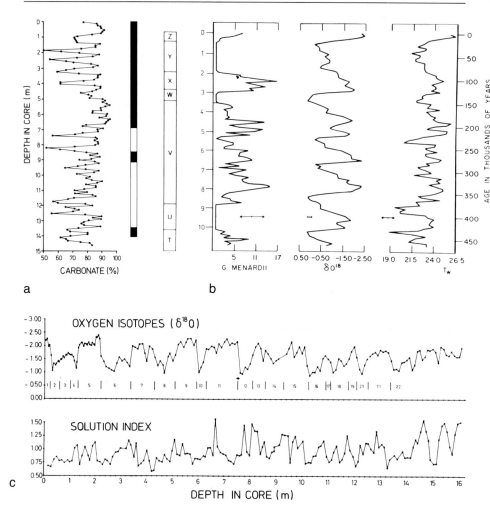

Fig. 9.4a–c. The Pleistocene record in deep sea sediments. **a** Carbonate cycles of the eastern Pacific, magnetically dated (J. D. Hays et al., 1969, Geol. Soc. Am. Bull. 80: 1481). **b** *G. menardii* pulses *(left)* and transfer temperature cycles *(right)* compared with oxygen isotope stratigraphy, Caribbean Sea (J. Imbrie et al., 1973, J Quat Res 3: 10). **c** Dissolution cycles compared with oxygen isotope stratigraphy, western Pacific. [P. R. Thompson, 1976, J. Foraminiferal Res. 6: 208; N. J. Shackleton and N. D. Opdyke, 1973, Quat Res 3: 39]

interglacials, the high sea level allows shallow water carbonates to build up – this carbonate is extracted from the ocean, and is no longer available for deposition on the deep-sea floor. Several other factors also play a role: changes in erosion rates on land and on shelves, growth and decay of forests which affects the abundance of CO_2 in the ocean – atmosphere system, changes in the temperature of the deep ocean, changes in the overall productivity of the ocean, and changes in deep-sea circulation. Deep-sea circulation governs much of the carbonate accumulation in the central

Pacific, through "basin-basin fractionation" with the Atlantic (Sect. 7.6.5). Whenever North Atlantic Deep Water (NADW) production is increased, the Atlantic traps the carbonate, leaving less for the Pacific.

On much of the Atlantic sea floor, carbonate cycles tend to swing counter to those in the central Pacific. The Atlantic carbonate cycles are largely *dilution cycles,* although effects from varying dissolution and production must also be considered. During glacials, the supply of terrigenous materials from the continents surrounding the Atlantic is greatly increased. In high latitudes, the moving ice grinds up enormous masses of rock. The plant cover, which protects soil from erosion, is less dense. In the subtropics, deserts are widespread, delivering dust. Flash floods in semi-arid regions are efficient conveyors of huge amounts of material. The tropical rainforest is much reduced, and semi-arid regions are expanded. The shelves are exposed and subject to erosion. All these factors contribute to the glacial increase of terrigenous deposition rates. As the supply of terrigenous material increases, of course, the proportion of carbonate in pelagic sediments decreases accordingly – the carbonate is diluted. By changing the degree of dilution, carbonate cycles are produced.

9.3.3 The Faunal (and Floral) Cycles can be expressed in various ways, most commonly as *warm-cold cycles.* Examples are the quantitative plots of Parker (1958) and Imbrie and Kipp (1971). Parker's method contrasts the relative abundance of warm-water and cold-water planktonic foraminifers as a function of core depth. The Imbrie-Kipp method calibrates the warm- and cold-water percentages against surface water temperature, by a statistical technique called factor regression. Based on this calibration, the most probable temperatures of surface waters are then calculated for samples deeper in the core. This technique was also used for constructing the 18 K map (Fig. 9.2). We see that there were considerable temperature variations in the Caribbean. We also see that the last change from cold to warm occurred about 11 000 years ago, with the beginning of the Holocene. The faunal cycles rather closely follow the associated oxygen isotopes, without, however, duplicating them. In principle, comparison of the two curves allows the separation of temperature effect from ice effect, in the oxygen isotope record, to be discussed next.

There is one other interesting fact that emerges from the faunal analysis: the present time is rather unusual in being so warm. For the last one half million years or so the climate was mostly much colder.

9.3.4 Oxygen Isotope Cycles. The fluctuations in the oxygen isotope composition of foraminiferal shells were first described by C. Emiliani in his classic paper *Pleistocene temperatures,* published in 1955. He analyzed the planktonic foraminifera from several long cores taken in the Caribbean and North Atlantic. He concentrated on those species with the lowest oxygen-18 *(Globigerinoides ruber* and *Globigerinoides sacculifer),* reasoning that these species must live in shallow water and therefore reflect surface water temperature. Since the temperature of growth affects the $^{18}O/^{16}O$ ratio (Sect. 7.3.2), the isotopic fluctuations reflect warm-cold cycles. In addition, the composition of seawater controls the $^{18}O/^{16}O$ ratios in the shells. The seawater composition fluctuates with the waxing and waning of the continental ice sheets. The reason is that ice is depleted in ^{18}O. During buildup of the ice, therefore, this heavier isotope stays preferentially in the ocean, increasing the ^{18}O/^{16}O ratio here. Shells grown in such water, then, are enriched in ^{18}O, and this enrichment is added to the enrichment caused by the lowered temperature of the water. When the glacial ice melts, the ^{18}O/^{16}O ratio in the ocean decreases again (see Fig. 5.15). Today, oxygen isotope stratigraphy forms the backbone of Pleistocene stratigraphy.

We use oxygen isotopes as a master template, to which other records must be compared to determine their significance. This is also true for nonmarine sequences, as in the interpretation of the loess record in China, for example (Fig. 9.5). It is seen (as has been well known for some time) that glacial intervals are characterized by eolian dust deposits ("loess"; numbered L1, L2, etc.) and interglacials by soil development ("soil", numbered S1, S2, etc.). This is a reflection, among other things, of drought-wet cycles. Detailed matching to the marine δ^{18}O record allows the observation that maximum wind supply occurred both during onset of glacials and during

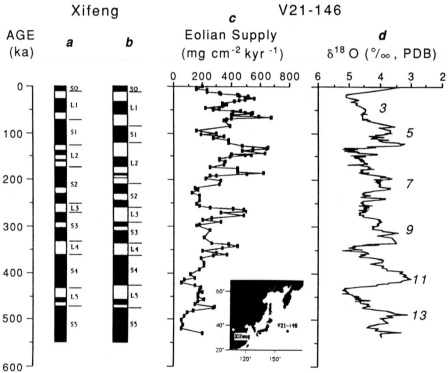

Fig. 9.5 a–d. Correlation of loess profiles from Xifeng (China, see *inset*) with marine δ^{18}O record during the last 500 000 years. **a** Loess profile, *L* loess; *S* soil; **b** Same profile with refined ages of loess sedimentation based on correlation to eolian flux record in (**c**); **c** Eolian flux record of the North Pacific, Core V 21–146 (mg/cm^2/kyr) from 3968 m water depth and with a distance of more than 3000 km from the loess source area; **d** δ^{18}O-Stratigraphy (‰, PDB) of this core, used for dating. [S. A. Hovan et al. 1989, Nature, 340: 296]

their maxima. Such supply of dust by wind has implications for deep-sea sedimentation – perhaps (as dust brings phosphate and iron) even for productivity in the ocean (and hence for atmospheric CO_2; see Sect. 9.4.2).

9.3.5 Milankovitch Cycles and Dating. As pointed out by Emiliani, the isotopic variations clearly indicate some sort of regular cycling such as could be produced by the *Milankovitch mechanism,* which invokes regular variations in the Earth's orbital parameters as a cause for the succession of ice ages separated by warm periods (Fig. 9.6). The hypothesis of Milutin Milankovitch (1879–1958) states that long-term fluctuations in the radiation received from the Sun, during summer seasons, in the high latitudes of the northern hemisphere, have controlled the occurrence of ice ages over the last 600 000 years.

For Emiliani's suggestion to be tested, a time scale for the isotopic variations was needed. Three dating methods are available: (1) C-14 dating of the uppermost portion of the record, and extrapolation downward. This method is not very reliable. (2) Uranium dating of corals which grew during the last high stand of sea level, a datum that can be correlated with the warm peak in the Isotope Stage 5 (Fig. 9.4 c). The best estimates for this age are near 124 000 years. From this we obtain an agerage sedimentation rate, which is more reliable than the one based on radiocarbon. (3) Magnetic reversals – the same that are recorded in the cooling basalt of the spreading sea floor (Sect. 1.8) – are recorded also in deep-sea sediments. The last major boundary between *magnetic epochs* (the *Brunhes* and the *Matuyama*) is dated at 790000 years ago. (An earlier date of 730 000 is now abandoned). This date can be recovered in very long cores only. It allows *interpolation* of the ages for the isotope variations.

The time scale derived from methods (2) and (3) is the one used in the comparison between isotope record and Milankovitch curve in Fig. 9.6. A certain similarity between the curves is obvious. Spectral analysis – that is, a search for the frequencies contained in the isotopic record – shows that the following periods are strongly represented: near 20 000 years, near 40 000 years, and near 100 000 years (see next Section). These are the main periods contained in the Milankovitch irradiation curve. Thus, the evidence is strong that irradiation of the northern hemisphere is the dominating factor in controllong the *frequencies* of the Pleistocene climatic fluctuations.

What are these frequencies? They describe the motions of the Earth's rotational axis and the change in shape of its path around the sun. The axis is not stationary in space, and does not always point to the North Star as in the present. Instead, it describes a circle, of which the North Star is one point. The circle is completed once in about 21 000 years. This is the *precession* (Fig. 9.6). Also, the inclination of the Earth's axis to the plane of its orbit changes through time. It is 66 1/2° at present, but varies between about 65° and 68° once in 41 000 years. This is the *obliquity* variation. Obliquity is very important, because high obliquity obviously means warm summers and cold winters, and vice versa. Finally, the Earth's orbit about the sun is not a circle but an ellipse (as Johannes Kepler, 1571–1630, showed). The ratio between the long and the short axis varies through time. This is the *eccentricity* variation. One cycle takes about 100 000 years. These, then, are the elements governing

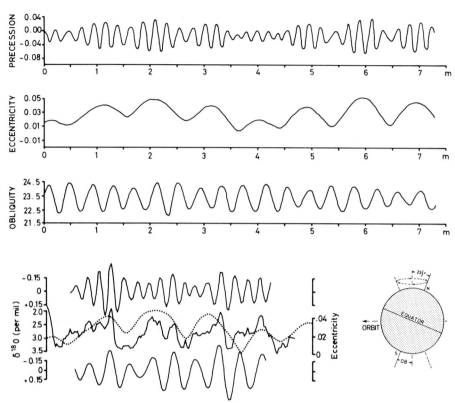

Fig. 9.6. The Croll-Milankovitch Theory of the ice ages, and its test. *Upper* 3 panels. Curves showing orbital parameters as calculated by A. L. Berger [in M. A. Kominz et al., 1979, Earth Planet Sci. Lett. 45: 394]. 1 m = 100 000 yrs. The eccentricity is the deviation of the orbit from a perfect circle, for which e=0. The eccentricity varies with a period near 100 000 years. The precession parameter is a function of both the position of perihelion (point of Earth's orbit which is closest to the Sun) with respect to the equinoxes (day = night positions) and the eccentricity of the orbit (which makes the difference of equinox and perihelion significant in terms of seasonal irradiation). The periodicity of the precession parameter is near 21 000 years. Note the small variations during times of low eccentricity. The obliquity is the angle of the Earth's axis, with respect to the vertical on the orbital plane *(inset, lower right)*. It varies with a period of 41 000 years. *Lower panel curves* showing the orbital periodicities contained in the oxygen isotope record in a subantarctic deep sea core. The *solid line in the middle* represents the original isotope record; superimposed (dotted line) is the eccentricity variation. *Top curve* 23 000-year component extracted from the isotope record by band pass filter (a statistical method); *bottom curve* obliquity component extracted in a similar fashion. *Inset right* illustrates obliquity. [J. D. Hays et al., 1976, Science 194: 1121] M. Milankovitch's classical paper was published in 1920: Théorie mathematique des phénomènes thermiques produits par la radiation solaire, Gauthier-Villars, Paris, 339 pp.

the changes in the seasonal irradiation of the northern hemisphere. Presumably, there is an especially sensitive latitudinal belt in the northern hemisphere where these changes in insolation are translated into varying amounts of snow cover, which in turn govern climate by the albedo feedback mechanism (Sect. 9.2.2).

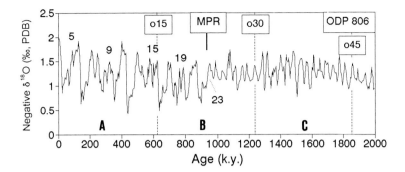

Fig. 9.7. Oxygen isotope record of *G. sacculifer,* ODP Site 806, western equatorial Pacific. Age scale based on counting obliquity cycles extracted from the record (and set at 41 000 years; see Fig. 9.6). *5, 9, 15, etc.* Isotope stages; *o15, o30, o45* position of crests of obliquity-related cycles; *MPR,* mid-Pleistocene Revolution (at 900 kyrs), showing a distinct change from obliquity-dominated to eccentricity-dominated- B/M, Brunhes-Matuyama bdry. at 790 kyrs. *A, B, C,* major cyclostratigraphic subdivisions, each with 15 obliquity cycles, where o15 = 625 000 yrs; *"Milankovitch Epoch"; o15–o30* "Croll Epoch"; o30–o45 "Laplace Epoch". [W. H. Berger and G. Wefer, 1992, Naturwisenschaften 79: 541; time scale adjusted according to Berger et al. (1994), Quaternary time scale for the Ontong Java Plateau: Milankovitch template for Ocean Drilling Program Site 806. Geology 22: 463–467.]

9.3.6 A Change in Tune. The Pleistocene record of $\delta^{18}O$ supports Milankovitch's propositions regarding the cause of ice age fluctuations. In particular, it is interesting how much the 100 000-year eccentricity cycle dominates the scene throughout the last 900 000 years. However, a surprise awaits us when we go further back than that: the lower half of the Quaternary essentially has no 100 000 year cycles at all! The $\delta^{18}O$ fluctuations (that is, the variations in ice mass) in that part of the record are entirely dominated by the obliquity cycle (Fig. 9.7). A little less than one million years ago, eccentricity acquired a strong voice in the mixture of tunes within the climatic record. (It is as though a strong bass joined the choir). We do not know the reason for this major change in the Earth's climate system (marked MPR, for *mid-Pleistocene revolution,* in Fig. 9.7). Apparently, there are no astronomical reasons: the orbital elements, as far as one can tell, were much the same before and after the event. Thus, we must assume that the response to forcing changed.

9.4 The Carbon Connection

9.4.1 A Major Discovery. One of the most exciting discoveries in the Earth Sciences, in the 1980s, was the reconstruction of the composition of the Pleistocene atmosphere from ice cores. Already in the last century, the physicist John Tyndall (1820–1893) suggested that glacial periods may have been caused by decreased atmospheric carbon dioxide. The famous Swedish chemist Svante Arrhenius (1859–1927) and the

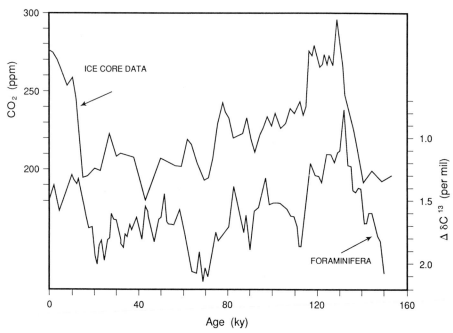

Fig. 9.8. Carbon dioxide concentrations in the Vostok ice core from Antarctica (J. M. Barnola et al., 1987, Nature 329: 408), compared with a productivity-related $\delta^{13}C$ signal from the eastern tropical Pacific (N. J. Shackleton et al., 1983, Nature 306: 319). The signal plotted is the difference in $\delta^{13}C$ of a planktonic and a benthic species, which constitutes a proxy for the nutrient content in deep waters. The time scale of the CO_2 record was adjusted using the deuterium record in the ice and the $\delta^{18}O$ record in the sediments. [W. H. Berger, V. S. Smetacek, G. Wefer, 1989, Productivity of the ocean: present and past. Wiley-Interscience, Chichester]

distinguished American geologist Thomas C. Chamberlin (1843–1928), both put forward hypotheses about how the concentration of CO_2 could be changed through time. The recent ice-core results prove that glacial CO_2 concentrations in air were lower than interglacial ones by a factor of 1.5 (Fig. 9.8, upper curve). The CO_2 fluctuations run parallel to the temperature at which the snow was deposited. Methane gas also was measured; the fluctuations are similar to those of carbon dioxide.

The observed amplitude of CO_2 variation corresponds to a greenhouse effect of between 1 and 2 °C, according to recent climate models. Changes are very rapid, making it very likely that the ocean is responsible: it has a large carbon reservoir (roughly 60 times that of the present atmosphere), and it has a "memory" of only about 1000 years (the average age of its deep water masses). Small changes in its overall chemistry, leading to a small proportional change of carbon dissolved in the ocean, can potentially have large effects on the atmosphere.

9.4.2 Paleoproductivity. There are a number of ways to affect the sharing of carbon dioxide between ocean and atmosphere (see Sect. 3.7.1). For example, if dissolved

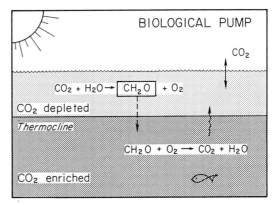

Fig. 9.9. Sketch of a mechanism linking productivity to the carbon dioxide content of the atmosphere. According to W. S. Broecker (Ocean chemistry during glacial time, Geochim Cosmochim Acta, 46, 1689–1705, 1982) when phosphate is added to the ocean a new equilibrium is established, with surface waters being more CO_2-depleted than before. This then reduces atmospheric CO_2. An increase in nutrient supply, through bio-pumping, also increases the $\delta^{13}C$ of the dissolved inorganic carbon in surface waters (by preferential removal of ^{12}C into organic matter which sinks). Thus, the biopump mechanism provides a simple conceptual model linking pCO_2 and the difference in $\delta^{13}C$ of planktonic and benthic foraminifers (as shown in Fig. 9.8)

inorganic carbon is preferentially stored at depth, and depleted in upper waters, the atmosphere should receive a smaller share of the total carbon available. By increasing the efficiency of *biologic pumping,* such increased vertical fractionation can be achieved (Fig. 9.9).

The bio-pump mechanism (first proposed by the American geochemist W. S. Broecker, in 1981) provides a simple conceptual model linking pCO_2 and $\delta^{13}C$ (that is, the records shown in Fig. 9.8). The reason for the link is that the pumping tends to remove ^{12}C more efficiently from the photic zone than ^{13}C, because ^{12}C is more readily incorporated into organic matter during photosynthesis. The effect is that ^{13}C is enriched in surface waters, relative to deep waters. The corresponding difference in $\delta^{13}C$, which is picked up in the shells of planktonic and benthic foraminifers, is a measure of the intensity of the fractionation resulting from the biological pumping. This intensity, in turn, depends entirely on the nutrient content of deep ocean waters.

The bio-pump model for changing CO_2 content of the atmosphere (which is controversial and probably can only account for a modest portion of the effect seen in the ice cores) implies higher productivity of the ocean during glacial time. There is indeed good independent evidence that productivity was higher during glacials. Off NW Africa, for example, the rate of burial of organic carbon during glacials is greatly increased (Fig. 9.10). The cause, presumably, is increased *eastern boundary upwelling* due to an increase in trade winds and monsoonal winds. Similar results have been reported from other coastal upwelling regions. Along the Equator, as well, productivity was higher during glacial time than now. The general increase in productivity may have resulted in the burial of sufficient additional carbon to pull down the CO_2 content significantly.

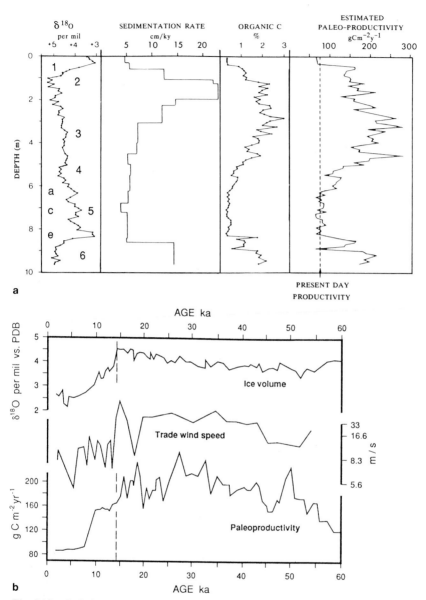

Fig. 9.10 a, b. Paleoproductivity reconstructions for the late Pleistocene, off NW Africa. **a** Oxygen isotopes, sedimentation rates, and organic matter content in *Meteor* Core 12392. *Right* resulting paleoproductivity estimate. [P. J. Müller and E. Suess, 1979, Deep-Sea Res. 26A: 1347, and P. J. Müller et al., 1983 in J. Thiede and E. Suess, Coastal upwelling, Part B. Plenum Press, New York], Isotope stages as in Figs. 9.7. and 9.4 c. (*1* Holocene; *2–4* last glaciation; *5a–e* last interglacial; *6* earlier glaciation). **b** Importance of wind in producing changes in upwelling, as seen in correlation with dust supply (from which the strength of the wind is calculated). [M. S. Sarnthein et al., 1987 in W. H. Berger and L. D. Labeyrie, eds. Abrupt climatic change – evidence and implications. Reidel Dordrecht]

Changes in the sites of *carbonate production* also must have been important. During times of low sea level, carbonate deposition would have been greatly decreased on the shelves, and this (together with erosion) would have increased the carbonate ion content of the ocean, and hence the "alkalinity" of the ocean. In turn, increased alkalinity allows the ocean to retain a larger share of the total CO_2 in the system.

9.4.3 The Long-Range View. On long time scales, not just the ocean's exchange with the atmosphere has to be considered, but also the weathering of silicates on land, and the input of CO_2 from volcanism. The weathering process can be summarily described by the formula

$$CO_2 + CaSiO_3 \rightarrow CaCO_3 + SiO_2 , \qquad (9.1)$$

which shows the long-term uptake of CO_2 from the atmosphere. The (volcanogenic) CO_2 on the left side, inasmuch as it evolved from the heating of subducted carbonates, is recycled on a very long-term scale (100 million to 1 billion years). Thus, within the Earth below subduction zones, the equation reads backwards.

From the *"Urey equation"* (9.1) we can deduce how to effect a long-term decrease of CO_2 in the atmosphere: by providing fresh $CaSiO_3$, through mountain building and deep erosion. Mountain-building and sea level drop proceeded all through the late Tertiary (Fig. 5.18), but especially in the last several million years, as shown in the changing ratio of strontium isotopes within marine carbonates (see Fig. 8.19). Here is one possible cause (of several) for the onset of the ice ages: a drop in atmospheric CO_2 as a long-term trend within the late Tertiary.

9.5 Tertiary Oceans: the Cooling Planet

9.5.1 Trends and Events: Oxygen Isotopes. Of the host of tools available to paleoceanographers and sedimentologists for studying the sediments newly recovered by deep-sea drilling, oxygen isotopes promised to yield the greatest return for investment, in the reconstruction of climatic change and the ocean's role in it. However, nothing much could be achieved without biostratigraphy. It delivers the time frame and also the all-important information on the response of plankton and benthos to physical change. Paleobiological studies bring ancient oceans alive. They established, for example, that "punctuations" in the history of evolution exist, and that they are a response to climatic change. Only the continuous record of deep-sea sediments, with its masses of microfossils, could establish such a pattern for sure. On land, hiatus-ridden records (that is, sequences with gaps) commonly do not allow a choice between gradual and jumpy evolution – as Charles Darwin pointed out long ago in his book, the *Origin of Species* (in 1859).

The central theme of climatic evolution in the Tertiary is an overall drop of sea level, which is accompanied by a general cooling (Fig. 9.11a). The intial work on the oxygen isotope record of the Tertiary deep ocean was done by C. Emiliani. In the 1950s he used samples of benthic foraminifera from the surface of the sea floor,

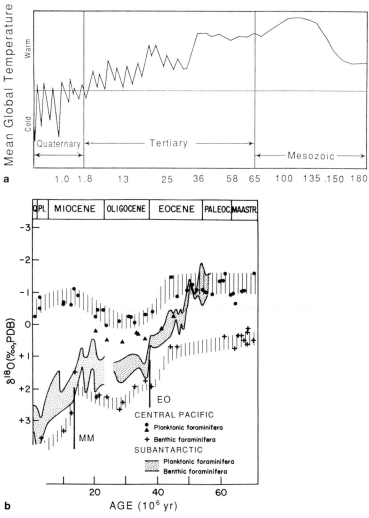

Fig. 9.11 a, b. Schematic diagram of the planetary cooling trend in the Cenozoic. **a** Generalized sea level curve, showing overall regression (J. Thiede et al., 1992, Polarforschung 61 [1]: 1; based on a compilation by L. A. Frakes). **b** Oxygen isotopic composition of planktonic and benthic foraminifera in deep-sea sediments. *Vertical lines* marked *MM* and *EO* show approximate position of major cooling steps, which involve high latitudes and deep ocean water masses. [W. H. Berger, E. Vincent, H. R. Thierstein, 1981, SEPM Spec. Publ. 32: 489, based on data of R. G. Douglas and S. M. Savin (1975), central Pacific, and N. J. Shackleton and J. Kennett (1975), subarctic Pacific]

where old sediments are laid bare by erosion, to identify an overall cooling trend since the Cretaceous, from the increase of oxygen-18 in their tests. His guess proved correct.

Seven years after the Deep-Sea Drilling Project started, two paleontologist-geochemist teams had worked out the essential trends (Fig. 9.11B). The stratigraphies of the oxygen isotopic composition of planktonic and benthic foraminifers shows separate trends in low latitudes, but similar trends in high latitudes. Thus, the overall cooling in the Tertiary is largely a high-latitude (and deep-water) phenomenon. In general, then, temperature gradients must have increased throughout the Tertiary, since the middle of the Eocene, some 40 million years ago. Wind speed depends strongly on temperature gradients. If so, winds and their offspring, the surface currents, greatly increased in the late Tertiary, as did coastal and mid-ocean upwelling. Direct evidence that this is true is found in the increasing diatom supply, both in the northern North Pacific and around the Antarctic, during the late Tertiary. Other evidence for fertility increases also exists. For example, radiolarian skeletons become continuously more delicate after the Eocene. Apparently the content of dissolved silica decreased through time, a sign that silica has been increasingly removed by diatom production in upwelling areas.

The high-latitude cooling indicated in the oxygen isotope trends shows two major steps: a more recent one in the middle of the Miocene, and an older one near the Eocene-Oligocene boundary. We do not know the cause of these steps, which simply enhance the overall trend of turning a warm ocean into a cold one. Perhaps, at some critical threshold of cooling around Antarctica, deep water formation off its shores was greatly enhanced, and this set off a chain reaction producing a permanent reorganization in the ocean's circulation patterns, making the change irreversible. Or perhaps the physical geography changed at that very time, barring or opening certain ocean passages and thus redirecting the heat carried by ocean currents. The two types of causes – *internal feedback* and *change of geography* – are not mutually exclusive. We know that both are at work. The question is, how important are they in each given situation.

9.5.2 The Great Partitioning. All through the Tertiary, changes in geography due to plate motions profoundly affect the configuration of exchange between ocean basins. A number of important valve points in the ocean's plumbing system (called "*gateways*") translate small motions of continents into large effects for ocean currents (Fig. 9.12). The dominant theme is a stepwise partitioning of the global ocean (originally connected by large circumtropical passages) into the cul-de-sacs we have today. In a sense, we trade northern and southern embayments along a broad world-encircling tropical ocean for three separate super-large basins connected through a ring current far in the south (the "*Southern Exchange*").

The gateways shown (Fig. 9.12) control access to the Arctic Ocean (east and west of Greenland), connect the global ocean along the Equator ("*Tethys Ocean*" between Africa and Eurasia, Panama Straits, Indonesian Seaway between Borneo and New Guinea), and control the evolution of the Circumpolar Current (Tasmanian Passage, Drake Passage). The difference from the present ocean to the Eocene ocean 45

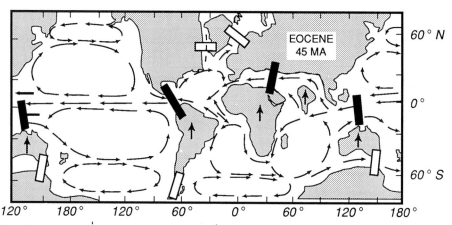

Fig. 9.12. Geography of the middle Eocene (ca. 45 Ma) and major critical valve points for ocean circulation. Tropical valves are closing *(filled rectangles)*, high latitude valves are opening up *(open rectangles)* throughout the Cenozoic. [Base map from B. U. Haq, Oceanologica Acta, 4 Suppl.: 71]

million years ago consists mainly in the closing of the tropical gateways, and the opening of the poleward ones. In detail, each of these closings and openings holds a wealth of interesting stories. One of the most fascinating, for example, is the drying out of the Mediterranean between about 6 to 5 million years ago, as discovered during Leg 13 of the Deep Sea Drilling Project (with W. B. F. Ryan and K. J. Hsü as Co-chief Scientists). Huge amounts of salt were deposited at that time in the Mediterranean basin, so much that the salinity of the global ocean may have been reduced by 6 %!

The overall trend towards isolation of the major ocean basins resulted both from the grand switching of the valves and a drop in sea level (Fig. 5.18). This fall – a long-term regression since the late Cretaceous – was briefly halted or reversed in the middle Eocene and middle Miocene. It is thought to have been caused by a decrease in spreading rates and by mountain building (in the late Tertiary). A decrease in spreading rates makes for a lesser rate of production of young (hence shallow) sea floor, which leads to deepening of the ocean. Mountain-building doubles up lithosphere, and leaves a hole to be filled. Both processes result in regression, therefore. Regression results in albedo increase (land reflects sunlight better than does the ocean), and CO_2 decrease (from the chemical weathering of exposed continental rocks). Hence the cooling trend is favored.

9.5.3 The Grand Asymmetries. What are the consequences for ocean circulation of the large-scale valve switch executed by continents moving away from Antarctica? There are implications for physical, chemical, and biological oceanography. Generally, we can say that increased compartmentalization allows for a broader range of temperatures and salinities in the global ocean. Water masses, then, become more different with time. Eventually, mixing of these masses is restricted to the Southern

Ocean, that is, to the "cold box" of the ocean modelers. Thus, the average ocean water has to be cold (which it is, at present). Since the cold water sphere becomes the global environment, while the tropical waters are isolated, deep-sea and cold-water faunas become global, while tropical ones become more provincial.

As concerns the heat budget, one leitmotif of Tertiary ocean history is the increased displacement of the intertropical convergence zone (ITCZ, the *heat equator*) to the north of the Equator. There are several causes for this. One is the whitening of the Antarctic continent. This has the effect of pushing climatic zones northward. Another is the northward movement of large continental masses which sets up monsoonal regimes favorable for northward heat transfer. The uplift of Tibet and the Himalayas as a consequence of the collision of the Indian with the Eurasian Plate had additional climatic consequences, for example, the strengthening of monsoons and an increase in weathering. A fourth factor is the peculiar geographic configuration in both major ocean basins, Atlantic and Pacific, which provides for deflection of westward-flowing equatorial currents, sending them northward to strengthen the Gulf Stream and the Kuroshio. The end result is that the southern hemisphere is robbed of heat by the northern hemisphere: glaciers in southern New Zealand are in walking distance from the seashore, at a latitude which corresponds to that of the vineyards of Bordeaux! The significance of this asymmetry for paleoclimatology was emphasized long ago by James Croll in his book *Climate and Time* (1875).

One important aspect of this planetary heat asymmetry is the fact that the North Atlantic tends to deliver deep water to the southern exchange around Antarctica. On the whole, it receives shallower water in return, not directly from the Antarctic, but via various routes along which this water was warmed. Thus, the North Atlantic uses deep water formation as a heat pump: warm water in, cold water out.

When was this *"Nordic heat pump"* first turned on? We do not know for sure, but there are certain clues from chemical asymmetries between Pacific and Atlantic (*basin-basin fractionation,* Sect. 7.6.5). Whenever the North Atlantic makes cold deep water, and sends it off to the southern mixing ring, it also sends nutrients and dissolved silica out of the basin. Also, in order to achieve the necessary density for making deep water, salinity needs to be increased by sending water vapor to the North Pacific. In turn, this sets up an estuarine deep circulation pattern there, with high nutrient and silica values at depth (Sect. 7.6.1). The result is that siliceous sediments become rare in the North Atlantic and abundant in the North Pacific, whenever the Nordic heat pump is in action. From the record of silica deposition, we see that the Pacific-Atlantic asymmetry greatly increased between 15 and 10 million years ago (*"silica switch"*). At that time the Nordic heat pump started working hard (Sect. 9.5.5).

9.5.4 The Onset of the (Northern) Ice Age. Once the Nordic heat pump was turned on, the north-to-south asymmetry was greatly strengthened – the ice age could proceed in Antarctica, while being kept out of northern latitudes. Eventually, however, the overall cooling caught up with the north as well, between 3.5 and 2.5 million years ago (Fig. 9.13). Northern glaciation set in, and the planetary climate moved into the glacial cycles which we discussed earlier. Again, the Plio-Pleistocene cooling was

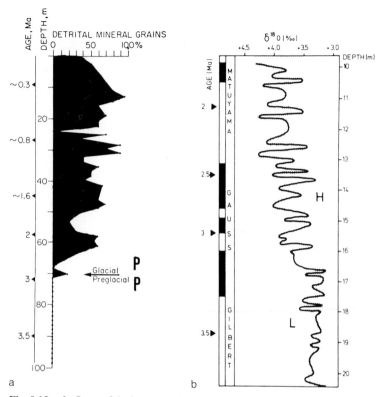

Fig. 9.13 a, b. Onset of the ice age. **a** Sudden increase in the abundance of detrital mineral grains at DSDP Site 116, NW Atlantic, indicating glacial activity on the adjacent continent. [W. A. Berggren, 1972, Init Rep Deep Sea Drilling Project 12: 953]. **b** Change from a low variability climate (*L*) to a high variability ocean climate (*H*), as indicated in the oxygen isotope stratigraphy of a piston core from the central equatorial Pacific. The isotopes refer to the benthic foraminifer *Globocassidulina subglobosa,* whose present composition is near +3.5 ‰ at this site. Magnetic stratigraphy controls the time scale. *PP* Panama Seaway closes gradually within the Gauss Epoch. [N. J. Shackleton, N. D. Opdyke, 1977, Nature 270: 216]

associated with regression. In addition, mountain building and the northward drift of land masses towards polar areas helped bring about a greater likelihood for snow to stay on the ground – the necessary condition for glaciation. The exact timing of the onset of ice age fluctuations may have to do with the closing of the *Panama Seaway* at the time. Atlantic and Pacific were now even more isolated from each other than before.

9.5.5 The Mid-Miocene Cooling Step. The "silica switch" mentioned above, that is, the large-scale transfer of dissolved silicate from the Atlantic to the Pacific between 15 and 10 million years ago, points to a major reorganization in the deep circulation of the ocean which preceded the northern ice age (Fig. 9.14). In the stable isotope

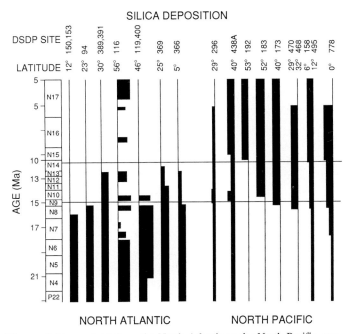

Fig. 9.14. The middle Miocene "silica switch" from the North Atlantic to the North Pacific, suggesting the turning on of the Nordic heat pump about 15 million years ago. Silica-rich sequences marked black. [F. Woodruff and S. M. Savin, 1989, Paleoceanog. 4: 87, based on a compilation by G. Keller and J. A. Barron, 1983, Geol Soc Am Bull 94: 590]

record, the crucial transition appears as a drastic increase in oxygen isotope values in benthic foraminifera, between 15 and 13 million years ago (Fig. 9.11). This increase reflects both growth of ice on Antarctica, and a world wide cooling of abyssal waters, which in turn reflects a fundamental change in deep circulation. A strong increase in the diversity of planktonic foraminifers at the time suggests the development of a strong thermocline below warm surface waters, providing for diversification of plankton habitats which resulted in diversification of plankton species.

The cooling step is preceded by a substantial carbon isotope excursion toward heavy values. This carbon isotope anomaly is seen in all oceans, at all depths. The excursion is synchronous with the onset of abundant sedimentation of organic-rich sediments in the margins of the Pacific. Presumably, the increase in organic deposition preferentially extracts ^{12}C from the ocean, so that the carbon isotope values of dissolved inorganic carbon in the ocean become enriched in the isotope ^{13}C, which increases the $^{13}C/^{12}C$ ratio as observed. In this manner, a link is established between margin sedimentation, and the isotope record of deep-sea sediments. There is a possibility that the increased burial of organic carbon was in part responsible for the mid-Miocene cooling that followed: the additional extraction of carbon from the ocean would have lowered the carbon dioxide content of the atmosphere, preparing the way for the subsequent build-up of ice.

Besides internal feedback mechanisms favoring cooling and ice growth on Antarctica (albedo increase and a lowering of atmospheric CO_2), we again have to consider rearrangements in geography. In the case of the middle Miocene, it is the *Indonesian Seaway* that closes some time between 20 and 10 million years ago. It is not clear, however, how the closing of this passage would have contributed to the cooling.

A much discussed question is to what extent the shift in $\delta^{18}O$ toward heavier values (Fig. 9.11b, MM) reflects cooling, and how much is due to ice buildup. Ice rafting off Antarctica (as seen in *"ice-rafted debris"*) apparently only sets in "in earnest" towards the end of the middle Miocene. At that time, one surmises, significant glacial ice growth began on Antarctica. Such buildup could have been responsible for roughly 50 m drop in sea level, beyond the general regression from uplift in Asia and elsewhere, which is reflected in strontium isotope stratigraphy.

9.5.6 The End-of-Eocene Cooling. Going back in time, we skip over the enigmatic Oligocene, with its low plankton diversity and its strange *Braarudosphaera* blooms, to take a closer look at the late Eocene and the Eocene-Oligocene boundary. On the very first leg of the Glomar *Challenger,* the Eocene posed a major puzzle: the drill bit hit chert in the western North Atlantic, at the level of a prominent *seismic reflector* (see Sects. 3.6.2 and 8.6.4). Later, similar beds were found in many parts of the ocean. In fact, only since the late Eocene (that is, the last 40 million years) do pelagic sediments look "modern". From then on they have a facies distribution similar to that outlined in Chapter 8.2.

In essence, Eocene sedimentary facies are less clearly differentiated from each other than post-Eocene ones. The chemical fractionation machine of the ocean was less efficient, both because ocean basins communicated more easily than now, and because there was less of a temperature gradient driving the machine (that is, the winds were weak, and hence the upwelling that concentrates organic-rich sediments, silica, and phosphate in the margins). This situation changed towards the end of the Eocene: the machine turned on. The polar regions cooled, presumably due to thermal isolation of the Arctic and Antarctic and due to positive feedback from snow cover and changing vegetation (increase of albedo, Sect. 9.2.2). The polar front migrated equator-ward, pushing a late Eocene rain belt (and thus the temperature regions) away from Antarctic shores. We see the effect on the oxygen isotope values in high latitudes: they quickly move to higher values during the late Eocene, with an especially dramatic change at the Eocene-Oligocene boundary (Fig. 9.11b, EO). This is the point at which the water in high latitude shelves is cold enough, and saline enough, to sink and fill the deep ocean basins.

Whatever the cause – most likely internal feedback processes set in motion by overall regression – the late Eocene cooling irreversibly initiated a new type of ocean, one where strong asymmetries began to develop between north and south, between Atlantic and Pacific, and between margins and deep sea.

9.6 The Cretaceous-Tertiary Boundary

9.6.1 The Evidence for Sudden Extinction. The Cretaceous-Tertiary boundary marks a major biostratigraphic break in the record. This break, of course, is the reason why the boundary between the Mesozoic and the Cenozoic was placed here in the first place. By convention, we count the time since as the age of the mammals (Appendix A3), which replaced that of the reptiles. In shallow marine sections exposed on land, the evidence for profound change is clear everywhere. However, one could never be quite sure to what degree the change observed was the result of superposition of sediments of greatly differing age. Pelagic sediments exposed on land gave the best evidence for a rapid and fundamental change in plankton biota at the "K/T"-Boundary, thanks to the studies carried out by a number of biostratigraphers, in the 1960's (e. g., W. A. Berggren, M. N. Bramlette and E. Martini, H. Luterbacher, and I. Premoli-Silva). Results from deep-sea drilling fully confirmed these early results, and demonstrated a large-scale, world-wide extinction event (Fig. 9.15). Careful scrutiny of the interval associated with extinction, and global correlation, showed that the catastrophe happened in a geologically short time, say, less than 100 000 years.

9.6.2 The Causes of Extinction. From the viewpoint of paleoceanography, there is little reason to suspect fundamental differences in circulation or heat budget between the latest Cretaceous and the early Paleogene, although a general regression apparently was in progress at the time. This observation, combined with the suddenness of the extinction, and the fact that it was highly selective (warm-water planktonic species were greatly reduced, but not deep-sea benthic species, for example) makes a buildup of gradual stress unattractive as a cause. Instead, we must look to a major unusual global disturbance.

The number of causes that have been proposed to explain the great extinctions at the end of the Cretaceous (including the demise of the dinosaurs) is truly remarkable. Radiation from a supernova, poison gas from a comet entering the atmosphere, climate catastrophe from large-scale volcanism or from enormous meteorites colliding with Earth are among the hypotheses testifying to the fertile imagination of geologists. However, the major cause of the sudden extinctions remained unknown, until L. and W. Alvarez and co-workers (1980) and J. Smit and J. Hertogen (1980) presented strong evidence for the *impact scenario,* that is, for collision with a large mass of rock coming in from outer space (e. g., a body ca. 10 km in diameter). The evidence rests mainly on the observation that *iridium* (a noble rare metal) is greatly enriched in the K/T boundary layer, and that *shocked quartz* is present, attesting to sudden high pressures. These findings made the topic of the K/T extinctions a dominant one all through the 1980s. Even a candidate for an impact crater has been found, on the Yucatán peninsula in Mexico, the 200-km-diameter Chicxulub structure.

Alternative scenarios are being debated as well, mainly those centered on volcanism. There is no reason, of course, why volcanism (or regression) could not have aided in bringing about unfavorable conditions for survival in the aftermath of an impact, or a series of impacts.

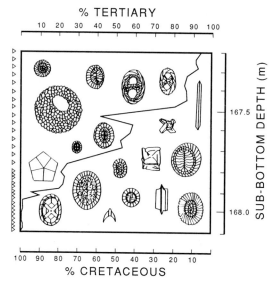

Fig. 9.15. Fundamental and rapid change in nannofossil content of deep-sea sediments from latest Cretaceous to earliest Tertiary ("K/T-Boundary"). A highly diversified tropical assemblage is replaced by a less diversified assemblage of opportunistic species, including dinoflagellate cysts *(large sphere with hole)* and the stress-tolerant form *Braarudosphaera (pentagon)*. Sample spacing at the *left* side. [DSDP Site 384 in the western South Atlantic studied by H. R. Thierstein and H. Okada, 1979, Init Reps Deep Sea Drilling Project 43: 601]

The exact manner in which the global environment was changed as a consequence of the impact is still a matter of lively discussion. A prolonged darkening of the sun (from particles in the stratosphere), acid rain (from the burning of molecular nitrogen when the air was heated), and drastic changes in temperature (from changes in greenhouse gases) are some of the possibilities that have been put forward.

9.7 Plate Stratigraphy and CCD Fluctuations

9.7.1 Backtracking and CCD Reconstruction. To interpret the meaning of a given sample from the sea floor correctly, the sediment has to be placed in its original latitude and depth, at the time of deposition. This is done by *"back-tracking"* the path that a site has taken as it aged. Without taking proper account of this path, the sediment sequences in drill cores cannot be interpreted. For example, when a site moves down from the East Pacific Rise, and northward across the equator, it collects a number of different sediments (calcareous ooze, siliceous clay, siliceous calcareous ooze, siliceous clay, red clay) as a result of plate motion only, and depending on its original position. To interpret such a facies sequence in terms of changes in the ocean system would be incorrect.

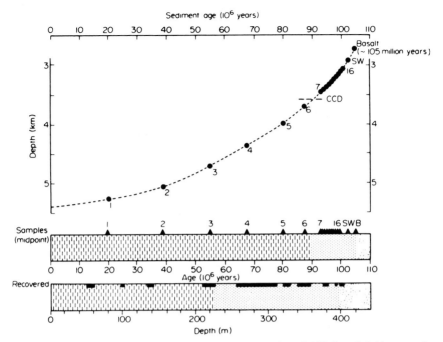

Fig. 9.16. Determination of paleodepth of the CCD by backtracking of drill sites. Subsidence track and sediment sequence of Site 137 Central Atlantic. Lithologic symbols *(from left to right at base): broken lines* clays; *dots* calcareous ooze; *random dashes* (below 400 m depth) basalt. CCD is reached near 3.6 km depth, 90 million years ago. [W. H. Berger, E. L. Winterer, 1974, Spec Publ Int Assoc Sedimentol 1: 11]

The changes of facies that are related to changes in oceanography emerge as a result of *"backtracking"* of sites, which brings the sediments accumulated on a piece of sea floor back to the original place of deposition, as a function of age. *Horizontal backtracking* is done by reversing the path of a drilling site, using the appropriate pole of rotation for the plate the site occupies. For some purposes it is sufficient to obtain the paleolatitude (that is, the distance to the equator). *Vertical backtracking* is necessary any time depth-sensitive properties are involved, such as carbonate content, sedimentation rate, or preservation of calcareous fossils.

The position of the CCD is, by definition, tied to depth of deposition, and the past positions of the CCD in any one region, therefore, have to be found by backtracking. How exactly does the CCD fluctuate? What can this tell us about the changing chemistry and fertility of the ocean? Before we can answer these questions, we must reconstruct the depth levels at which the CCD was crossed, for a sufficient number of drilling sites. To do this we use the generalized subsidence curve (Fig. 1.9a) and proceed as follows:

First, we identify the point which corresponds to the basement age of a given deep-sea record, on the subsidence curve of Fig. 1.9. Then we walk back up the

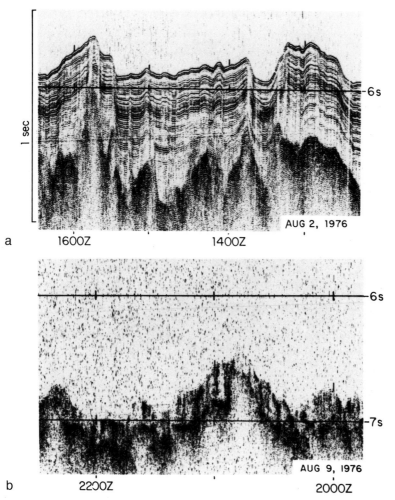

Fig. 9.17 a, b. Effect of sea-floor subsidence on carbonate sedimentation. **a** Seismic profile of well-layered, well-preserved calcareous sediments, western flank of the East Pacific Rise, near the equator. *6s* is the echo-return time and corresponds to 4500 m depth. **b** Similar profile, further west, at greater depth (*7s* 5250 m). Topography and sediment cover are irregular, and are interpreted as deep-sea karst, produced by dissolution of carbonate which migrated into greater water depth with the subsiding sea-floor. [W. H. Berger et al., 1979, Mar Geol 32: 205]

curve, the distance corresponding to the age of the sediment for which we wish to determine the depth of deposition. We obtain a paleodepth estimate, as a difference to present water depth, for the time of deposition. To refine this estimate, we subtract one half the thickness of the sediment accumulated since that time of deposition, to account both for the buildup of the sea floor and for isostatic subsidence. The paleo-CCD emerges as the paleodepth at which the facies changes from carbonate to noncarbonate (Fig. 9.16). In the example, the site crossed at a paleodepth of 3.6 km, 90 million years ago.

To obtain a number of crossings, which would allow reconstruction of fluctuations, we apply the same procedure to a number of sites in the same region. In an age-depth diagram, each drill site will appear as a subsidence curve, on which the sedimentary facies (and rates of sedimentation) can be plotted. The changes in CCD will appear as changing depths of the carbonate-to-noncarbonate boundary. Conditions do not always allow positioning of the CCD by backtracking. Upon moving down the ridge flank, calcareous sediment can be deeply eroded, wherever bottom currents remove the protective clay cover, so that *"deep-sea karst"* can form (Fig. 9.17). When this happens, backtracking will not find the depth of the CCD crossing; the path of the site will show a hiatus extending above and below that depth level.

9.7.2 Atlantic and Pacific CCD Fluctuations. Reconstructions of CCD fluctuations provide useful information on deep circulation, on overall and regional productivity, and on the sharing of carbonate between shelf and deep-sea floor. Quite generally, the CCD is one of the important proxies for the chemical state of the ocean that bears on atmospheric CO_2 content.

Reconstructions differ somewhat depending on the available datings of basement and sediments, and on the assumptions regarding the rates of subsidence. Nevertheless, the general trends have been well known since the early 1970s (Fig. 9.18): the CCD stood high in the late Eocene, dropped near the Eocene-Oligocene boundary, rose in the Miocene when it reached a peak between 10 and 15 million years ago, and then fell to its present depth near 4.3 km. The most dramatic changes occurred at the end of the Eocene, and within the last 10 million years, in both oceans. The drastic drop at the end of Eocene is global, and hence reflects changes in the global carbon

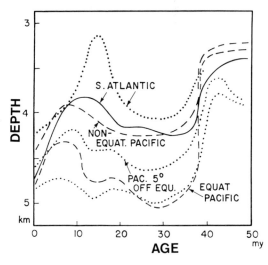

Fig. 9.18. Reconstructions of CCD fluctuations for various oceanic regions. *Solid* and *dashed lines.* Reconstructions of Tj. H. van Andel et al., 1977, J Geol 85: 651. *Dotted line* Reconstructions of W. H. Berger, P. H. Roth, 1975, Rev Geophys Space Phys 13: 561. The reconstructions agree in the general patterns of the fluctuations, which appear correlated with sealevel changes on the whole.

cycle. The sharp drop in the Atlantic after 10 million years ago has a strong regional component, and reflects (among other things) the increase in production of North Atlantic Deep Water discussed in connection with the "silica switch" (Sect. 9.5.5).

The overall similarity in the CCD fluctuations of Pacific and Atlantic indicates that the chemical climate of the ocean is changing on a global scale throughout. Comparison with sea level reconstructions and $\delta^{18}O$ stratigraphy suggests that periods of high sea level are characterized by a shallow CCD and by warm high latitudes; periods of low sea level by a deep CCD and cold high latitudes (and cold deep waters). Why should a relatively warm ocean have a *shallower* CCD than a cold one? Is not cold water *less* favorable to the preservation of carbonate than warm water?

9.7.3 Possible Causes of CCD Fluctuations. A simple hypothesis linking sea level to CCD fluctuations is the concept of *"basin-shelf-fractionation"*. The shelf, being shallow, is the favored place for carbonate to accumulate, because of the correlation between solubility and pressure. Flooded shelves, then, are carbonate traps, and remove $CaCO_3$ from the ocean so that the deep-sea floor starves. Conversely, bared shelves supply carbonate to the deep sea. Thus, it is not the temperature which is important, but the sea level! Temperature happens to be correlated with sea level for various reasons (decrese of albedo [Sect. 9.2.2] and also increase in pCO_2 during flooding of shelves).

There is good reason to believe that the mass balance hypothesis of CCD fluctuations does not suffice. We must, in addition, employ a more subtle argument, based on internal cycling of carbonate within the deep ocean basins. Remember that the biological productivity of the ocean is responsible for precipitating carbonate. Removal of solids from a solution lowers saturation. High productivity, then, results in an undersaturated ocean with a shallow CCD. In this model, we can read the CCD fluctuations as productivity fluctuations, with fertility high in the Eocene and Miocene, low in the Oligocene. There is other independent evidence that productivity was low in the Oligocene, supporting this concept.

CCD fluctuations also are closely associated with the history of *erosion* on the deep sea floor. The dissolution of carbonate itself is a form of erosion, of course. Compilations of sedimentation rates and *hiatuses* (gaps in the record) show that both these stratigraphic parameters fluctuated consideraby through time. However, the relationships between these fluctuations and those of the CCD are still obscure. One problem is that hiatuses may be "produced" during drilling, whenever recovery is difficult. This happens, for example, in the chert-rich sediments of Eocene Age.

9.8 Cretaceous Oceans: a Question of Oxygenation

9.8.1 A "Stagnant" Ocean? When the Deep Sea Drilling Project recovered mid-Cretaceous sediments in the Atlantic Ocean, it was found that some of these were very rich in organic matter, indicating either high organic supply to the sea floor, or reduced losses of organics due to low oxygen content in deep water at the time, or

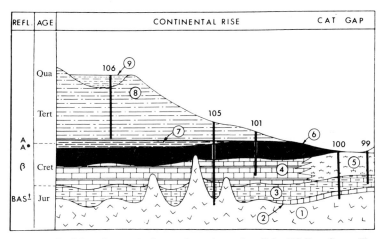

Fig. 9.19. Cretaceous organic-rich sediments in the continental margin off the US East Coast. The diagram is highly schematic and is based on both drilling and seismic profiler records. *1* Basalt; *2* greenish-gray limestone; *3* red clayey limestone; *4* white and gray limestone; *5* calcareous ooze and chalk; *6* black clay; *7* multicolored clay; *8* hemipelagic mud; *9* terrigenous sand and clay. Age: *JUR* Jurassic; *CRET* Cretaceous; *TERT* Tertiary; *QUA* Quaternary. [Y. Lancelot et al., 1972, Init Rep Deep Sea Drilling Project 11: 901]

both (Fig. 9.19). Regionally, the oxygen content apparently dropped low enough to prevent burrowing organisms from establishing a mixed layer on the sea floor. Thus, laminations are common in sediments of this time, especially in the margins.

The various indicators of low oxygen conditions are concentrated during certain well-defined periods which may have been quite brief (*"anoxic events"*). The occur typically in the Atlantic, although some have been found in the Pacific also. What is the significance of these anaerobic and near-anaerobic deposits? What are the implications for the chemistry and fertility of the ocean at the time? For climate?

To answer these questions, we need to do several things: (1) Search for modern analogues of deposition under low oxygen conditions (see Sect. 7.6.3). This will give us an appreciation for the type of environment that might have prevailed in the middle Cretaceous, or during parts of it. (2) Consider the physical principles of oxygenation in a warm ocean. (3) Analyze the material to determine its origin (how much is washed in from land?). (4) Note the stratigraphic details associated with wide-spread, well-defined anoxic events, to obtain clues about possible specific causes for specific events. We next take up these points in sequence.

9.8.2 Modern Analogs. There are two situations where sediments with high organic content are being deposited today: the partially restricted basins with estuarine circulation, and the open ocean continental slope, where it is intersected by a strong oxygen minimum (Fig. 9.20). The Baltic and the Black Sea are examples for the first, the Gulf of California and the Indian continental slope of the Arabian Sea for the second. Common to both is a high supply of organic matter relative to the oxygen

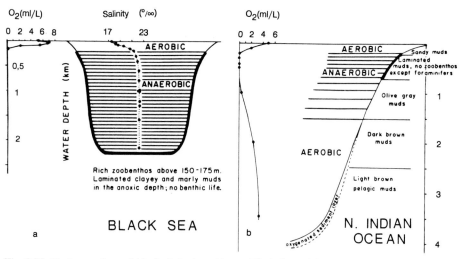

Fig. 9.20. Modern analogs of black shale deposition. **a** Black Sea, with strong salinity stratification preventing vertical overturn and hence blocking supply of oxygen to the deep layers. **b** Intersection of an oxygen minimum layer with a continental margin, in the northern Indian Ocean. The minimum layer is a worldwide phenomenon; it is reinforced in areas of high production. [J. Thiede, Tj. H. van Andel, 1977, Earth Planet Sci Lett 33: 301]

supply from the deep water which bathes the sea floor onto which the organic matter falls. To expand the likelihood of obtaining "black" deposits, then, we can increase productivity, or decrease the oxygen supply, or do both.

Which of these conditions applies to the mid-Cretaceous? There is no evidence for high productivity in the mid-Cretaceous, either from sedimentation rates or from the type of sediment delivered. In fact, productivity may well have been *decreased* compared to typical present values. Thus, it has to be a low supply of oxygen which is the crucial factor.

9.8.3 Oxygen in a Warm Ocean. We can readily appreciate why the oxygen content of deep waters was low: the temperature was relatively high. At present, the ocean has a temperature of between 0 and 5 °C, except for a thin upper layer. Saturation values for oxygen are near 7.5 ml/l. Typical actual values in the deep ocean are between 3 and 5 ml/l, that is, about 3.5 ml/l lower, because of oxygen consumption through decay. For the Cretaceous ocean, isotopic measurements suggest deep water temperatures close to 15 °C. For water temperatures between 15 and 20 °C, the oxygen saturation is near 5.5 ml/l; that is, 2 ml/l less than for the present cold deep water. We might expect, then, a typical value near 2 ml/l for the deep Cretaceous ocean, after subtracting a loss to account for decay. Under these conditions, any above-average loss of oxygen in an ocean basin with estuarine-type circulation (Fig. 7.12) would make the basin susceptible for developing regional oxygen deficiency. Unusually high supply of *terrigenous* organic matter to such a basin could further reduce the oxygen supply. This is a common factor today, along continental margins (but is unimportant in the open ocean).

One scenario that also needs to be considered is the formation of heavy warm saline water masses originating from high evaporation shelf seas. Pulsed input from such sources to the deep ocean could conceivably produce stagnant bottom water layers in somewhat restricted areas, as existed in the early Atlantic.

9.8.4 Milankovitch in the Cretaceous? Many pelagic sediments of Cretaceous age – some now exposed in the mountains of Italy, others in the deep ocean – show distinct cyclic deposition. Alternations of organic-rich and carbonate-rich facies are typical. The detailed mathematical analysis of numerical sequences of such alternations (by Fourier expansion) indicates that orbital cycles were important in producing them. What might be the mechanism translating orbital information into sediment cycles, during ice-free times? We do not know – presumably these mechanisms have to do with fluctuations in productivity, perhaps linked to oxygen supply to the deep sea (which releases nutrients) and to a changing rate of overturn, from fluctuations in deep-water formation, and from a slight change in the latitude where deep waters formed. It is unlikely that we shall know about these mechanisms soon – after all, we do not yet understand the cycles of the Quaternary very well, even though they govern the period we live in.

9.8.5 "Anoxic Events" and Volcanism. The widespread occurrence of Cretaceous organic-rich sediments was referred to above as something that is not entirely unexpected, given the fact that deep waters were much warmer than today. However, the question remains why such black deposits occur globally during well-defined time intervals (leading S. O. Schlanger and H. C. Jenkyns to propose the term *Oceanic Anoxic Events,* or OAE, in 1976). That there were such times when an unusual amount of organic matter was deposited is strongly supported by the $\delta^{13}C$ stratigraphy in

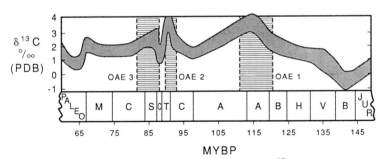

Fig. 9.21. "Oceanic Anoxic Events" of Schlanger and Jenkyns, and the $\delta^{13}O$ record of pelagic marine limestones, showing coincidence of the "events" with positive $\delta^{13}C$ excursions (generated through the lock-up of ^{12}C-rich carbon). Width of black band reflects uncertainty in the value of $\delta^{13}C$. *Lower part*: Cretaceous stages from Berrassian (right) to Maastrichtian (left). OAEs are centered in the Aptian, Cenomanian/Turonian and Santonian/Campanian [M. A. Arthur, W. E. Dean, S. O. Schlanger, 1985, AGU Geophys Monogr 32: 504; modified]

marine carbonates (as emphasized by P. A. Scholle and M. A. Arthur, in 1980). During the crucial intervals, one observes distinct positive excursions in the $\delta^{13}C$ record, which result from the preferential extraction of ^{12}C into organic matter, whose rate of deposition was greatly increased at the time (see Fig. 9.21).

One fascinating scenario, which is quite plausible, is the idea that extensive volcanism was responsible for the origin of the OAEs. There is evidence for enormous outpourings of basaltic lava in the South Pacific around 120 million years ago, in the Aptian. One surviving witness of this enhanced volcanic activity is the Ontong Java

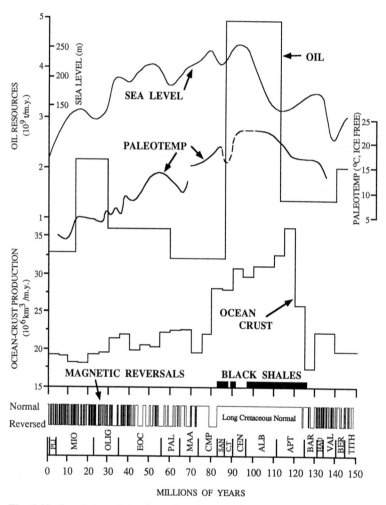

Fig. 9.22. Correlation of the rate of production of ocean crust with sea level and temperature changes, black shale and oil formation, and the abundance of magnetic reversals: basis of the superplume hypothesis of black shale formation. [R. L. Larson, 1991, Geology 19: 963, courtesy of the author]

Plateau east of New Guinea, which is some 40 km thick and has the size of Texas. A "superplume" event may be responsible for its origin. According to recent theories, the basalt forming the plateau derived from a great hot blob of magma that rose from the core-mantle boundary during the early Cretaceous, and arrived at the surface some time in the Aptian. The release of carbon dioxide to the atmosphere (and possibly other disturbances related to hydrothermal reactions and heat input) may be ultimately responsible for anaerobism in the deep sea. The coincidence of major volcanism, anaerobism (and petroleum formation), and sea level rise, certainly is striking (Fig. 9.22). In addition, as R. L. Larson points out, the frequency of magnetic reversals is affected: delivery of a plume of magma extracts energy from the core-mantle boundary, which halts the processes responsible for reversals.

The last illustration in this chapter emphasizes the complexity of factors that are important in ocean history and paleoceanography. Especially the influence of mantle processes on climate evolution are being hotly discussed in the geologic community. Is the formation of black shales and related petroleum source rocks really determined by volcanism? At this time we cannot say. In the next chapter we address the more descriptive aspects of resources from the ocean floor, including hydrocarbons.

Further Reading

Emiliani C (1955) Pleistocene temperatures. J Geol 63, 538–578

Parker FL (1958) Eastern Mediterranean foraminifera. Reports of the Swedish Deep-Sea Expedition, 1947–1948, 8, 217–283

Imbrie J, Kipp NG (1971) A new micropaleontological method for quantitative paleoclimatology: application to a late Pleistocene Caribbean core. In: Turekian KK, ed. The Late Cenozoic Glacial Ages. Yale University Press, New Haven, p 71–181

Schlanger SO, Jenkyns HC (1976) Cretaceous oceanic anoxic events: causes and consequences, Geol Mijnbouw, 55, 179–184

Scholle PA, Arthur MA (1980) Carbon isotope fluctuations in Cretaceous pelagic limestones: potential stratigraphic and petroleum exploration tool. Am Assoc Pet Geol Bull 64: 67–87

Berggren WA (1962) Some planktonic Foraminifera from the Maestrichtian and the Danian Stages of southern Scandinavia. Stockholm Univ Contr Geol 9: 1–106

Bramlette MN, Martini E (1964) The great change in calcareous nannoplankton fossils between the Maestrichtian and Danian. Micropaleontology, 10: 291–322

Bramlette MN (1965) Massive extinctions in biota at the end of Mesozoic time. Science, 148: 1696–1699

Luterbacher HP, Premoli Silva I (1964) Biostratigrafia del limite Cretaceo-Terziario. Riv Ital Paleontol 70: 67–128

Alvarez LW, Alvarez W, Asaro F, Michel HV (1980) Extraterrestrial cause for the Cretaceous-Tertiary extinction. Science, 208, 1095–1108

Smit J, Hertogen J (1980) An extraterrestrial event at the Cretaceous-Tertiary boundary, Nature, 285, 198–200

Catastrophism and Earth History, The New Uniformitarianism, edited by WA Berggren and JA Vancouvering, Princeton University Press, Princeton, NJ (1984)

CLIMAP Project Members (1981) Seasonal reconstruction of the Earth's surface at the last glacial maximum. Geol Soc Am Map Chart Ser MC-36, Boulder, Colo

Warme JE, Douglas RG, Winterer EL (eds) (1981) The Deep Sea Drilling Project: a decade of progress. SEPM Spec Publ 32, Soc Econ Paleontol Mineral, Tulsa, Okla

Berger A, Imbrie J, Hays J, Kukla G, Saltzman B (eds) (1984) Milankovitch and climate, 2 vols. Reidel, Dordrecht

Kennett JP (ed) (1985) The Miocene Ocean: paleoceanography and biogeography. Geol Soc Am Mem 163, Boulder, Colo

Sundquist ET, Broecker WS (eds) (1985) The carbon cycle and atmospheric CO_2: natural variations Archean to present. Geophys Monogr 32, Am Geophys Union, Washington DC

10 Resources from the Ocean Floor

10.1 Types of Resources

The ocean floor contains energy sources (petroleum and gas) and raw materials (sand and gravel, phosphorite, corals and other biogenic carbonates, heavy metal ores). Also, the sea floor is used as a dump site for waste, which represents a considerable economic value. In terms of dollars and cents, energy (hydrocarbons) is the most important resource, while (at present) raw materials are of regional importance only. Nothing as yet has been gained from deep sea ores, although they are of great scientific interest and are *potentially* valuable. The various resources are summarized in Table 10.1. Many of the figures given are rather crude guesses: resources within the ground are difficult to quantify. We shall briefly treat the geologic background for seafloor resources here, with some mention of the economic and the political problems associated with the use of the sea floor.

10.2 Petroleum Beneath the Sea Floor

10.2.1 Economic Background. The present high standard of living of the industrial countries, unprecedented in all of history, is largely dependent on the availability of large amounts of cheap energy. In the USA and elsewhere, petroleum delivers about one half of this vital resource. In the USA, 54 % (1991) of the oil is imported, in Germany and France around 97 %, and in Japan even 100 %. In the long run, these imports will become more and more expensive, independent of political ups and downs, reflecting the strong and growing demand, as well as the realization by the sellers that their goods cannot last forever. In 1991, world production was about 3.15 billion tons of petroleum and 2.1 trillion cubic meters of natural gas. Around 30 % of the oil and 20 % of the gas came from offshore sources.

Total potential resources and even reserves (that is, producible petroleum under prevailing economic and technological conditions) are hard to estimate. World *resources* may be 200 to 300 billion tons of oil and 200 to 250 trillion cubic meters of gas, and world *reserves* 135 billion tons and 124 trillion cubic meters. Expressed in static life expectancy, reserves for oil are good for 43 years (135 billion tons/3.15 billion tons per year) and for gas 60 years (124 trillion cubic meters/2.1 trillion cubic meters per year). US oil reserves were estimated at around 4.4 billion tons in 1975 and recently at around 3.5 billion tons only. Hence, static life expectany would be

Table 10.1 Sea-bed Resources (After J.M. Broadus, 1987, Science 235: 853, simplified)

Sea-bed deposits (10^6 metric tons)	Sea-bed production	World production	Sea-bed share of world re-venues (%)	Sea-bed reported potential resources	World onshore resources	Sea-bed comparison to world resources (%)
Crude oil (metric tons of oil equivalent)	789	2789	28	> 61 430	181 860	34
Natural gas (metric tons of oil equivalent)	247	1296	19	> 60 000	228 210	26
Sand and gravel	112	7802	1	> 660 000	Very large	Small
Shell carbonates	17	1667	1	90 000	Very large	Small
Phosphorites	–	159	–	7 940	129 500	6
Mineral placers Tin	0.028	0.2	14	Co 2.5 / 6–24	34.5 / 11	7 / 55–220
Nodules and Crusts	–			Ni 35–131 / Mn 706–2600 / Cu 29–108	130 / 10 900 / 1 600	27–100 / 6–24 / 2–7
Massive sulfides				??		

some 8 years only, if all US consumption were to be supplied from US reserves as defined in available statistics.

So far new discoveries have kept up with increasing use. At some point, of course, new discoveries will fall behind. Some experts think that this is happening right now.

10.2.2 Origin of Petroleum. Petroleum is a complex mixture of hydrocarbons and other organic compounds originating from organic matter produced both on land and in the sea, but largely from marine plankton. Therefore, essentially, petroleum is stored fossil solar energy; but this storage system is extremely inefficient. At present, only 0.23 % of solar radiation is used for photosynthesis (the basis of organic production). During Earth's history, less than 0.1 % of the organic production was preserved in sediments as organic carbon compounds. Only about 0.01 % of this organic matter in sediments became concentrated in oil and gas fields. Why so little?

The reasons are manifold. (1) The organic matter captured within the sediment must be converted to fluid petroleum by thermochemical processes, requiring a blanket of sediments more than 1000 m thick and temperatures of 50 to 150 °C. This is,

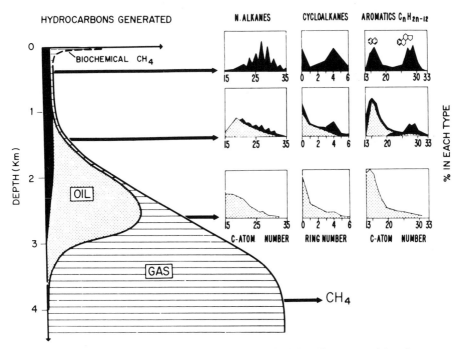

Fig. 10.1. Origin of petroleum. After burial of the organic-rich sediment containing the source material *(black)*, carbon compounds (chains and rings) are produced. In the zone above 1 km depth, the numbers of C-atoms tend to be large within the organic molecules. In the oil zone, compounds with fewer C-atoms (and more H-atoms) are produced by a cracking process, due to elevated heat and pressure. Under a large overburden, gas is produced (*CH4* = methane). [B. P. Tissot, D. H. Welte, 1978, Petroleum formation and occurrence. Springer, Heidelberg]

of course, very schematic. Long reaction times may compensate for low temperatures. Also, high temperatures may convert organic matter into petroleum very rapidly, even with only a modest sediment cover. This is demonstrated in the Guaymas Basin in the southern Gulf of California, where hydrothermal activity produced oil from sediments less than 5000 years old. If temperatures become too high, oil is cracked to natural gas (Fig. 10.1). (2) Petroleum must migrate from organic-rich source-rock sediments to porous and permeable reservoir rocks such as sandstones or vuggy limestones in response to compaction pressure and gravity (oil is lighter than interstitial waters) (Fig. 10.2). (3) Reservoirs must be big enough to be of interest, and they must trap petroleum with impermeable cover rocks such as thick shales or evaporites. Otherwise, the more volatile hydrocarbons escape to the surface. Examples of lost petroleum are the pitch lakes in Trinidad and Iraq, where oil has leaked through the surface, and the more volatile parts have evaporated into the atmosphere. The La Brea tar pits in Los Angeles are a familiar illustration of the process. About 200 natural marine oil seeps are reported worldwide. Petroleum seepage into the marine environment has been estimated as 0.6 million tons per year (a little less than the amount of oil estimated to be spilled at sea). (4) These several petroleum-forming processes must take place within the correct time frame: each process needs to complete its turn in the proper sequence. When everything looks just right, the drill hole may yet be "dry", because the *timing* in the interplay of the natural processes was wrong.

10.2.3 Where Offshore Oil is Found. The classic area of offshore oil fields is the *Gulf of Mexico*. The conditions of entrapment are much like those onshore. Marine sediments overlie salt, which is gravitationally unstable, being less dense than the overburden. It pushes up in plumes, making the so-called *salt domes*. The upturned sedimentary strata butting against the salt provide traps for petroleum (Figs. 10.2 and 10.3).

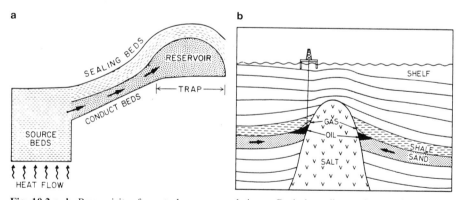

Fig. 10.2 a, b. Prerequisites for petroleum accumulation. **a** Basic ingredients of a petroleum reservoir system, showing migration from source beds (shales rich in organic matter) into reservoir (porous rock, e. g., vuggy reef limestone or sandstone). **b** Salt dome tectonics as an example for trapping conditions.

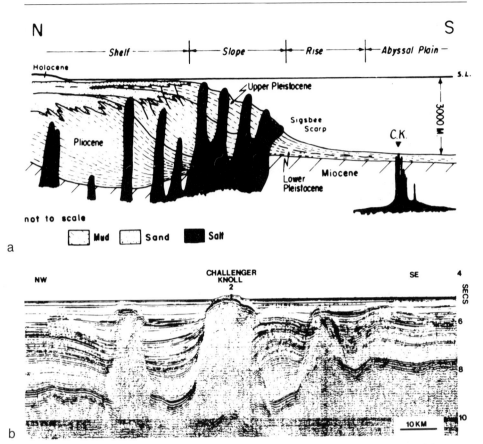

Fig. 10.3 a, b. Salt plumes in the Gulf of Mexico. **a** Schematic profile of Pliocene-Pleistocene sediment wedge intruded by early Mesozoic salt deposits. Intrusion of salt in the Sigsbee Abyssal Plain produced the Challenger Knoll. (CK). [C. J. Stuart, C. A. Caughey, 1977. Am Assoc Pet Geol Mem 26: 249: modified]. **b** Seismic profiler record from Sigsbee Knolls region. Challenger Knoll was drilled in Hole 2 of Leg 1 of the Deep Sea Drilling Project, establishing its nature as a salt dome feature. Time scale is two-way reflection time (sea floor = 4.5 s ≙ 3400 m). [J. L. Worzel, C. A. Burk, 1979, Am Assoc Pet Geol Mem 29: 403]

Another area from which offshore oil has been produced for a long time is the *Continental Borderland off Southern California*. There, oil originates from Miocene organic-rich marine strata originally formed under conditions of upwelling and oxygen deficiency. The oil is trapped in sandy layers abutting faults, hence there is considerably natural seepage into the ocean, as at Coal Oil Point near Santa Barbara and at the shores of Santa Monica Bay.

The newest highly promising areas for offshore petroleum are on the *Arctic shore of Alaska*, where large-scale onshore recovery is already proceeding (Prudhoe Bay). The drawback to offshore exploitation here is that climatic conditions are highly inclement.

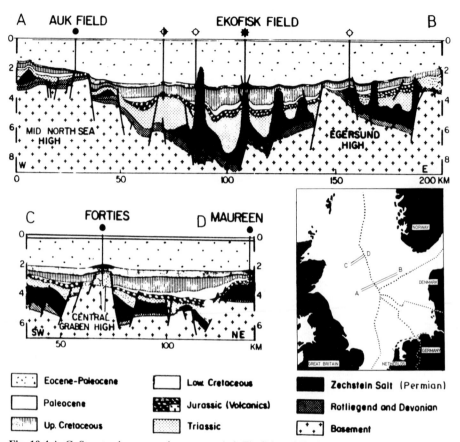

Fig. 10.4 A–C. Structural cross-sections across Auk-Ekofisk and Forties-Maureen fields, North Sea vertical exaggeration 6.7x. *Insert* Map with national shelf boundaries. [P. A. Ziegler, 1977, Geo. Journal 1: 7]

In recent years, large reserves of oil and gas were discovered in the thick sedimentary deposits of the *North Sea*. Its central part is a rift system now inactive. Rifting began during earliest Triassic, peaked in late Jurassic/earliest Cretaceous, and terminated during Paleocene, i. e., after some 160 million years. High rates of subsidence and sedimentation rates favored oil formation. The North Sea is the most important hydrocarbon province of Western Europe (Fig. 10.4). Recoverable petroleum reserves are estimated at more than 2 billion tons (United Kingdom 0.6, Norway 1.5 and Denmark 0.1 [1989]) – just about two thirds of one year's present world consumption. Natural gas reserves are about 4000 billion cubic meters. In 1989, production of oil was some 175 million tons (92 from United Kingdom and 75 from Norwegian waters) and 150 billion cubic meters of natural gas (45 United Kingdom, 31 Norway, 72 Netherlands, including mainland).

Exploration in the North Sea started in the early 1960s. In 1969 the Ekofisk Field was discovered. Offshore production began in 1967 for gas and in 1971 for oil. The technical difficulties in recovering hydrocarbons in this "fiercest of all seas" are enormous. Blinding fog, swift and drastic changes in weather, wind speeds of more than 160 km per hour, and waves frequently 30 m high, are dangerous even for the huge platforms. In March 1980 in the Ekofisk field, the Norwegian platform Alexander L. Kielland collapsed in stormy weather, a catastrophe leaving over a hundred people dead. Subsidence of the sea floor due to extraction of oil from overpressured chalks in this field made it necessary to elongate the legs of several platforms by several meters – a major engineering endeavor in the midst of the North Sea.

In the south of the hydrocarbon-rich region, a broad belt of gas fields stretches east-west from Germany to southern England (Fig. 10.4). Gas migrated from Carboniferous coal measures to porous Lower Permian sandstones and is sealed by overlying Zechstein (upper Permian) evaporites. Oil fields, however, are concentrated around the north-south rift in the northern North Sea mentioned above, which originated in the first stages of opening of the northern North Atlantic. The major faults bordering the rift lie on either side of the political median line between the United Kingdom and Norway. In the south, the oil occurs in Cretaceous chalk reservoirs, next to salt domes. Some fields are developed on Jurassic sandstones on the crests of horsts and tilted blocks. The Forties and other fields get oil from basal Tertiary sands.

10.2.4 Present Oil Exploration. As already mentioned, one of the prerequisites of petroleum formation and migration is an elevated temperature. A cover of roughly 1 to 2 km of sediment is required with some possible exceptions in areas with high heat flow or in sediments where large time spans for conversion are available. Most of the ocean bottom, roughly 80 or even 90 %, offers no chances for exploration because it

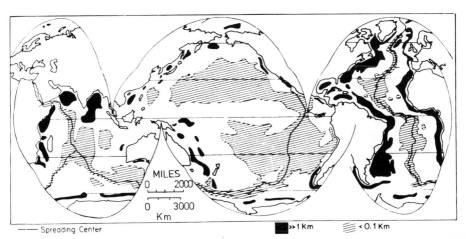

Fig. 10.5. Regions where Cenozoic sediments are well over 1 km thick. Only areas where source beds are deeply buried will produce petroleum (see Fig. 10.1). [W. H. Berger, 1974, in C. A. Burk, C. L. Drake, eds. 1974, The geology of continental margins. Springer, Heidelberg]

is too young and the sediment carpet is too thin (Fig. 10.5). The most promising prospects for petroleum concentrations are continental shelves and slopes, and small ocean basins with thick sediments produced by high accumulation rates and which are rich in organic matter (e. g., the Gulf of Mexico, the Caribbean, the Mediterranean including the Black Sea, the Bering Sea, Okhotsk, Japan, South China Seas, and the Indonesian Archipelago). The enormous sediment wedges of the continental rises bordering the large ocean basins, although rich in organic matter in places, may or may not contain suitable reservoir rocks; this question is being explored.

The shelf off the US East Coast has become the site of intensive exploration, and leases have been granted in places, after much opposition from citizen's groups concerned about the possibility of oil pollution. As shown by occurrences off California (Santa Barbara oil spill, 1969), in the North Sea (Ekofisk oil spill 1977), or in the Gulf of Mexico (Ixtoc-1 oil spill 1979), such concern is not without basis. The problem is a political one of profit and risk distribution; effects of spills on the ocean as a whole cannot be shown to be damaging (but neither is it known whether spills are innocuous). Regional damage, of course, can be extensive.

Some years ago, off the US East Coast, the Baltimore Canyon Trough area (Fig. 10.6) was regarded as the best prospect of the outer continental shelf, because of a sediment thickness of more than 10 km and favorable structural and other conditions. However, extensive drilling did not result in the discovery of oil fields.

In addition to the shelves, the upper *continental slopes* are interesting areas for petroleum exploration. Their sediments were formed in the realm of fertile coastal waters (including deltaic conditions). Hence, carbon content of the sediments is high (Fig. 6.4). Also, an oxygen minimum develops in many areas on the upper slope (see

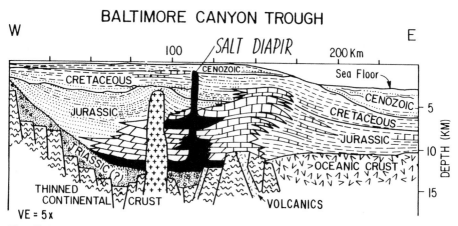

Fig. 10.6. Schematic section through the Baltimore Canyon Trough off New Jersey. Based on seismic profiles by the US Geological Survey. In the center, the "Great Stone Dome", probably an area of volcanic sills and dykes, intruding the sedimentary sequence. The oceanic/continental crust boundary is highly speculative. [L. R. Jansa and J. Wiedmann. In: von Rad et al., 1982, Geology of the Northwest African Continental margin: 225, Springer, Heidelberg, modified]

Fig. 9.20), and favors burial of organic matter. We have briefly mentioned the Meso-zoic structures identified as salt domes in the eastern South Atlantic, off Africa (Sect. 2.3, Fig. 2.4) where organic-rich Cretaceous sediments were found by deep drilling. Clearly, from the Gulf of Mexico experience, petroleum may be expected in such a situation.

Regardless of the promise of certain continental slope regions, activity on the shelves, even in regions with heavy seas or in polar areas, is going to dominate petroleum recovery from the sea floor in the foreseeable future. Expenses of drilling increase rapidly beyond the shelf break because platforms either have to have ex-tremely long legs, or they have to be positioned dynamically, that is, by constant calculated application of power to several thrusters. Also, safety problems rapidly increase with depth. Beyond the continental slopes, in the deep sea proper, where sedimentation rates are low, the chances of finding oil rapidly diminish anyhow. This may be one reason that the Exclusive Economic Zone extends "only" to 200 nautical miles (371 km) offshore.

10.3 Raw Materials from Shelves

10.3.1 Phosphorites. The naturalist on HMS *Challenger*, John Murray, became a wealthy man (unlike his modern counterparts) as a result of joining an oceanographic expedition. He acquired an interest in a mining venture which resulted from the discovery of phosphorite on Christmas Island in the western equatorial Pacific. What is more, the British Crown's Treasury eventually recovered the expenses of the ex-pedition, through the taxes derived from the phosphorite produced.

Phosphorites were briefly mentioned in Section 3.8.2. They are cryptocrystalline apatites and vary in composition. A general formula is Ca_{10} $(PO_4, CO_3)_6$ F_{2-3}, with increases in carbonate running parallel to increases in fluoride (or hydroxide). Mo-dern phosphorites typically occur in areas of high productivity – off Southern and Baja California, off Peru, off South Africa, for example (Fig. 10.7). They occur as black or brown nodules from pellets up to head size, and as irregularly shaped cakes. They are interpreted as replacement products of fine-grained lithified carbonates, as mineralization of pre-existing organic matter, as precipitates from microorganisms filling cavities within carbonates, or as direct precipitates from interstitial waters.

Off California such deposits contain about 25 to 30 % P_2O_5 and 40 to 45 % CaO, but these values vary in other regions. The common depth of deposition is on the shelf and upper slope. Fossil phosphorites are abundant in Florida and Georgia. Miocene phosphorites are mined on a large scale; in West Africa, Eocene deposits are being exploited. Seamounts bearing Cretaceous carbonates commonly carry phos-phorites.

The association of geologically young phosphorite deposits with present-day re-gions of *upwelling* (Fig. 4.19) suggests that the source of the phosphorus is organic matter. Apparently, the algae growing in these regions, in the surface waters, extract the phosphorus from the water. Also, crustaceans and fish concentrate it further in their bodies and excrement. During decomposition of organic debris on the sea floor,

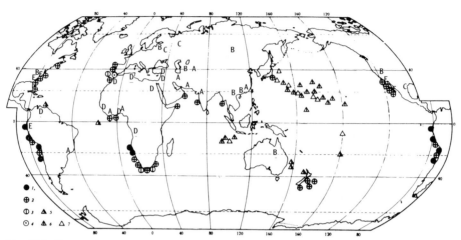

Fig. 10.7. Phosphorites. *Circles* Shelf phosphorites *1* Holocene *2* late Tertiary *3* early Tertiary *4* Cretaceous; *triangles* sea mount phosphorites *5* late Tertiary *6* early Tertiary *7* Cretaceous. [G. N. Baturin, P. L. Bezrukov, 1979, Mar. Geol. 31: 317]. Continental phosphate deposits: *A* Precambrium; *B* Paleozoic; *C* Jurassic to Lower Cretaceous; *D* Senon to Eocene; *E* Miocene to Pleistocene. [M. Slansky, 1980. Mém. BRGM, France, 114]

much phosphate is released to the interstitial water (and also to seawater). Thus, interstitial waters right below the sea floor may become saturated with the phosphate mineral apatite. Precipitation of apatite, replacement of pre-existing carbonate minerals, and impregnation of sediment can then proceed. Off Peru, nodules with diameters of several centimeters grow within soft sediments with a rate of some mm/1000 years, indicating diffusive upward flux of dissolved phosphorus. The resulting phosphatic concretions are resistant to transport by currents; they can be mechanically exhumed and concentrated during periods of sediment reworking. Indeed, concretions are commonly associated within the sediment with hiatuses or other discontinuity surfaces with drastically reduced net deposition rates.

Phosphorites are largely used for fertilizers, but also as a source of phosphorus in the chemical industry. Marine phosphorites are enriched in some trace elements compared with marine shales (Ag 20x, U 30x, Cd 50x). Offshore phosphorites typically contain less than 25 % P_2O_5.

Weathering processes on land concentrate P_2O_5 up to 30 to 40 %. Such residual phosphorites are mined, for example, in Florida and Morocco (Fig. 10.7), surface concentrations within marine P-rich deposits from off California, western South America, South Africa, or on the Chatham Rise (Fig. A6.1), east of New Zealand, are up to 80 kg/m^2 phosphorites with up to 25 % P_2O_5. They occur over large areas and in water depths of less than 400 m. Exploitation will depend on the prices charged by producers on land and on the domestic demand and supply situation. In any case, the consumption of phosphorites from land increased from 3.5 million t P in 1900 to 140 million t in 1971 and presumably will continue to increase in the future.

10.3.2 Shell Deposits. Calcareous shell deposits were or are dredged in places as raw material for calcium carbonate, and also for building roads. For example, oyster shells have been mined in San Francisco Bay for use in making cement, and in Galveston Bay in the Gulf of Mexico for calcination and reaction with seawater, to extract magnesium. Dredging of shells hinders the growth of benthic organisms and adversely affects the productivity of the sea bed. Conflicts commonly arise where dredging and fishing activites overlap.

With the advent of worldwide souvenir markets and of face-mask diving, the collecting of shells and corals has become an important source of income for islanders of the Pacific and other coastal peoples. Not surprisingly, attractive and rare species suffer considerable depredation from such collecting.

10.3.3 Placer Deposits. Concentrations of heavy minerals and ore particles on beaches and in estuaries are locally mined for metals such as titanium, gold, platinum, thorium, zirconium, and valuable minerals such as diamond. Seventy percent of the world production of zirconium is being extracted from placer deposits off East Australia. Diamonds are found in beach deposits of Southwest Africa, as well as offshore. Magnetite is being mined from beach placers in certain areas in Japan and New Zealand. In the USA, gold has been mined from beach deposits near Nome, Alaska; thousands of tons of ilmenite ($FeTiO_3$) were extracted at one time from

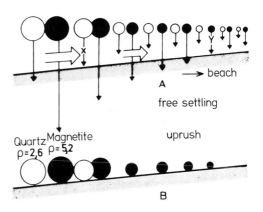

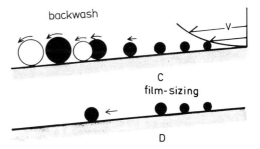

Fig. 10.8 A–D. Origin of heavy mineral placers, schematic. **A** Suspended sediment particles are washed onshore, settle to the floor, large and heavy ones (shown *solid*) first. **B** Resulting grain association shows enrichment of large and heavy particles on the beach face. **C** Backwash rolls away large particles (greater velocity away from sediment interface, *V*, to the *right*). **D** Resulting association consists of medium-size, well-sorted heavy minerals. [E. Seibold, 1970, Chem Ing Tech 42: A 208l]

Redondo Beach, California; beaches in western Oregon yield chromite and other heavy minerals as well as gold and platinum; titanium minerals are taken from beach sands along the eastern Florida coast.

How do placers originate? The process of concentrating heavy particles on the beach has much in common with panning for gold – water motion works on the different settling velocities of the particles and on their different sizes, to separate heavy from light, large from small.

Terrigenous beach sands commonly consist of more than 95 % quartz, tropical ones of calcareous grains. Volcanic rocks and other igneous sources supply minerals which are considerably heavier than quartz or calcite (density of 2.65 and 2.70). These minerals have densities of greater than 2.85 g/cm^3. The "heavy minerals" are ubiquitous, usually making up a few percent of the sand. The back-and-forth movement of the waves washing over the beach face can concentrate them greatly (Figs. 10.8 and 10.9). Under the microscope, one recognizes minerals such as ilmenite (iron-titaniumoxide), rutile (titaniumoxide), zircon (zirconiumsilicate), and monazite (phosphate, containing cerium and thorium).

Thick placers of heavy particles form only in the beach zone. Offshore placers, therefore, were formed when the sea level stood lower. The gold deposits off Nome, Alaska, are a case in point. Glaciers of the last ice age brought gold-containing debris from the hinterland and dropped it on the shelf. This morainal material was then worked up by the surf, as the sea level rose.

River deposits may contain heavy mineral concentrations, too, such as cassiterite (SnO_2) in Thailand, Malaysia, or Indonesia. In these areas, former river beds on the shelf, which are now submerged down to about 100 m water depth, are of potential economic interest and have been exploited offshore in shallower water for nearly a century. At present, the offshore revenue reaches some 14 % of that of worldwide tin exploitation on land.

10.3.4 Sand and Gravel. Considerable amounts of sand are taken locally for building roads and houses, as well as for coastal protection (dams) and for fill. However, the main use of beach sand is for building sand castles and for plain enjoyment. Recreation is one of the most important uses of beaches – by far exceeding mineral extraction in business value. Beaches are commonly eroded by winter storms (Sect. 4.2.2); in some areas sand is brought in from offshore to replace the eroded material, usually at considerable expense.

Gravel only rarely reaches the sea, unless high mountains are near, or unless glaciers bring morainal debris. Such debris, when washed by waves or rivers on the bare ice age shelves can then yield gravel. In the Baltic and in the North Sea, for example, gravel is mined as a filler for concrete.

In summary, the economic importance of raw materials from shelves is very modest: they contribute less than 1 % of the more conventional onshore production, with the exception of tin (not to mention sunken treasures from antiquity, and from Spanish pirates, licensed and otherwise). For economic impact of marine resources on a global scale, one must look to hydrocarbons, fisheries, recreation, and waste diposal.

Fig. 10.9. Beach placers. The beach south of Quilon (SW India) has a heavy mineral deposit on the top part (where the boat rests). During SW monsoon, high waves sort the sand as shown in Fig. 10.8., and produce layers of "black sand" *(inset)*. The inset profile is 20 cm high. [Photo E. S.]

10.4 Heavy Metals on the Deep-Sea Floor

10.4.1 Importance of Manganese Deposits. The rich metal deposits on the deep-sea floor have been a favorite topic of discussion among marine geologists ever since the *manganese nodules* (better: *ferromanganese concretions*) were discovered by the *Challenger* Expedition a century ago and were shown to be rich in copper, cobalt, nickel, and other heavy metals (Fig. 10.10). The abundances of these metals in seawater are extremely low, largely due to the low solubility of their oxides and hydroxides. Also, they are readily extracted by biological processes, and sent to the sea floor within organic matter.

The amount of nodules on the Pacific Ocean floor seems to be in the neighborhood of 100 to 200 billion tons. What is the economic value of these deposits? Right now, close to zero. It is too expensive and (because of complex international legal problems) too risky to mine them, transport the material back to shore, extract the wanted heavy metals, market them, and still make a profit. Nevertheless, there is a *potential* value, if prices of copper, nickel, and cobalt rise high enough. This potential value is a bone of contention among UN members: those nations technically unable to mine the material wish to make sure that those able to do so must share the profits, if any should materialize. In addition, metal-exporting countries are apprehensive

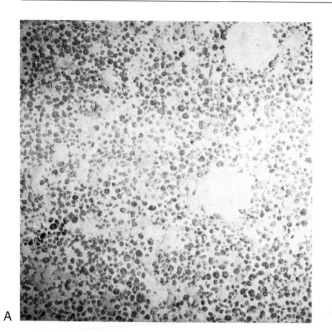

A

a

B

1cm   b

Fig. 10.10 A, B. Manganese nodules. **A** View ($\sim 10\,m^2$) of nodule-covered deep sea floor, central tropical Pacific (Photo Metallgesellschaft Frankfurt). **B** Manganese nodules recovered on *Challenger* stations in the Central Pacific. *a* Nodule with upper smooth and lower uneven surface (Station 274 5000 m); *b* Sliced section (parallel to the sea bottom) showing internal layering and nuclei (Station 254, 5700 m). [J. Murray, A. F. Renard, 1891, Report on deep-sea deposits. H M S *Challenger*, 1873–1876. Reprinted 1965 by Johnson Reprint, London]

about the potential competition. Many of them belong to the developing countries. The resulting concept of the deposits as a "common heritage of mankind" has slowed development of the resource, since venture capital is hesitant to take the risk both of failure and of an unknown degree of taxation in case of success. Eventually, however, the resource will probably be exploited. Mn, Co, Cu, and Ni are so-called strategic minerals. For example, Co is used for dense and strong alloys in jet engine parts. Because cobalt production is concentrated in a few African countries, the hostilities between Angola and Zaire caused an increase in Co prices from US $ 3–6 per pound during 1960–1977 to more than $ 22 in 1979. Cu, Mn, and Ni prices remained stable – an illustration of difficulties in attempting to make economic forecasts in the mining industry.

10.4.2 Nature of Manganese Deposits. What are the ferro-manganese deposits like? Where do they occur? How much of the valuable trace metals do they contain? How did they originate?

The *appearance* of the manganese deposits varies. Nodules come in sizes of 1 to 10 cm and look much like small potatoes except for being black. The surface can be smooth or rough. Not all deposits are nodules: some are crusts several centimeters thick, others are pavements covering the sea floor in areas of active currents. The nodules are rather porous. They are easily crushed, which should facilitate on-board processing and chemical extraction of metals when the time comes. On cutting the nodules, one notes a concentric structure. In the center there is commonly a core of altered volcanic material. Fragments of older nodules, bones, or shark teeth can also serve as core. From the age of such cores it is quite obvious that the deep-sea nodules must grow very slowly, a few millimeters per million years, at most (Fig. 10.10 Bb).

The manganese nodules occur in areas of low sedimentation rate: because of their slow growth they would soon be covered up in regions of high sediment supply (Fig. 10.11).

Calcareous ooze accumulates at about 10 m per million years, 1000 to 10 000 x faster than nodules; hence no nodules develop (excepting "micronodules" and encrustations on shells of foraminifera in many areas). Ferromanganese is deposited here also, but is greatly diluted with carbonate. Brown pelagic clay (Red Clay) accumulates at less than 1 m up to about 2.5 m per million years, the higher values applying in the Atlantic Ocean. Surprisingly, nodules have time to grow at the surface under these conditions. Apparently they are moved about sufficiently, through the activity of benthic organisms, to stay on top of the clay. In fact, such movement may be necessary to keep them round rather than mushroom-shaped, and to prevent them from coalescing into a pavement. If we could mount a time-lapse camera on the sea floor, taking a picture every 100 years for 10 000 or 100 000 years, what a spectacle of dancing nodules we might see!

In places, bottom currents prevent clay deposition or even cause erosion over large areas. Indeed, in sediment cores nodules can be seen to be concentrated at horizons of Tertiary hiatuses.

The distribution of nodules is patchy (Fig. 10.10 A). For example, in the eastern central Pacific, records were taken with bottom-near television cameras, for hundreds

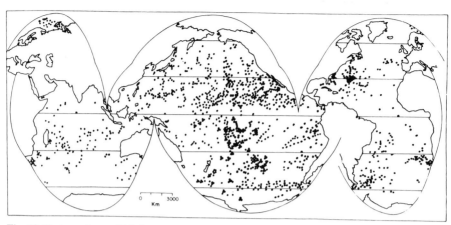

Fig. 10.11. Areas from which ferromanganese concretions have been reported. The richest areas are in the north-equatorial Pacific. [W. H. Berger, 1974, in C. A. Burk, C. L. Drake [eds], 1974. The geology of continental margins. Springer, Heidelberg; mainly after D. R. Horn [ed] 1972, Ferromanganese deposits on the ocean floor. Harriman, New York]

of kilometers. For 5 % of the records the sea floor was covered to more than 50 % by nodules, that is, up to 25 kg of ore per square meter. Conversely, 5 % of the records showed sea floor free of nodules. Elsewhere the cover varied, sometimes considerably, even over distances of only 50 m. The reasons for patchiness are not clear; in part it may be due to alternating burial and uncovering of nodules below slowly moving clay carpets, somewhat akin to dune migration.

10.4.3 Origin of Manganese Nodules. Questions about the origin of ferromanganese concretions commonly address themselves to the following aspects of ferromanganese distributions. (1) The ultimate source; for example, weathering of continental or oceanic rocks or sediments or exhalations from volcanic or hydrothermal vents. (2) The immediate source; that is, the surrounding seawater or the mobilization of ferromanganese and other metals (by reduction) within sediments and diffusion to the interface. (3) The mode of transport into or within the ocean; that is, whether delivery is as solute, mineral, or coating on mineral.

The first of these problems, that of ultimate sources, was the one originally raised by Murray and Renared (1891), and it has dominated much of the subsequent research. With the discovery of widespread hydrothermal activity, the idea of a "volcanic" origin of the nodules has gained in acceptance. It is now believed that both hydrothermal release of metals and terrigenous input are important sources.

Ferromanganese crusts derive most of their metal content from dissolved and particulate matter (able to scavenge metal ions) in ambient bottom water. Nodules lying on sediments also grow by input from below, i. e., from metal ions released by mobilization in interstitial waters. This is illustrated by uneven lower nodule surfaces (Fig. 10.10 Ba). High organic matter contents in the sediment promote this mobilization and may also promote bioturbation, thus preventing burial of the nodules.

Table 10.2. Metal contents of manganese nodules, crusts, and hot vent sulfides.

%	Manganese nodules[a]					Manganese[b]		Hot vent sulfides[c]		
	Average Pacific	Indian O.	Atlantic	max.	min.	Crusts	Hydrothermal oxides	Average	max	min
Mn	17.2	14.9	13.6	34.00	5.41	21.6	27.3	–	–	–
Fe	11.8	14.6	15.5	26.32	4.36	16.5	11.6	19.1	34.0	2.5
Ni	0.63	0.38	0.33	2.00	0.13	–	–	–	–	–
Co	0.36	0.31	0.24	2.57	0.045	0.63	0.0023	–	–	–
Cu	0.36	0.17	0.16	2.5	0.028	–	–	2.9	9.2	0.2
Pb	0.047	0.053	–	0.51	0.046	–	–	0.17	12.1	0.03
Zn								16.9	36.7	4.0
SiO$_2$								10.2	28.1	1.2

[a] Mainly after J. S. Tooms, 1972, Endeavour 31: 113, [b] After F. T. Manheim and C. M. Lane-Bostwick, 1989, Nature, 335: 59, [c] After S. D. Scott, 1991, in: K. J. Hsü and J. Thiede, 1992.

In the various discussions about the origin of ferromanganese, the *trace element* content, as well as the *iron-to-manganese* ratio, play an important role. Typical values for Pacific, Atlantic, and Indian Ocean are given in Table 10.2. Note that the Mn/Fe ratio is greater in the Pacific than in the Atlantic. Also, the content of trace elements in Pacific nodules, on average, is about twice that of Atlantic ones.

What factors control the Mn/Fe ratio? Which control the trace element content?

The Mn/Fe ratio, on the whole, increases with the degree of oxidation and with depth – the more "deep-sea" character the nodules have, the more manganese they contain. Shallow water ferromanganese concretions (e. g., on continental slopes) are generally iron-rich. In the economically interesting manganese zone north of the central equatorial Pacific, Mn/Fe ratios are as high as ten. High manganese content appears to be favored by both high biogenous sediment supply and low rates of accumulation. Under these conditions, presumably, the carrier material dissolves but leaves its content of trace elements. Also, in the "manganese zone" in the Pacific, the underlying sea floor consists of dissolving biogenous sediment rich in trace metals. The sediment was originally formed underneath the Equator, then moved northward and downward due to plate motion. This motion brought the calcareous sediment from a zone of accumulation into a zone of dissolution. Thus, one way to concentrate manganese and trace elements is to have organisms precipitate the metals (which they do very efficiently), bring them to the sea floor in shells and fecal matter, and dissolve or oxidize these carriers to obtain a more nearly pure concentrate.

Other mechanisms for concentrating the metals also must exist. For example, the element cobalt tends to be high (> 1 %) within ferromanganese accumulating on seamounts under highly oxidizing conditions. Here the manganese precipitates extremely slowly out of seawater (~ 1 mm/million years), at the same time scavenging (co-precipitating catalytically) the chemically similar cobalt. Indeed, cobalt contents of crusts near hydrothermal vents are two orders of magnitude lower, because they are diluted by high Mn and Fe precipitation, as indicated by growth rates of more than 1000 mm/million years.

10.4.4 Ores from Spreading Axes. For some of the ferromanganese deposits on the deep-sea floor, the origin is hardly in doubt: those on the crest of active spreading ridges. Here, seawater circulates through cracks in the newly formed crust, reacting with the hot basalt (Fig. 10.12). Precipitation of sulfides (Fe, Mn, Cu, Zn, etc.) originating there may become economically important in the distant future. Up to now some 100 localities were discovered, by far the most in the Pacific. There is a great variability in metal contents (Table 10.2).

Seawater penetrates the hot basalt and reacts with it. It takes up SiO_2 and metals, gives up Mg to alteration products (smectites and other clay minerals) and gains Ca. Seawater sulfate is stripped of its oxygen by the reactions with reduced iron in the basalt, and sulfides precipitate accordingly. The hydrothermally active zone above the magma chambers feeding the central rift is 3–5 km thick, and temperatures within the fluids issuing from the vents reach some 350 °C. The high temperatures, of course, greatly accelerate all rates of reaction.

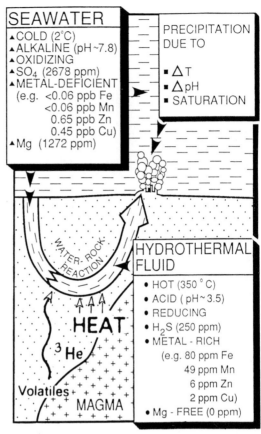

Fig. 10.12. Hydrothermal circulation model for a sediment-starved spreading ridge.The more important factors for reactions between seawater and hydrothermal fluids are illustrated. In addition, precipitates near vents act as scavengers – as iron oxides do for vanadium. [Modified from S. D. Scott, 1991, in: K. J. Hsü and J. Thiede, 1992]

On exiting, the hot acid waters (pH about 3) mix with the cold, slightly alkaline seawater, so that a great number of minerals, largely sulfides, must precipitate. The *precipitates* form crusts and spectacular chimneys several meters high (over 10 m in cases), with growth rates up to 1 m/year (Fig. 6.20). Mounds, miniature spires, and various baroque edifices can be built. In places, anhydrite ($CaSO_4$) forms, due to high concentrations of sulfate when the exiting H_2S is re-oxidized. A chimney made of anhydrite in the North Fiji basin (back-arc spreading) was baptized "La Dame Blanche", a reference to its ghost-like appearance (Fig. 10.13), and in contrast to the more common "*black smokers*", made of metal sulfides and oxides. Amorphous silica can also form chimneys. In a broad zone around the vents, iron and manganese oxides are precipitated, as the reduced Fe^{2+} and Mn^{2+} meets normal seawater, oxidizes, and forms hydroxides.

1 m

Fig. 10.13. "La Dame Blanche", a chimney made of anhydrite at an active hydrothermal site with fluids of 285 °C and active animal colonies. North Fiji Basin axis, 1900 m water depth. [J. M. Auzende et al., 1989, C R Acad Sci Paris, 309, II, 1787]

As the ores and other minerals precipitate within cracks in the basalt, they tend to clog the hydrothermal system, eventually turning it off. Ores near the surface may be brecciated by hydrofracturing and may be covered by new lava in several cycles. Fossil massive sulfide ore deposits – as in Southeast Japan (Kuroko) or Cyprus (Troodos) – permit the study of these complicated relationships on land.

Metalliferous hydrothermal systems were recently also discovered in *back-arc basins:* Manus basin north of Papua New Guinea, Lau basin west of the Tonga Plate boundary (Fig. Appendix A6.1) and others with oceanic crust. Vents in the Okinawa Trough issue from continental crust.

Of course, hydrothermal processes are everywhere associated with volcanic activity – Santorini off Greece, Ebeko in the Kurile Islands, Loihi seamount off Hawaii (the latest creation of the fire goddess Pelé) are examples.

Hydrothermal activity is of fundamental importance for *plate tectonics* (heat budget of the lithosphere) and for *geochemistry* (origin of seawater, fractionation of metals from basalts, silica cycle). The heat transfer by this process is of the order of one-fourth of the total global heat loss. Every 8 to 10 million years the volume of the entire ocean passes through hydrothermal systems – conditioning the seawater and resetting its chemistry. Ultimately, this process also stabilizes the chemistry of the atmosphere and climate. It should be kept in mind, that seawater/basalt interaction is not restricted to volcanically active regions – the entire seafloor is young and cools continuously; thus weak hydrothermal flow can occur almost everywhere.

Economically, the Ridge Crest deposits and other deep-sea hydrothermal ores are as yet not important, despite their presumed large volume throughout the oceanic crust. So far, it is much easier to mine the ores on land, many of which (especially in active margins) may owe their existence to preconcentration by deep-sea hydrothermal activity (Fig. 1.20).

10.4.5 Red Sea Ore Deposits. A rather specialized case of Ridge Crest accumulation is represented by the heavy metal deposits in the Red Sea, which are of considerable economic interest. Promising metalliferous deposits occur in the Atlantis-II Deep, named after the Woods Hole research vessel. The basin, which lies offshore of Mecca, was found in 1963 by the British vessel *Discoverer; Atlantis II* of Woods Hole explored it in 1964 and 1965; *Meteor* and other German research vessels were there in 1965 and afterwards.

The *Glomar Challenger* paid a visit in 1972, drilling in this area.

What is it that attracts all this attention to the Red Sea?

In the central Red Sea – an active spreading center which opened only a few million years ago – there are several enclosed basins, the "deeps". The Atlantis II Deep is more than 2000 m deep, and only 6 by 15 km in area. The bottom is filled with a hot salt brine with a temperature of about 60 °C and a salinity of 25 %, seven times that of seawater. Iron is 8000 times more concentrated in the brine than in seawater, zinc 500 times, copper 100 times. The sediment below the brine is incredibly colorful, brick-red layers alternating with ocher, white, black, greenish. A variety of minerals provides the coloring; economically the most important are the sulfides in the dark layers. Zinc contents of up to 10 %, copper contents of 3 % or even 7 % were measured. Unfortunately, the minerals are extremely fine-grained, which will make extraction difficult.

How did these deposits originate? Apparently, two kinds of processes are important. First, we have to apply the hydrothermal mechanisms summarized in Fig. 10.12. Second, thick *sedimentary* deposits of Tertiary age are nearby, abutting the newly forming sea floor. These contain several-hundred-meter-thick salt and gypsum layers. Hot water circulating through such sediments can dissolve out metals and salt and hence produce metalliferous brine issuing into the brine pools. The metals precipitate upon cooling, and when oxygen is supplied, by mixing with normal seawater. However, such mixing is greatly obstructed by the high density of the brine, and can occur only at the very tops of the brines. Thus, the metals are trapped in the brine.

Just how rich are the deposits? In the Atlantis-II Deep alone there are supposedly 3.2 million tons of zinc, 0.8 million tons of copper, 80 000 tons of lead, 4500 tons of silver, and even 45 tons of gold. Whether the *net worth* of this ore is significant (after accounting for recovery, processing, and transport) remains to be seen. However, it is the most studied and most promising occurrence of deep sea metals found so far.

Initial rifting of continental margins produced polymetal sulfides on other areas, also. The Guaymas basin in the Southern Gulf of California is an example.

10.5 Waste Disposal and Pollution

10.5.1 A Change in Pace. Man's impact on the marine environment spans a wide range of interference with natural processes. One of the oldest, one may assume, is the increased delivery of sediments to estuaries and harbors caused by increased erosion on land from working the soil, and from deforestation. Basically, this should accelerate the filling-in of estuaries and affected lagoons. Conversely, the building of large dams upriver has led to interception of sediments, and the starvation of beaches, which promotes the erosion of the coastline. For example, since the construction of the Aswan High Dam in 1964 in the uplands of Egypt, shoreward erosion around the Nile delta region has increased to over 150 m annually near headlands.

Coastal cities have contributed nutrients for algal growth along their shores for centuries, and this process has continued and accelerated with population growth. In places where waste amounts are sufficiently large and water bodies restricted, bays are fouled, and sensitive ecosystems (such as coral reefs) are damaged.

What is new in our time is a massive increase of *waste* off rapidly growing cities, with unhealthy environmental effects well away from the sources of pollution. Such effects may include dangerous levels of pathogenic bacteria, and bursts of growth ("blooms") of noxious dinoflagellates. Also new is the production of potent and long-lived industrial poisons which are effective at low concentrations. One prime example is the accumulation of pesticides in certain seabirds, through the food-chain effect, which can wreak havoc with their reproductive success. Oil spills and the introduction of radioactive materials likewise are modern developments, as is the large-scale addition of chemical-industrial waste.

Because of the economic interests involved, the study of problems connected with disposal and pollution has become an important subject in environmental marine geology, and it is a field that is likely to grow. Studies involve the cooperation of many scientists (e. g., chemists and biologists), commonly from different countries, as pollution crosses boundaries. How can we monitor environmental changes and detect possible causes? A promising method is biological monitoring (Fig. 10.14). The observed changes set a warning signal, but it is commonly difficult to define the – natural or anthropogenic – factors responsible for the changes.

10.5.2 Sewage and Sludge. Industrial man uses energy and raw materials and produces waste. Most of the waste, such as *sewage* sludge, is relatively innocuous provided it is without potent poisons (certain metals, for example, or biocides). Of course, the amounts introduced must not exceed what the ocean can absorb by dilution and bacterial action (which makes estuaries more vulnerable than open ocean sites). Under these conditions sludge mainly acts as fertilizer.

When dumping of *sludge* proceeds on a grand scale in a limited area, problems arise. For example, several years ago foul bottom conditions as well as diseased benthic organisms were reported from shelf areas off New York in regions of dumping. Fishing activities, especially lobster fishing, are adversely affected under such circumstances. This has led to restrictions on dumping of sludge close to the coast, or on the shelf; but even "deep" dumping can cause problems. Disposal of municipal

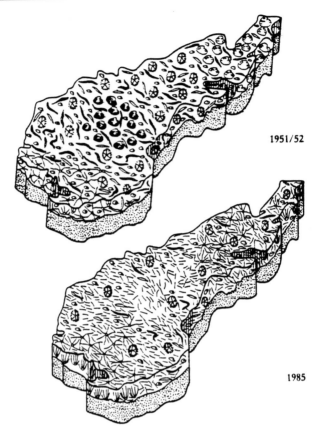

1951/52

1985

Fig. 10.14. Benthic biota changed drastically between 1951/52 and 1985 in the Central North Sea (Doggerbank about 20 m waterdepth). Mussels and sea urchins were increasingly replaced by the less sensitive brittle stars and worms. This type of monitoring is useful in identifying change in stressed environments. [J. Lohse et al., 1989, Die Geowissenschaften, 7.6, 155, Weinheim]

sewage sludge in moderately deep water as far as 185 km off the coast of New Jersey from 1986 to 1992 at a rate of 8–9 million wet metric tons per annum affected the benthic food web some 2500 m deeper. It had been hoped that dispersal and dilution of sewage particulates in surface waters would prevent waste from entering the food web at such depths.

Much greater problems arise in confined water bodies. In the Baltic Sea, the influx of sewage and agricultural fertilizer results in large-scale *eutrophication,* which leads to a greatly increased oxygen demand at the sea floor. A shortage of oxygen, sporadically exacerbated by climatic conditions, can produce massive fish kills.

Where sewage contains dangerous bacteria and where solids settle out in bays close to cities, one potential problem is the stirring up of such material by storms, and the *fouling* of heavily used recreational areas. Seaside recreation is big business in many coastal cities. Thus, the pressure for "clean" disposal becomes intense in situations where tourism is at stake. A prime example is the Mediterranean, where there are as many tourists as inhabitants along the shores, and where up to 25 % of coastal waters is now deemed unfit for swimming (depending on criteria used).

10.5.3 Oil Spills. Certain kinds of pollution and dumping events have high media visibility. Among these are *oil spills*. Such reports greatly raise the awareness level of the public, regarding the potential problems stemming from misuse of the ocean, and from irresponsible behavior. In most cases, however, even massive spills (Torrey Canyon, 1967; Amoco Cadiz, 1978; Exxon Valdez, 1989) have had limited impact in terms of area and time period affected (although that is of little comfort to those whose livelihood is at stake).

One possible exception with potentially long-lasting regional effects may be the deliberate release of oil in the upper Persian Gulf, from wells in Kuwait, during the hostilities in 1991. More than half a million tons of crude oil entered Gulf waters from from the Mina Al-Ahmadi oil terminal, polluting 770 km of Saudi coastline. Significant impacts from direct oiling are reported for the intertidal habitats (mangroves, sandy and rocky beaches, and mudflats). Long-term damage is expected especially for the upper tidal zone.

10.5.4 Chronic Hydrocarbon Pollution. In constrast to catastrophic spills, the steady leakage of hydrocarbons from shipping and other industrial activities (including oil change in automobiles) receives much less attention. In 1985 the annual petroleum hydrocarbon input to the sea was estimated to reach about 3 million tons – with 12.5 % of it from tanker accidents. Additional risks arise from the operation of more than 600 offshore oil production platforms globally.

10.5.5 Radioactive Pollution. The principal source of artificial *radionuclides* in the marine environment used to be nuclear weapon-testing. This source produced a very low-level contamination with no obvious ill effects. Scientists use this contamination for tracing water masses (e. g., by noting the tritium content of deep waters), and for dating purposes in skeletal parts (e. g., a 14^C spike in the 1960s, in coral and shell). A relatively smaller source of such nuclides are fission products introduced from military and commercial reactors. Dumping of radioactive waste has been prohibited among the signatories to the *London Dumping Convention* (1972) since 1983. However, it may be more widespread than realized, as has recently become obvious (e. g., weapons disposal by the former USSR in the Barents Sea). Disposal of radioactive materials (albeit of low-level activity) is also known to have occurred several decades ago off California and off the US East coast. The impact, if any, is unknown.

The massive release of radioactive materials to the atmosphere at *Chernobyl* (April 1986) provided a unique opportunity to study the pathways of a number of dangerous radionuclides, within the marine environment. Results (from trapping and coring in the Black Sea and elsewhere) suggest that many, perhaps most, of the metals involved (e. g., ^{137}Cs) are rapidly removed from surface waters and transported to the sea floor within biogenic debris. Similar results are known from studying the effluent from the Hanford reactor, in the Columbia river mouth.

Problems of an entirely different order are posed by *high-level radioactive waste*. Given that large amounts of such waste exist on land now, and that these pose a considerable hazard for present and future generations, the disposal of such materials through burial in the sea floor is an option that has received some discussion and

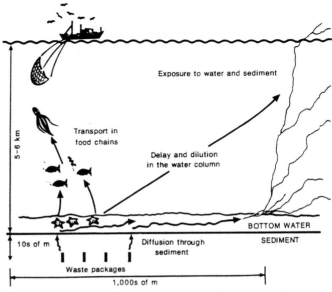

Fig. 10.15. Investigation of risks associated with sub-seabed disposal of high-level nuclear waste. The diagram shows hypothetical pathways of contaminants from postulated leaks from waste repositories. These pathway (and others) have to be assessed before subsea disposal can be considered as low-risk option. Vitrification of canister packages, circulation driven by high temperatures generated by radiation, diffusion through sediments or erosion by deep-sea storms, all these might provide for escape into bottom waters and uptake by benthic organisms, resulting in physical or biological transport to surface waters. [M. F. Kaplan et al., 1984, Sandia Natl Lab Rept Sand 83–7106]

study. One type of proposal envisions canisters placed into deep-sea sediments (Fig. 10.15).

Geologic questions which must be considered in connection with such contemplated emplacement are as follows: (1) how stable is the area in which disposal is to take place (i. e., might the sediment be eroded before the radioactivity has died down sufficiently?); (2) what type of circulation will develop around the containers (which are hot), and where would leaked materials end up? (i. e., would they be adsorbed on sediment particles, or reach the sea floor?); (3) what kind of reactions might leaked material undergo, and how would this influence mobility and bio-activity? Of special interest are possible pathways into the food web of marine organisms that might be eaten by humans.

There is no evidence that *deep-sea disposal* by burial would be infeasible given reliable containers and well-chosen areas. The main problems may not be geologic at all: the issue is safety during transportation and emplacement. Also, there is the ethical problem of the concept of producing highly dangerous substances only to unload them in "no-man's land" at the risk of future generations.

10.5.6 Assessment of risk. Clearly, it is desirable to assess the potential impact of dumping of hazardous materials *before* such dumping takes place. In any case, one would hope that existing (if unfortunate) instances of hazardous dumping are used to study the processes which determine the fate of the materials in question. These include *dispersion* of the substances, and alteration through *chemical reactions*. Also, the *bio-activity* is of vital importance (Fig. 10.15).

Dispersion is controlled by bioturbation, resuspension, and transport by currents, which oppose the effects from natural burial. Chemical reactions on and within the sediment can change the toxicity of materials, and the ease with which they migrate through the sediment, or are taken up by organisms. A well-known example with dire consequences is the mobilization of mercury by methylation within anaerobic sediments. Methyl-mercury can end up in fish which then become poisonous, causing "*Minamata Disease*", if consumed in sufficient quantitiy. The disease is named after a bay in Japan where contamination of fish occurred from release of mercury, leading to much suffering and death.

An important aspect of the waste disposal problem is the scarcity of acceptable dumping sites on land. One reason for this is the subjective desire of most communities to site dumps outside their own jurisdiction where possible; another is the objective danger of groundwater contamination, which rules out many otherwise suitable areas. Incineration (which can help heat a city) is rarely favored because of its effects on air quality. Thus, the price of dumping has risen sharply. This makes the ocean attractive as a dumping ground.

Deep-sea dumping is a classic case of cost externalization. The communities doing the dumping reap the benefit of being rid of the noxious material which arose from their activities, while the risk is spread evenly to all using the oceans, including – in the case where the damage is irreversible – future generations.

One important task of marine geologists studying the sea floor – and perhaps the most difficult yet – is to help assess the risks with which we burden the future.

Further Reading

Rona PA, Boström K, Laubier L, Smith KL (eds) (1983) Hydrothermal processes at sea floor spreading centers. Plenum Press, New York

Tissot B, Welte DH (1984) Petroleum formation and occurrence, 2nd edn. Springer, Berlin New York

Notholt AJG, Jarvis I (1990) Phosphorite research and development. Geol Soc Lond Spec Publ 22, London

Cronan DS (1992) Marine mineral in exclusive economic zones. Chapman and Hall, London

Hsü KJ, Thiede J (eds) (1992) Use and misue of the seafloor. Dahlem Konferenzen. Wiley, New York

Epilog

We have attempted, in this brief survey of sea floor studies, to show where we have been and where we are now. But where are we going? An answer can only be tentative, of course. Just like other fields in the earth sciences, marine geology has experienced an exponential expansion of knowledge in the last two decades. This knowledge explosion is largely driven by technological developments: satellites, submarines, deep-sea drilling, all sorts of remote sensing devices, highly sophisticated laboratory equipment, and electronic information handling. Also, large-scale integration of geology with physics, chemistry, and biology has continued at a rapid pace, with new specialities arising from this cross-fertilization in quick succession, and recombining across disciplines to attack common targets: messages from the mantle, ridge-crest processes, fluid circulation in the margins, large-scale extinction, global environmental change.

The safe prediction is to postulate a continuation of these trends resulting in ever-greater specialization in technological skills, and in increased cooperation across disciplines when focusing these skills on studying sea-floor processes and on reconstructing sea-floor history, as well as the history of ocean and climate.

What were the "hot" topics of research and discussion in the last decade? How do they stack up against the topics of the two preceding decades?

Sea-floor studies of the 1960s were characterized by the overwhelming success of *geophysics:* discovery of the grand motions of the continents and the sea floor by continuous echo sounding (for bathymetry), by the mapping of earthquake foci and heat flow, by seismic profiling, and especially by the mapping of paleomagnetism.

The 1970s saw a great expansion of geologic knowledge based on testing and applying the new paradigm of *plate tectonics,* the crowning achievement of marine geophysics. This expansion owed much to *deep-sea drilling,* which provided the "ground truth" for sea-floor spreading, and made available an enormous amount of rocks and sediments, the likes of which had not been seen before, or only in bits and pieces. In rapid succession we gained an understanding of the main structural features of the sea floor, of processes involved in creating and destroying sea floor, and of the growth of continents through mountain-building in subduction zones. *Geochemistry* took the lead in starting to unravel messages from the mantle, in the shape of isotopic composition and rare earth content of volcanic products in different settings, from basement rock to island arcs to mantle plumes. These messages, combined with large-scale mapping of heat flow, magnetism, and gravity, and seismologic modeling, are providing glimpses of the structure of the mantle, and the character of the convection processes responsible for sea-floor spreading and continental drift. *Ridge-crest processes* took center stage as it became clear that the interaction between seawater and basalt is a determinant factor in ocean chemistry. *Deep-sea biology* made headlines with the discovery of an entirely new kind of deep-sea community, the hot vent assemblage on the East Pacific Rise. These communities represent a food web whose base is *chemosynthesis:* the oxidation of sulfide issuing from the vents by archaic bacteria specially adapted to high pressure and high temperature. The fact that sunlight is unnecessary to sustain primitive life forms has led to new speculations about the conditions surrounding the origin of life. Was Earth's heat the energy source nourishing the first living things?

Also in the 1970s, *stratigraphy* and its offspring *paleoceanography* received a tremendous boost from deep-sea drilling. The main features of the Cretaceous and Cenozoic history of the ocean were established, using *biostratigraphy, magnetostratigraphy,* and *chemostratigraphy* (mainly the *isotopic* ratios in the elements oxygen, carbon, and strontium, as recorded in calcareous microfossils, and also the *preservational* state of the fossils, which records saturation and reactivity of deep waters). It was found that the great *evolutionary breaks* near the epoch boundaries were not an artifact of poor preservation of the record, which is so common in the land sections. Deep-sea sediments are made mainly of fossils, so neither was there any question of missing the message due to lack of messengers. What emerged is that in the ocean evolution proceeds mainly in distinct steps rather than gradually, and that major discontinuities are tied to rapid climatic change.

Drilling in the margins discovered their architecture and established the course and effects of continental break-up (in the Atlantic) and of continental accretion (in the Pacific). *Seismic stratigraphy,* in conjunction with drilling results, greatly expanded our knowledge about margin structure and margin processes, and allowed the (tentative) reconstruction of sea level variation on a global scale.

In the meantime, Quaternary research on the youngest deep-sea sediments established the cyclic nature of the record, and the dominance of *astronomic forcing* within the background climatic fluctuations which characterize the "normal" situation between the climate steps.

The discoveries of the 1970s provided the base for the search for mechanisms in the 1980s. The new catch phrases are *"earth dynamics"* and *"global systems"* and they will continue to be most important. How does mantle convection control tectonic or geochemical processes at the Earth's surface, and how is it tied to the coevolution of climate and life? How is the chemistry of the ocean and of the atmosphere determined, and what is the role of the circulation of seawater through the ocean crust and through the ocean margins, and how is this circulation affecting crustal materials? How do *hot vents* (on ridge crests) and *cold seeps* (along the margins) control the abundance and diversity of chemosynthetic-based communities? How do ocean processes (productivity, sedimentation) codetermine climate? What exactly happened at those times when extinction (or speciation) greatly accelerated in the ocean?

Significant contributions to these major topics came from all disciplines, but several major discoveries and developments set the tone. For the study of mantle processes (degree of mixing, nature of layering), three-dimensional *seismic tomo-*

graphy (permitting the detection of inhomogeneities) and the discovery of basin-size *isotopic provinces* proved crucial. Vigorous discussion about the style of mantle convection (e. g., the relative importance of shallow and deep convection and the role of mantle plumes) continues.

The proper description of ridge-crest processes depends on mapping the *magma chambers* which underlie the crest like large pearls on a string, and on measuring the emissions of gases and fluids. Details of topography become available as *acoustic swath-mapping,* using multiple source and microphone arrangements, becomes almost routine. Geologists now "look" at the sea floor here at the ridge crest and elsewhere with giant "acoustic searchlights", opening up new frontiers of marine geomorphology.

For understanding processes within the subduction zone, the detection of fluid overpressure along overthrust faults, and of seeps along the continental slope, were very important. Thus, the *role of fluid circulation* within the crust has become a topic of intensive research.

Regarding the debate about the processes controlling evolution, the *iridium spike* at the Cretaceous-Tertiary boundary (reported in 1980) provided a rallying point. This debate, bolstered by the study of pelagic sediments on land and on the deep-sea floor, has led to a re-evaluation of the tenets of classic uniformitarianism, and has emphasized the crucial importance of episodic events in Earth history, especially concerning the role of impacts of extraterrestrial bolides. It is as though geologists suddenly realized that Earth is not an isolated fortress in space, but is vulnerable to bombardment – as illustrated by the surface of our Moon.

For Quaternary climate history, the discovery of ice age *carbon dioxide fluctuations* in polar ice (likewise reported in 1980) was of primary importance. The tie-in to the deep-sea carbonate record proved to be highly instructive regarding the ocean's role in controlling atmospheric CO_2. This topic has become the subject of intense research efforts.

In the continental margins, especially in the passive ones, *sea level fluctuations* are recorded in the geometry of sediment bodies of vast volumes. Some of these bodies contain petroleum and other resources, which has greatly stimulated four-dimensional basin analysis using the new insights gained from plate tectonics.

History, it is said, is a guide to destiny. First marine geophysics and, more recently, marine geology and stratigraphy have led the way back to a global view of the evolution of Earth and its life. The next step in this development is to achieve detailed correlation between the record in the sea and that on land. Sequence-stratigraphy, magnetic stratigraphy, and bio- and chemostratigraphy will have to be greatly refined for the purpose, and accurate dating of events becomes crucial. Only with such tools in hand can we attempt synchronous global reconstructions of paleo-environments, to elucidate with some exactness the complex interactions of the many processes governing climate change (Fig. E.1). For geologists, the answer to questions of cause and effect must be sought by determining "before" and "after" across oceans and continents, with ever-increasing resolution. As we expand our knowledge of rates of changes in Earth history and the forces controlling them, we gain fun-

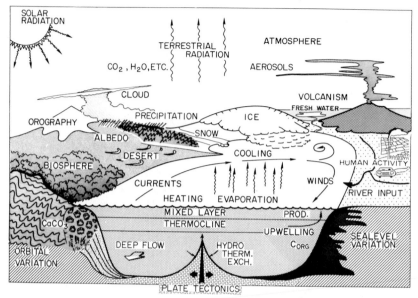

Fig. E.1. Major components of the climate system and central role of the ocean as mediator between the various elements. (W. H. Berger and L. A. Mayer, 1987, Paleoceanography 2: 613)

damentally new insights on how the face of the planet evolves, and the living things on it.

During the 1990s and beyond, human interactions with the sea will increase, leading to intensified investigation of *human impacts*. Mankind's activities now have reached truly global dimensions, outcompeting natural factors in many arenas of geologic change. This trend began in the Neolithic age, with agriculture and deforestation gradually and regionally changing the rates of erosion and affecting soil formation. It is now becoming a factor for profound and rapid alteration of the global environment.

The knowledge gained in these various researches will help define our place in nature, and will (in principle) allow us to use Earth's resources more wisely. One area where there is a pressing need for wisdom is in the disposal of waste, especially of dangerous chemical and radioactive waste. Another is the assessment of hydrocarbon resources. Yet another is the entire complex of forecasting, including earthquakes and volcanism, and especially the effects of the large-scale climatic change which is expected for the next centuries, as a result of human activities.

List of Books and Symposia

(additional entries at chapter ends)

Aigner T (1985) Storm depositional systems. Springer, Berlin Heidelberg New York

Andersen NR, Malahoff A (eds) (1977) The fate of fossil fuel CO_2 in the oceans. Plenum Press, New York

Anderson DL (1989) Theory of the Earth. Blackwell Scientific, Oxford

Bally AW et al. (eds) (1979) Continental margins – geological and geophysical research needs and problems. Natl Acad Sci, Washington DC

Bally AW et al. (eds) (1981) Geology of passive continental margins. AAPG Educational Course Notes 19, Amer. Assoc. Petrol. Geol., Tulsa, Okla.

Barth MC, Titus JG, Ruckelshaus WD (1984) Greenhouse effect and sea level rise. Van Nostrand Reinhold, New York

Bascom W (1964) Waves and beaches. Doubleday, Garden City, New York

Barthurst RGC (1975) Carbonate sediments and their diagenesis, 2nd edn. Elsevier, Amsterdam

Baturin GN (1982) Phosphorites on the seafloor. Elsevier, Amsterdam

Bentor YK (ed) (1980) Marine phosphorites – geochemistry, occurrence, genesis. SEPM Spec Publ 29, Soc Econ Paleontol Mineral Tulsa, Okla

Berger WH, Labeyrie LD (eds) (1987) Abrupt climatic change – evidence and implications. Reidel, Dordrecht

Berggren WA, van Couvering JA (eds) (1984) Catastrophism and Earth history, the new Uniformitarianism. Princeton Univ Press, Princeton NJ

Berner RA (1971) Principles of chemical sedimentology. McGraw-Hill, New York

Berner RA (1980) Early diagenesis. Princeton Univ Press, Princeton N J

Biddle KT (ed) (1991) Active margin basins. AAPG Mem 52. Am Assoc Petrol Geol, Tulsa, Okla

Boillot G (1981) Geology of continental margins. Longman, London (translated fr. French edn 1978)

Bolin B, Degens ET, Kempe S, Ketner P (eds) (1979) The global carbon cycle. SCOPE Rep 13. Wiley, Chichester

Bolin B, Döös BR, Jäger J, Warrick RA (eds) (1986) The greenhouse effect, climatic change, and ecosystems. SCOPE 29. Wiley, Chichester

Bott MHP (1982) The interior of Earth, its structure, constitution and evolution, 2nd edn. Edward Arnold, London

Bouma AH, Brouwer A (eds) (1964) Turbidites. Elsevier, Amsterdam

Bradley RS (1985) Quaternary paleoclimatology: methods of paleoclimatic reconstruction. Allen and Unwin, Winchester Mass

Bradley RS (ed) (1991) Global changes of the past. Office for Interdisc. Earth Studies, Boulder Colorado

Brenchley P (ed) (1984) Fossils and climate. Wiley, New York

Brenchley PJ, Williams BPJ (1985) Sedimentology – recent developments and applied aspects. Blackwell Scientific, Oxford

Briggs JC (1974) Marine zoogeography. McGraw-Hill, New York

Broecker WS, Peng T-H (1982) Tracers in the sea. Lamont-Doherty Geol Obs, Palisades, New York

Broedehoeft JD, Norton DL (eds) (1990) The role of fluids in crustal processes. Studied in Geophysics. National Academy Press, Washington DC

Bromley RG (1990) Trace fossils – biology and taphonomy. Hyman, London

Brooks J, Fleet AJ (eds) (1987) Marine petroleum source rocks. Geol Soc Spec Publ 26, Geol Soc, London

Broussard JM (ed) (1975) Deltas. Houston Geol Survey

Bruland KW (ed) (1984) Global ocean flux study. Proc worksh. National Acad Press, Washington DC

Burnett WC, Froelich Ph (eds) (1988) The origin of marine phosphorite. Geol 80 (Spec Issue)

Burnett WC, Riggs SR (1989) Phosphate deposits of the world, vol 3. Neogene to modern phosphorites. Cambridge Univ Press

Cande SC (1988) Magnetic lineations of the world's ocean basins. AAPG Map Ser Am Assoc Petrol Geol Tulsa, Okla

Charnock N, Edmond JM, McCave IN, Rice AL, Wilson TRS (eds) (1990) The deep sea bed: its physics, chemistry and biology. The Royal Society, London

Chisholm SW, Morel FMM (eds) (1991) What controls phytoplankton production in nutrient-rich areas of the open sea? Limnol Oceanogr 36 (8) (Spec Issue)

Cline RM, Hays JD (eds) (1976) Investigation of late Quaternary paleoceanography and paleoclimatology. Geol Soc Am Mem 145, GSA, Boulder

Collison J, Thompson D (1988) Sedimentary structures. Unwin Hyman, London

Condie KC (1982) Plate tectonics and crustal evolution. Pergamon Press, Oxford

Cooper AK, Davey FJ (eds) (1987) Antarctic continental margin: geology and geophysics of the western Ross Sea. Am Assoc Petrol Geol, Tulsa, Okla

Cox A, Hart RB (1986) Plate tectonics – how it works. Blackwell Scientific, Palo Alto

Cronan DS (1980) Underwater minerals. Academic Press, London

Crowley TJ, North GR (1991) Paleoclimatology. Oxford Univ Press, New York

Cushing DH, Walsh JJ (eds) (1986) The ecology of the seas. Blackwell Scientific, Oxford

Das S, Boatwright J, Scholz CH (eds) (1986) Earthquake source mechanics. AGU Maurice Ewing Ser Vol 6. Am Geophys Union, Washington DC

Davies RA Jr (ed) (1978) Coastal sedimentary environments. Springer, Berlin Heidelberg New York

Deep-Sea Drilling Project, Initial Reports (1969–1987) Government Printing Office, Washington DC, vols 1–96

Degens ET (1989) Perspectives on biogeochemistry. Springer, Berlin Heidelberg New York

Degens ET, Ross DA (eds) (1969) Hot brines and recent heavy metal deposits in the Red Sea. Springer, Berlin Heidelberg New York

Degens ET, Ross DA (eds) (1974) The Black Sea – geology, chemistry and biology. AAPG Mem 20. Am Assoc Petrol Geol, Tulsa, Okla

Delaney J (ed) (1988) The Mid-Oceanic Ridge – a dynamical global system. Proc worksh. National Academy Press, Washington DC

Dickinson WR (ed) (1974) Tectonics and sedimentation. SEPM Spec Publ 22. Soc Econ Paleontol Mineral, Tulsa, Okla

Drooger CW (ed) (1973) Messinan events in the Mediterranean. K Ned Akad Wet, North-Holland, Amsterdam

Duedall IW, Kester DR, Ketchum BH, Park PK (eds) (1983) The dumping of wastes at sea. Wiley-Interscience, New York, 3 vols

Edwards JD, Santogrossi PA (eds) (1990) Divergent/passive margin basins. AAPG Mem 48. Am Assoc Petrol Geol, Tulsa, Okla

Ehlers J (1988) The morphodynamics of the Wadden Sea. Balkema, Rotterdam

Einsele G, Seilacher A (eds) (1982) Cyclic and event stratification. Springer, Berlin Heidelberg New York

Eittreim SL, Hampton ML (eds) (1987) Antarctic continental margin: geology and geophysics of offshore Wilkes Land. Am Assoc Petrol Geol, Tulsa, Okla

Emery KO (1960) The sea off southern California. Wiley, New York

Emery KO, Uchupi E (1972) Western North Atlantic Ocean: topography, rocks, structure, water, life, and sediments. AAPG Mem 20, Am Assoc Petrol Geol, Tulsa, Okla

Emery KO, Uchupi E (1984) The geology of the Atlantic Ocean. Springer, Berlin Heidelberg New York

Fairbridge RW (ed) (1966) The encyclopedia of oceanography. Reinhold, New York

Fairbridge RW, Bourgeois J (eds) (1978) The encyclopedia of sedimentology. Dowden Hutchinson & Ross, Stroudsburg, Pa

Falkowski PG, Woodhead AD (eds) (1992) Primary productivity and biogeochemical cycles in the sea. Plenum Press, New York

Faure G (1977) Principles of isotope geology. Wiley, New York

Fischer AG, Judson S (eds) (1975) Petroleum and global tectonics. Princeton University Press, Princeton

Frakes LA (1979) Climates throughout geologic time. Elsevier, New York

Freeman TJ (ed) (1989) Disposal of radioactive waste in seabed sediments. Society for Underwater Technology. Graham and Trotman, London

Frey, RW (1975) The study of trace fossils. A synthesis of principles, problems, and procedures in ichnology. Springer, Berlin Heidelberg New York

Frost SH, Weiss MP, Saunders JP (eds) (1977) Reefs and related carbonates – ecology and sedimentology. AAPG Stud Geol 4, Tulsa, Okla

Füchtbauer H (ed) (1988) Sedimente und Sedimentgesteine, 4 Aufl. Schweizerbart, Stuttgart

Funnell BM, Riedel WR (eds) (1971) The micropaleontology of oceans. Cambridge Univ Press

Gage JD, Tyler PA (1991) Deep-sea biology: a natural history of organisms at the deep-sea floor. Cambridge Univ Press

Garrels RM, Mackenzie FT (1971) Evolution of sedimentary rocks. Norton, New York

Gass IG, Lippard SJ, Shelton AW (eds) (1984) Ophiolites and oceanic lithosphere. Blackwell Scientific, Oxford

Gerdes G, Krumbein WE (1987) Biolaminated deposits. Springer, Berlin Heidelberg New York

Geyer RA (ed) (1980/1981) Marine environmental pollution, I: Hydrocarbons; II: Dumping and mining. Elsevier, Amsterdam

Ginsburg RN (ed) (1975) Tidal deposits, a casebook of Recent examples and fossil counterparts. Springer, Berlin Heidelberg New York

Ginsburg RN, Beaudoin B (eds) (1990) Cretaceous resources, events and rhythms. NTO ASI Ser. Kluwer Academic, Dordrecht

Glennie K (1990) Introduction to the petroleum geology of the North Sea. Blackwell Scientific, Oxford

Glynn PW, Wellington GM, Wells JW (1983) Corals and coral reefs of the Galapagos Islands. University of California Press, Berkeley

Gray J, Boucot AJ (eds) (1979) Historical biogeography, plate tectonics and changing environments. Oregon State Univ Press, Corvallis

Hansen JE, Takahashi T (eds) (1984) Climate processes and climate sensitivity. Am Geophys Union Geophys Monogr 29

Haq BU, Milliman J (eds) (1984) Marine geology and oceanography of Arabian Sea and coastal Pakistan. Van Nostrand Reinhold, New York

Harland WB, Armstrong RL, Cox AV, Craig LE, Smith AG, Smith DG (1990) A geologic time scale 1989. Cambridge Univ Press

Hart SR, Gülen L (eds) (1989) Crust/Mantle recycling at convergence zones. Kluwer Academic, Dordrecht

Hay WW (ed) (1974) Studies in paleo-oceanography. SEPM Spec Publ 20. Soc Econ Paleontol Mineral, Tulsa, Okla

Heezen BC (ed) (1977) Influence of abyssal circulation on sedimentary accumulations in space and time. Elsevier, Amsterdam

Heezen BC, Tharp M (1964) Physiographic diagram of the South Atlantic Ocean. Geol Soc Am, New York

Heezen BC, Tharp M (1964) Physiographic diagram of the Indian Ocean. Geol Soc Am, New York

Heezen BC, Hollister CD (1971) The face of the deep. Oxford Univ Press, New York

Ruddiman WF, Wright HE (eds) (1987) North America and adjacent oceans during the last deglaciation. The geology of North America, K-3, Geol Soc Am Boulder, Colo

Rumohr J, Walger E, Zeitzschel B (eds) (1987) Seawater-sediment interactions in coastal waters. Springer, Berlin Heidelberg New York

Sawkins FJ (1990) Metal deposits in relation to plate tectonics, 2nd edn. Springer, Berlin Heidelberg New York

Saxon S, Nieuwenhuis J (1982) Marine slides and other mass movements. Plenum Press, New York

Scholl D, Grantz A, Vedder J (eds) (1987) Geology and resource potential of the continental margin of western North America and adjacent ocean basins – Beaufort Sea to Baja California. Am Assoc Petrol Geol, Tulsa, Okla

Scholle PA, Spearing D (eds) (1982) Sandstone depositional environments. AAPG Mem 31. Am Assoc Petrol Geol, Tulsa, Okla

Scholle PA, Bebout DG, Moore CH (eds) (1983) Carbonate depositional environments. AAPG Mem 33. Am Assoc Petrol Geol, Tulsa, Okla

Schopf, TJM (1980) Paleoceanography. Harvard Univ Press, Cambridge Mass

Schroeder JH, Purser BH (eds) (1986) Reef diagenesis. Springer, Berlin Heidelberg New York

Schwartz ML (ed) (1982) The encyclopedia of beaches and coastal environments. Hutchinson Russ, Stroudsburg, Pa

Schwarz HU (1982) Subaqueous slope failures, experimental and modern occurrences. Contrib Sedim 11. Schweizerbart, Stuttgart

Scoffin TP (1987) An introduction to carbonate sediments and rocks. Blackie, Glasgow

Scott DB, Pirazolli PA, Honig CA (eds) (1989) Late Quaternary sea-level correlation and applications. Kluwer Academic, Dordrecht

Scrutton RA (ed) (1982) Dynamics of passsive margins. AGU Geodyn Ser 6. Am Geophys Union, Washington DC

Selley RC (1976) An introduction to sedimentology. Academic Press, London

Selley RC (1988) Applied sedimentology. Academic Press, London

Seyfert CK, Sirkin LA (1979) Earth history and plate tectonics: an introduction to historical geology, 2nd edn. Harper and Row, New York

Shea JH (ed) (1985) Plate tectonics. Van Nostrand Reinhold, New York

Shepard FP (1963) Submarine geology, 2nd edn. harper and Row, New York

Shepard FP, Marshall NF, McLoughlin PA, Sullivan, GG (1979) Currents in submarine canyons and other sea valleys. AAPG Stud 8. Am Assoc Petrol Geol, Tulsa, Okla

Sheridan RE, Grow JA (eds) (1988) The Atlantic continental margin: U.S. The geology of North America, vol I-2. Geol Soc Am, Boulder

Silver LT, Schultz PH (eds) (1982) Geological implications of impacts of large asteroids and comets on the Earth. Geol Soc Am Spec Pap 190

Slansky M (1986) Geology of sedimentary phosphates. Elsevier, Amsterdam

Sliter WV, Bé AWH, Berger WH (eds) (1975) Dissolution of deep-sea carbonates. Cushman Found Foram Res, Spec Publ 13

Snelling NJ (ed) (1985) The chronology of the geological record. Geol Soc Lond, London

Stanley DJ, Moore GT (eds) (1983) The shelf break: critical interface on continental margins. SEPM Spec Publ 33, Soc Econ Paleontol Mineral, Tulsa, Okla

Stein R (1991) Accumulation of organic carbon in marine sediments. Lecture Notes in Earth Sciences. Springer, Berlin Heidelberg New York

St. John B (1984) Sedimentary provinces of the world. AAPG Map Ser Am Assoc Petrol Geol, Tulsa, Okla

Stow DAV, Piper DJW (eds) (1984) Fine-grained sediments: deep-water processes and facies. Geol Soc Lond Spec Publ 15

Sutton GH, Manghnani M-H, Moberly R (eds) (1976) The geophysics of the Pacific Ocean basin and its margin. AGU Geophys Monogr 19. Am Geophys Union, Washington DC

Swift DJP, Palmer HD (eds) (1978) Coastal sedimentation. Benchmark Papers in Geology. Dowden Hutchinson Ross, Stroudsburg Pa

Talwani M, Pitman WC (eds) (1977) Island arcs, deep sea trenches, and back-arc basins. AGU Maurice Ewing Ser, vol 1. Am Geophys Union, Washington DC

Talwani M, Harrison CG, Hayes DE (eds) (1979) Deep drilling results in the Atlantic Ocean: ocean crust. AGU Maurice Ewing Ser 2. Am Geophys Union, Washington DC

Talwani M, Hay W, Ryan WBF (eds) (1979) Deep drilling results in the Atlantic Ocean: continental margins and paleoenvironment. Maurice Ewing Ser 3 Am Geophys. Union, Washington DC

Tarling DH (1983) Paleomagnetism. Chapman and Hall, London

Tarling DH, Tarling MP (1971) Continental drift: a study of the Earth's moving surface. Bell, London

Taylor B, Exon NF (eds) (1987) Marine geology, geophysics, and geochemistry of the Woodlark Basin – Solomon Islands. Am Assoc Petrol Geol, Tulsa, Okla

Thiede J, Suess E (eds) (1983) Coastal upwelling – its sediment record. Part B: Sedimentary records of ancient coastal upwelling. Plenum Press, New York

Tillman RW, Ali SA (eds) (1982) Deep water canyons, fans, and facies: models for stratigraphic trap exploration. AAPG Reprint Ser 26. Am Assoc Petrol Geol, Tulsa, Okla

Trabalka JR (ed) (1986) Atmospheric carbon dioxide and the global carbon cycle US Dept Energy, US Gov Printing Office, Washington DC

Tucholke B, Fry V (1985) Basement structure and sediment distribution in the NW Atlantic Ocean. AAPG Map Ser Am Assoc Petrol Geol, Tulsa, Okla

Turcotte DL, Schubert G (1982) Geodynamics. Wiley, New York

Turekian KK (ed) (1971) The late Cenozoic glacial ages. Yale Univ Press, New Haven, Conn

van Andel TjH, Heath GR, Moore TC (1975) Cenozoic tectonics, sedimentation and paleooceanography of the central Pacific. Geol Soc Am Mem 143 Boulder

van der Zwaan GJ, Jorissen FJ, Zachariasse WJ (eds) (1992) Approaches to paleoproductivity reconstructions. Mar Micropalentol 19 (1/2). Elsevier, Amsterdam

van Straaten LMJU (ed) (1964) Deltaic and shallow marine deposits. Elsevier, Amsterdam

Vogt PR, Tucholke BE (1986) The western North Atlantic region. The geology of North America, vol M. Geol Soc America, Boulder

von Rad U, Hinz K, Sarnthein M, Seibold E (eds) (1982) Geology of the northwest African continental margin. Springer, Berlin Heidelberg New York

Vorren TO, Bergsager E, et al. (eds) (1992) Arctic geology and petroleum potential. Elsevier, Amsterdam

Walsh JJ (1988) On the nature of continental shelves. Academic Press, Orlando Fla

Watkins JS, Mountain GS (eds) (1990) Role of ODP drilling in the investigation of global change in sea level. JOI/USSAC Office, Washington DC

Watkins JS, Montadert L, Dickerson PW (eds) (1979) Geological and geophysical investigations of continental margins. Am Assoc Petrol Geol Mem 29, Tulsa, Okla

Weaver CE (1989) Clays, muds, and shales. Elsevier, Amsterdam

Wiens HJ (1962) Atoll environment and ecology. Yale Univ Press, New Haven Conn

Winterer EL, Hussong DM, Decker RW (1989) The eastern Pacific Ocean and Hawaii. The geology of North America, vol N. Geol Soc America, Boulder

Wolf KH, Chilingarian GV (eds) (1992) Diagenesis III. Elsevier, Amsterdam

Yarborough H, Emery KO, Dickinson WR, Seely DR, Dow WG, Curray JR, Vail PR (1977) Geology of continental margins. AAPG Course Notes 5. Am Assoc Petrol Geol, Tulsa, Okla

Ziegler PA (1988) Evolution of the Arctic-North Atlantic and western Tethys. AAPG Mem 43. Am Assoc Petrol Geol Tulsa, Okla

Important Research Journals in English that Regularly Carry Articles on Subjects Related to Marine Geology

Published by the American Geophysical Union (Washington, D. C.): Journal of Geophysical Research; EOS, Transactions; Paleoceanography; Maurice Ewing Series (monographs).

Published by the Geologic Society of America (Bouler, Colorado): Bulletin of the G. S. A.; Geology; G. S. A. Memoir.

Published by the American Association of Petroleum Geologists (Tulsa, Oklahoma): AAPG Bulletin; AAPG Memoir.

Published by the Society of Economic Paleontologists and Mineralogists (Tulsa, Oklahoma): J. Sedim. Petrol.; SEPM Special Publ.

Published by the American Association for the Advancement of Science (Washington, D. C.): Science

Published by the Royal Astronomical Society, Deutsche Geophysikalische Gesellschaft and European Geophysical Society (Blackwell, Oxford): Geophysical Journal International (formerly Geophys. J. of the Roy. Astr. Soc.).

Published by the European Union of Geosciences (Blackwell, Oxford): Terra Nova.

Published by Macmillan Magazines Ltd., London: Nature.

Published by Elsevier, Amsterdam: Earth and Planetary Science Letters; Marine Geology; Palaeogeography, Palaeoclimatology, Palaeoecology; Marine Micropaleontology.

Published by Pergamon Press, Oxdford: Geochimica et Cosmochimica Acta; Deep-Sea Research.

Appendix

A1 Conversions Between Common US Units and Metric Units

Temperature
C = (F-32) · 5/9
F = C × 9/5 + 32 where C is degrees Celsius or "centigrade" and F is degrees Fahrenheit
Freezing point of water: 0 °C, 32 °F
Boiling point of water: 100 °C, 212 °F
Typical room temperature: 20 °C, 68 °F
Degrees Kelvin (= absolute scale) = Celsius + 273.2°

Length

1 cm = 0.394 inch	1 inch = 2.54 cm
1 m = 3.281 feet	1 ft = 0.305 m
1 km = 0.621 miles	1 mile (statue) = 1.609 km
1 cm = 10 mm (millimeter)	1 mile (nautical) = 1.852 km
	1 mm = 1000 μm (micrometer, microns)

Volume
1 liter = 1000 milliliters = 0.264 US gallons
1 US gallon = 3.781 l
1 barrel (oil) = 42 gallons

Mass
1 kilogram = 1000 grams = 2.205 pounds
1 pound = 0.454 kilograms
1 metric ton = 1000 kg = 1.102 short tons

A2 Topographic Statistics

Earth
Equatorial radius = 6378 km
Polar radius = 6356 km
Area = 510×10^6 km^2
Volume = 1.083×10^{12} km^3
Northern hemisphere: 61 % oceans
Southern hemisphere: 81 % oceans
Sea floor = 71 % of Earth's surface

Oceans
Atlantic Ocean (incl. Arctic and marginal seas)
  Area = 107 million km^2
  Volume = 351 million km^3
Indian Ocean
  Area = 74 million km^2
  volume = 285 million km^3
Pacific Ocean
  Area = 181 million km^2
  Volume = 714 million km^3
Total ocean
  Area = 362 million km^2
  Volume = 1347 million km^3
Depths: for depths see Table 2.2

Source for A2, H. U. Sverdrup et al. (1942). The oceans, Prentice Hall, Englewood Cliffs, New Jersey.

Shelves: for shelf areas see Table 2.1

A3 The Geologic Time Scale

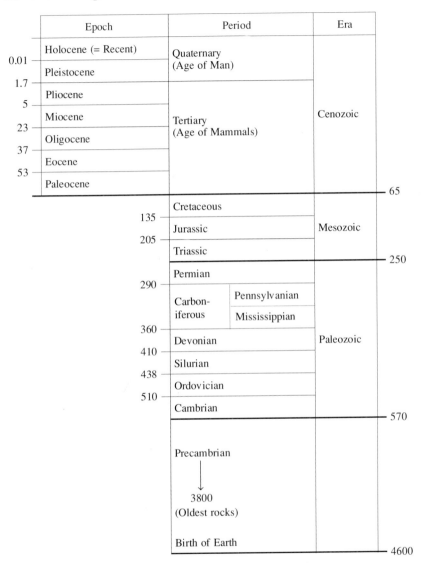

	Epoch	Period	Era
0.01	Holocene (= Recent)	Quaternary (Age of Man)	Cenozoic
1.7	Pleistocene		
5	Pliocene	Tertiary (Age of Mammals)	
23	Miocene		
37	Oligocene		
53	Eocene		
	Paleocene		
65			
135		Cretaceous	Mesozoic
205		Jurassic	
		Triassic	
250			
290		Permian	Paleozoic
360		Carboniferous — Pennsylvanian / Mississippian	
410		Devonian	
438		Silurian	
510		Ordovician	
570		Cambrian	
3800 (Oldest rocks)		Precambrian	
4600		Birth of Earth	

Numbers indicate boundary ages in millions of years. Note that the age of fossils, the *Phanerozoic* is only one eighths of Earth's history. All other time belongs to the Precambrian. The deep-sea record begins in the Jurassic (see Chap. 9). The Holocene is the youngest and the shortest epoch; it begins with the disappearance of the large ice caps from Canada and Scandinavia, 10 000 years ago.

Sources for A3, Episodes, 1989, 12,2. International Union of Geological Sciences.

A4 Common Minerals

Silicate Minerals

Most abundant, crust-forming minerals. Basic building blocks. SiO_4-tetrahedra (see graph).

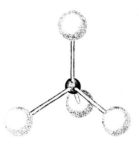

Fig. A4.1 SiO_4-tetrahedron, a silicon atom surrounded by four oxygen atoms

Quartz SiO_2. Framework structure: each tetrahedron shares all oxygens with other tetrahedra. Found in granitic rocks and in sediments. Opal = disordered hydrated silica ($SiO_2 \cdot nH_2O$)

Feldspars. Framework structure as in quartz. Replacement of Si by Al in some tetrahedra results in negative charge which is balanced by cations. Found in virtually all rocks. Examples:

Albite	Na Al Si_3O_8
Anorthite	Ca $Al_2Si_2O_8$
Microcline	K Al Si_3O_8

Micas. Sheet structure: each tetrahedron is tied to others by three oxygens. Sheets are bonded by cations. Ubiquitous. Examples:

Muscovite	K Al_2(Al Si_3)O_{10} (OH)$_2$
Biotite	K Mg_3(Al Si_3)O_{10} (OH)$_2$ (also usually contains iron)

Clay minerals. Similar to mica, but deficient in cations. See Fig. 8.7 for clay mineral structure. Abundant in sediments and in lowgrade metamorphic rocks. Produced during chemical weathering of rocks with feldspar and mica. Examples:

Montmorillonite (= smectite)	$(Al,Fe^{3+},Mg)_3$ (OH)$_2$[(Si, Al)$_4O_{10}$]Na $\cdot$ nH_2O
Illite	$K,H_3O)Al_2(H_2O, OH)_2[Al\ Si_3O_{10}]$
Chlorite	$(Al,Mg,Fe)_3(OH)_2(Al,Si)_4O_{10}Mg_3(OH)_6$
Kaolinite	$Al_4(OH)_8[Si_4O_{10}]$

Hornblende-type minerals. Double chain structure. Abundant in metamorphic rocks, and in many igneous rocks. Example:

Amphibole	$(Ca_2Mg_5)[Si_8O_{22}]$(OH)$_2$

Augite-type minerals. Single chain structure. Abundant in basalts. Example:

Pyroxene	$(Mg,Fe)SiO$

Olivine series. $(Mg,Fe)SiO4$ Isolated tetrahedra bonded via cations. Abundant in basalts.

Zeolites. Feldspar-like structure but with large interstices containing water. Common in volcanogenic sediments and in deep-sea clays. Examples:

Phillipsite	$(1/2Ca,NA,K)_3[Al_3Si_5O_{16}] \cdot 6\ H_2O$
Clinoptilotite	$(CaNa_2)[Al_2Si_7O_{18}] \cdot 6H_2O$

Nonsilicate Minerals

Carbonates. Bulk of biogenous sediments. Examples:

Calcite	$Ca\ CO_3$ shells and skeletons, see Table 3.3
Aragonite	$Ca\ CO_3$ ditto
Dolomite	$Ca\ Mg\ (CO_3)_2$ diagenesis, see Chapter 3.

Evaporite minerals. Precipitated from seawater. Examples:

Anhydrite	$Ca\ SO_4$
Gypsum	$Ca\ SO_4 \cdot 2H_2O$
Halite	$Na\ Cl$
Epsomite	$Mg\ SO_4 \cdot 7H_2O$
Bischofite	$Mg\ Cl_2 \cdot 6H_2O$
Carnallite	$K\ Mg\ Cl_3 \cdot 6H_2O$

Iron oxides and sulfides. Ubiquitous in igneous and sedimentary rocks. Examples:

Goethite	$\alpha\text{-}FeOOH$
Lepidocrocite	$\gamma\text{-}FeOOH$
Limonite	$FeOOH \cdot nH_2O$
Hematite	$a\text{-}Fe_2O_3$
Magnetite	Fe_3O_4
Pyrite	FeS_2
Marcasite	FeS_2
Hydrotroilite	$FeS_2 \cdot nH_2O$

Heavy minerals (see Chap. 3.5.2)

Apatite	$Ca_5(PO_4)_3(F,Cl,OH)$
Augite	Ca,Mg,Fe – silicate
Barite	$BaSO_4$
Epidote	Ca,Fe,Al – silicate
Garnet	Mg,Ca,Fe,Al – silicates
Glauconite	K,Fe – mica
Hematite	iron oxide
Hornblende	Ca,Fe,Na,K,Mg,Al – silicates (common in andesites)
Ilmenite	$FeTiO_3$
Limonite	iron (hydr)oxide
Marcasite	iron sulfide
Magnetite	iron oxide
Olivine	Mg,Fe,Ca – silicates (common in oceanic crust)
Pyrite	iron sulfide
Pyroxene	Mg,Fe,Ca – silicates (common in basalt)
Rutile	TiO_2
Titanite	Ca,Ti – silicate
Zircon	Zr – silicate

Source for A4, any mineralogy textbook

A5 Grain Size Classification for Sediments

Boulder		Gravel		Sand		Silt		Clay
mm	256		2		0.063		0.004 (a)	

(a) Some classifications use 0.002 mm

A6 Common Rock Types

Igneous Rocks. Mineral assemblages (dominated by silicates) which crystallized from hot, fluid magma. Slow cooling produces coarse crystals, typical for intrusive rocks. Fast cooling produces small crystals, typical for extrusive rocks (volcanic lava). Classification is by mineral types (from high-silica to low-silica minerals) and by crystal size.

Intrusive rocks	*Extrusive rocks*	
(Coarse crystals)	(Fine-grained)	Decreasing
Granite	Rhyolite	proportion
Granodiorite	Dacite	of SiO_2, increase
Diorite	Andesite[a]	in Mg and Fe
Gabbro	Basalt[b]	
Peridotite	Basalt[b]	$\downarrow$
Dunite	Basalt	

[a] Typical volcanic rock in island arcs and landward of trenches (Andes!). Partial melting of down-going oceanic crust appears to be an important process in generating andesite (see Fig. A6.1)
[b] Typical basement rock for deep sea floor, that is, ocean crust below sediments.

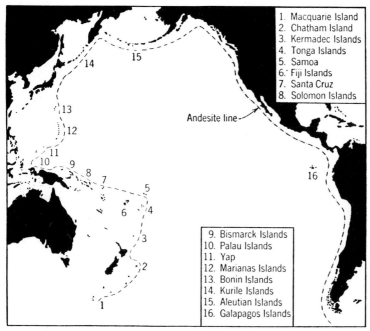

Fig. A6.1 Andesite line. The "andesite line" separates andesitic volcanoes from basaltic ones. The recognition of this boundary predates plate tectonics by many decades (see Fig. 1.2). Graph from Kuenen Ph H (1950) Marine Geology. John Wiley and Sons, New York

Sedimentary Rocks. Consolidated sediments. Particle assemblages consisting of minerals, rock fragments, shell, and amorphous substances in various proportions. Produced by weathering, transportation, deposition, alteration, and cementation. Classification is by grain size and composition (see Chap. 3).

Detrital or Clastic Rocks

(Unconsolidated)	(Consolidated)	
Gravel	Conglomerate	Decreasing
Sand	Sandstone	grain-size
Silt	Siltstone	
Clay	Claystone	↓
Mud[a]	Mudstone	
	Shale[b]	

[a]*Mud* is a mixture of silt and clay.

[b]*Shale* is the most common sedimentary rock; it is siltstone, claystone, and mudstone which is laminated and parts along bedding planes.

Examples of common sand- and siltstones are:

Quartz sandstone: produced through recycling of sedimentary rocks (quartz is resistant to weathering)

Arkose: feldspar-rich sandstone, produced from erosion of granite and granodiorite

Graywacke: cemented "dirty" sand, that is, with abundant clay and rock fragments. Composition is similar to that of sediments accumulating on active margins (see Fig. 2.7).

Chemical Rocks. Assemblages of minerals produced by precipitation from watery solution, either by organisms or inorganically. Classification is by origin and composition.

(Biogenous)	(Nonbiogenous)
Reef-limestone	Dolomite
Bedded limestone	Evaporite
Chert[a]	
Coal	
Phosphorite	

[a] *Chert* denotes a series of rocks ranging from silicified mudstones to very fine-grained quartz rock. The origin of the silica (quartz) is biogenous opal (at least in Mesozoic and younger rocks).

Metamorphic Rocks. Mineral assemblages produced by deformation and recrystallization of igenous rocks or sedimentary rocks through elevated temperature and pressure, with or without addition of material through injection of solutions. Classification is by degree of foliation, types of minerals present, and grain size. Metamorphic rocks occur at active margins (see Fig. 2.7).

(Nonfoliated)
Hornfels (from shale, tuff, or lava)
Marble (from limestone or dolomite)
Quartzite (from quartz sandstone)
Amphibolite (from basaltic material or corresp. sediments)
Granulite (from shale, graywacke, or andesite)

(Foliated)
Slate (from shale, tuff)
Schist (from basalt, andesite, tuff, or shale)
Gneiss (from granite, diorite, schist, shale, etc)

Source for A6, any geology text book

A7 Geochemical Statistics

Chemical Composition of Crust, Ocean, and Atmosphere

	Cont. crust	Ocean crust	Seawater	Atmosphere
O	46.3	43.6	85.8	21.0
Si	28.1	23.9	–	–
Al	8.2	8.8	–	–
Fe	5.6	8.6	–	–
Ca	4.2	6.7	0.04	–
Na	2.4	1.9	1.1	–
Mg	2.3	4.5	0.14	–
K	2.1	0.8	0.04	–
Ti	0.6	0.9	–	–
H	0.14	0.2	10.7	–
P	0.10	0.14	–	–
Cl	–	–	2.0	–
N	–	–	–	78.1
Ar	–	–	–	0.9
CO_2	–	–	tr	0.03
	100.0	100.0	99.8	100.0

The crust is mainly made of alumuno-silicates, with the alkaline and earth-alkaline elements as cations, and with iron inside and outside of the silicate minerals. Continental crust has a composition resembling that of granodiorite, ocean crust that of basalt. Seawater is largely a solution of sodium chloride. The atmosphere is a mixture of nitrogen and oxygen, both intimately tied to biological cycling.

Composition of Igneous and Sedimentary Rocks

	Igneous (continent)	Sedimentary (continent)	Tholeiitic basalt (deep sea)	Alkali olivine basalt (ocean volcanoes)	Red clay (deep sea)
SiO_2	59.1	57.9	50.2	48.2	53.7
Al_2O_3	15.3	13.3	16.2	16.5	17.4
FeO	3.8	2.1	7.1	7.6	0.5
Fe_2O_3	3.1	3.5	2.6	4.2	8.5
CaO	5.1	5.9	11.4	9.1	1.6
Na_2O	3.8	1.1	2.8	3.7	1.3
MgO	3.5	2.7	7.7	5.3	4.6
K_2O	3.1	2.9	0.2	1.9	3.7
H_2O	1.1	3.2	$<1^a$	$<1^a$	6.3
TiO_2	1.1	0.6	1.5	2.9	1.0
MnO	–	0.1	0.2	0.2	0.8
P_2O_5	0.3	0.1	0.1	0.5	0.1
CO_2	0.1	5.4	–	–	0.4
C(org)	–	0.7	–	–	0.1
SO_3	–	0.5	–	–	–
	99.4	100.0	99.8	99.9	100.0

a)Assumed as zero in the calculation.

On the whole, igneous rocks and the sediments derived from them (mainly shales) have very similar composition (columns 1 and 2). Note the depletion, in the sediments, of sodium (Na_2O), much of which ends up in seawater. Note also the high CO_2 content in sediments (carbonates), the high carbon content (coaly matter), and the sulfur (evaporites). The sea floor basalts differ from continental igneous rocks in the abundance and oxidation state of iron, in the calcium content, in magnesium, and especially in potassium. Red clay has a composition much like sedimentary rock in general, except for depletion in calcium carbonate (dissolution!), and enrichment in manganese. Also note the difference in sulfur (no evaporites) and in carbon (high degree of oxidation). (Sources for A7: Fairbridge R. W. ed. 1972 The encyclopedia of geochemistry and environmental sciences, Van Nostrand Reinhold Co. New York; A. E. J. Engel and C. G. Engel, 1971 in A. Maxwell [ed] The Sea vol. 4 pt. 1, 465–519, John Wiley and Sons, New York

A8 Radioisotopes and Dating

General. Certain atoms have a nucleus which emits radiation, whereupon the nature of the atom changes. Atoms of the same kind are called *isotopes* and those which emit radiation from the nucleus are *radioactive* isotopes. Many elements have both *stable* and *unstable* (= radioactive) *nuclides* (= isotopes), but elements heavier than lead (see A4) generally tend to be unstable. The radiation emitted from a nucleus may be α, β, or γ-radiation. An alpha particle consists of two neutrons and two protons (= helium nucleus); a beta particle is a high-speed electron, and gamma rays are like X-rays except more energetic.

The probability that a given atomic nucleus emits radiation (or "decays") is independent of its age and of environment (pressure, temperature, chemical conditions). From this it follows that the number of atoms in a given set of radioactive isotopes decreases according to a simple exponential law, which may be written:

$$N/N_0 = e^{-\lambda t},$$

where N/N_0 is the proportion of remaining atoms after time t, and λ is the *decay constant*. Solving the equation for $N/N_0 = 1/2$ we obtain a value for t which is called the *half-life*

$$t_{1/2} = \ln 2/\lambda.$$

The decaying isotope is called *parent* and the resulting new isotope is called *daughter*. Geochronological dating is done by assessing the abundance of parent and/or daughter elements. When N/N_0 is known, and λ, then t can be calculated.

Types of Radioactivity. Three kinds of radioactivity are useful in measuring time in the geologic record.

The first is *primary* radioactivity from radioisotopes with very long half-lives, which have been around since the Earth was formed. These isotopes are potassium-40, rubidium-87, thorium-232, uranium-235, and uranium-238 (^{40}K, ^{87}Rb, ^{232}Th, ^{235}U, ^{238}U). The decay of ^{40}K to ^{40}Ar has been used to date volcanic rocks *(potassium-argon dates)*. Such dating delivered the time scale for magnetic lineations on the sea floor (Chap. 1). The decay of ^{87}Rb to ^{87}Sr has been used to date igneous rocks and meteorites, and to establish the timing of metamorphism *(rubidium-strontium method)*. The method is also used to study inhomogeneities in the upper mantle, as reflected in different $^{87}Sr/^{86}Sr$ rations in oceanic basalts on the seafloor. ^{232}Th decays to ^{208}Pb in a series involving six alpha steps (each decreasing the atomic weight of the preceding parent by 4). The ^{232}Th decay has been used to date the time of crystallization of certain minerals. The two *uranium decay series* are especially important in dating Pleistocene deposits.

Another kind of radioactivity is called *secondary;* it consists in the alpha, beta, and gamma emissions from the daughter elements along a decay series. The third kind of radioactivity is *cosmic-ray-induced.* Certain radioisotopes are continually being produced by cosmic ray bombardment of the atmosphere and the ocean. Their abundance represents an equilibrium between new production on Earth, and the decay by radioactivity. Examples are carbon-14, hydrogen-3 (= tritium), beryllium-7, beryllium-10, and silicon-32.

A fourth kind of radioactivity – the man-made type – is useful for measuring the rate of rapid geologic processes, such as water movements, dispersion of sediments, sediment mixing on the sea floor, and growth rate of skeleton-producing organisms. The degree to which bomb-produced tritium and carbon-14 are dispersed in the ocean, for example, contains clues about rates of deep water formation and rates of carbon cycling. The depth to which plutonium penetrates sediment on the sea floor tells something about the activity of benthic organisms. Some short-lived isotopes in the natural uranium decay series also are used for the study of such processes.

Uranium Decay Series. There are two such series, one for ^{238}U and for ^{235}U; the heavier uranium is the more abundant of the two isotopes (99.27 % versus 0.72 %). The two series proceed by alpha and beta decay, as follows (half-life below each arrow):

$$^{238}U \xrightarrow[4.49 \cdot 10^9 \text{ yrs}]{\alpha} {}^{234}Th \xrightarrow[24.1 \text{ ds}]{\beta} {}^{234}Pa \xrightarrow[1.18 \text{ min}]{} {}^{234}U \xrightarrow[2 \cdot 48 \cdot 10^5 \text{ yrs}]{\alpha} {}^{230}Th \xrightarrow[7.5 \cdot 10^4 \text{yrs}]{\alpha} {}^{226}Ra$$

$$^{226}Ra \xrightarrow[1622 \text{ yrs}]{\alpha,\alpha,\alpha,\beta,\beta,\alpha} {}^{210}Pb \xrightarrow[22 \text{ yrs}]{\beta,\beta,\alpha} {}^{206}Pb \text{ (stable)}$$

$$^{235}U \xrightarrow[7.13 \cdot 10^8 \text{ yrs}]{\alpha} {}^{231}Th \xrightarrow[25.6 \text{ hrs}]{} {}^{231}Pa \xrightarrow[3.25 \cdot 10^4 \text{ yrs}]{\alpha} {}^{227}Ac \xrightarrow[22 \text{ yrs}]{\beta} {}^{227}Th \xrightarrow[18.6 \text{ ds}]{\alpha} {}^{223}Ra$$

$$^{223}Ra \xrightarrow[11.1 \text{ ds}]{\alpha,\alpha,\alpha,\beta,\alpha,\beta} {}^{207}Pb \text{ (stable)}$$

The ratios between start and end elements, and the ratio between the stable endproducts *(lead-lead-method)* have been used for geochronometry on long time scales. The relatively short-lived intermediate products offer opportunities for dating on time scales of the Pleistocene. Both sediment accumulation and the age of raised coral terraces have been dated by uranium series analysis.

Sediment dating proceeds from the observation that uranium is much more soluble than thorium (Th) or protactinium (Pa), daughter products in the decay series. The decay products thorium and protactinium enter the sediment at the stages shown in Fig. A8.1. At this point the particular elements form insoluble compounds and they also exist long enough to reach the sea floor as precipitates within particles. Decay then continues within the sediment at a rather slow rate: ^{230}Th has a half-life of 75 000 years, and ^{231}Pa one of 32 500 years.

From Fig. A8.1 it is obvious that we should find, within the sediment, the thorium-230 and the protactinium-231 in *excess* of what is expected to be delivered by the uranium-238 and the uranium-235 present in the sediment. Also, since these daughter isotopes decay slowly in their turn (to radium-226 and actinium-227, respectively), this excess abundance should decrease with the age of the sediment, that is, downward in a core. These expectations are fulfilled, and the decreasing abundance of excess thorium-230 (or of excess protactinium-231) downcore yields a measure for the sedimentation rate (Fig. A8.2).

In the figure, measurements on thorium activity are plotted against depth in core. Thorium activity is shown as the difference (in disintegrations per minute) between the activity of ^{230}Th and of ^{234}U, its immediate parent. This difference is the *unsupported* or *excess* thorium-230. It is assumed to be the remnant of thorium originally entering the sediment at its former surface. The broken line shows where one would expect to plot the measurements if the original activity of excess thorium had been 10 dpm, and the sedimentation rate had been unchanged at 2.4 cm/1000 years. The fit is reasonable but not perfect: the thorium activity is too high near the top of the core and too low between 2 and 4 m depth-in-core.

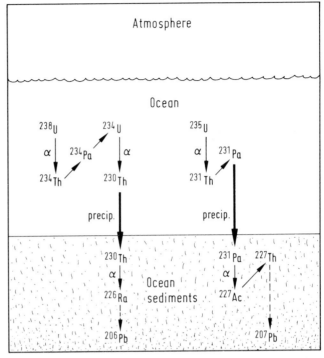

Fig. A8.1. Behavior of uranium series isotopes in the ocean. Uranium is much more soluble than thorium and protactinium. These elements, therefore, precipitate when formed by decay of their immediate parent. ^{230}Th and ^{231}Pa are the first long-lived isotopes in their respective series. Thus, they have time to reach the sediment. (W. M. Sackett, 1964, Ann N Y Acad Sci 119: 340)

Carbon-14-Dating. Carbon-14 arises in the atmosphere from the interaction of nitrogen-14 (the most abundant isotope in the air) with slow moving neutrons which owe their origin to cosmic ray bombardment of the atmosphere. About 10 kg of carbon-14 is produced per year, in this fashion. The half-life of ^{14}C is 5700 years. Thus, the proportion of radiocarbon which disintegrates each year (back to nitrogen-14) is one hundredth of 1.2 % of the total radiocarbon present. For steady state, the amount formed equals the amount decaying, and the total mass of radiocarbon present, therefore, should be near 80 000 kg. Each living organism (and each carbonate shell abuilding) has its share of this radiocarbon within it. Bomb-produced ^{14}C has considerably increased the total amount (and the share).

For dating purposes, the condition before the introduction of man-made radiocarbon is taken as the normal state (A. D. 1950). In reality there is not a steady state, as was shown by tree-ring dating for the last 8000 years. Radiocarbon dates may deviate from true ages by as much as 10 % in this period (and for earlier periods as well; see Fig. 5.6).

The maximum range of the radiocarbon method in sediments is about 40 000 years, which corresponds to seven half-lives or 1/128 of the original activity. One measures the ratio between radiocarbon and the total carbon present. The original ratio of radiocarbon to stable carbon (carbon-12 and a minor amount of carbon-13) is about one in 10^{12} (one in one thousand billion). The age is

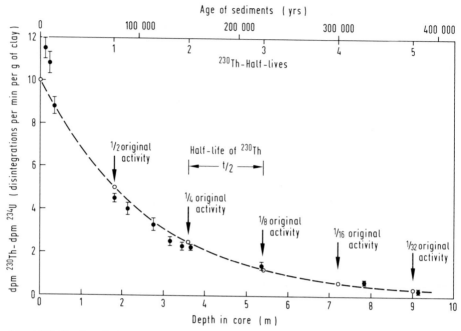

Fig. A8.2. Thorium-dating of deep-sea core V 12-122 raised from the Caribbean deep sea floor (17°00' N, 74°24' W, 2800 m, core length 10.95m). The part of thorium-230 activity which is unsupported by uranium-234 decay in the sediment is plotted as disintegrations per minute, on a carbonate-free basis. The dashed line and hollow circles show the decrease of activity expected if the initial value is 10 dpm, and the rate of sedimentation is 2.4 cm/1000 yrs, and assuming no variation of the supply of ^{230}Th through time. (Data from T. L. Ku and W. S. Broecker, 1966, Science 151: 448)

calculated from the equation

$$(^{14}C/\text{total C})_{present} = (^{14}C/\text{total C})_{initial} \cdot e^{-\lambda t},$$

where the decay constant is 1/8200.

From this follows

$$t = 82000 \ln \frac{(^{14}C/C)_{initial}}{(^{14}C/C)_{present}}.$$

The initial ratio is assumed to be that of living matter in A. D. 1950. There also is a correction for differential incorporation of ^{14}C and ^{12}C into various types of living and skeletal matter. This correction is based on the measurement of $^{13}C/^{12}C$ ratios.

Either carbonate or organic carbon may be dated. In sediments, contamination from redeposition can be a problem in fine-grained carbonate, as well as in organic carbon, especially near continents. Also, the ^{14}C content of organic carbon can change through exchange with the atmosphere, during improper storage of sediment. Sediment mixing is an important problem in interpreting ^{14}C dates of marine deposits. This problem does not arise in the dating of single large shells, of course.

The amount of carbon necessary for routine dating is 10 g of carbon (or 100 g of carbonate). Refined methods can cope with amounts much smaller than this.

A9 Systematic Overview for Major Groups of Common Marine Organisms Important in Sea-Floor Processes

The number of species of living organisms on Earth is not known; estimates range from 3 to 10 million. As far as animals (including protozoans, i. e., eukaryotic unicellular organisms without chlorophyll) are concerned, there are well over a million species, of which 16 % (160 000) live in the ocean. On land, some 75 % of the animal species are insects. Hence, the number of *noninsect* species in the sea is two thirds of the total noninsect species. Benthic species outnumber pelagic ones by 50 : 1.

On the basis of similaritiy (and presumed common origin, therefore) *species* are grouped into *genera* (sing. *genus*), genera are grouped into *families*, these into *orders*, these into *classes*, and classes finally into *phyla* (sing. *phylum*) (see Table A9.1).

Table 9.1. Taxonomic classification: four examples. (J. L. Sumich, 1976, An introduction to the biology of marine life. WC Brown Co., Dubuque, Iowa)

Taxonomic categorie	Blue whale	Common dolphin	Purple sea urchin	Giant kelp
Kingdom	Animalia	Animalia	Animalia	Plantae
Phylum/ division	Chordata	Chordata	Echinodermata	Phaeophyta
Class	Mammalia	Mammalia	Echinoidea	Phaeophycae
Order	Mysticeti	Odontoceti	Echinoida	Laminariales
Family	Balaenopteridae	Delphinidae	Strongylocentrotidae	Lessoniaceae
Genus	*Balaenoptera*	*Delphinus*	*Strongylocentrotus*	*Macrocystis*
Species	*musculus*	*delphis*	*purpuratus*	*pyrifera*

The highest taxonomic category is the *kingdom*. Traditionally two kingdoms have been recognized, the "animals" and the "plants", a division which goes back to the time of the 18th century naturalist C. Linné. The biologist E. Haeckel (in 1866) recognized the *protists* (unicellular organisms) as a separate kingdom, and H. F. Copeland (in 1938) elevated the *monera* (bacteria and blue-green algae) to this rank. A five-kingdom system is now widely adopted (R. H. Whittacker, 1966, Science 163: 150). Such a system is shown in the *phylogenetic tree* (Fig. A9.1) which gives a summary of organism classification. "*Archaeobacteria*", many of them methanogenic, somewhere between prokaryotes and eukaryotes, may constitute a separate sixth kingdom.

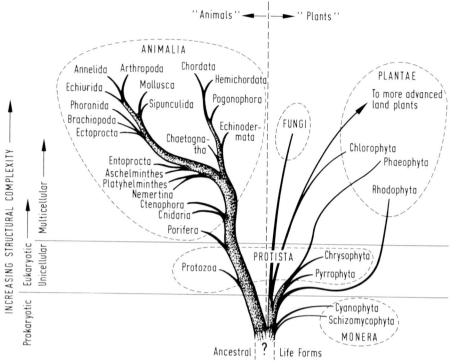

Fig. A9.1. Evolutionary tree. Hypothetical evolutionary tree showing probable relationships of the major groups of organisms. Kingdoms are outlined with *dashed lines*. (J. L. Sumich, 1976, An introduction to the biology of marine life. W. C. Brown Co., Dubuque, Iowa)

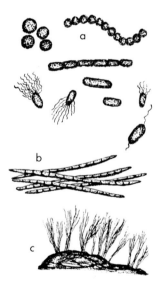

Kingdom Monera. Prokaryotic cells: lacking nuclear membranes.

Bacteria. Responsible for remineralization of organic matter and regeneration of nutrients in the water column and on the sea floor. Also for nitrogen fixation, sulfate and nitrate reduction, sulfide precipitation (pyrite), ferromanganese precipitation (manganese nodules). Photosynthetic bacteria in the sunlit zone add to primary production. Chemosynthetic bacteria form the base of the food chain at hydrothermal vents (**a**)

Blue-Green Algae (Cyanophyta). Photosynthesis in plankton and benthos. Stromatolite formation. Nitrogen fixation. Endolithic forms in calcareous shells. Chemosynthetic forms. Algalmats in lagoons (**b, c**)

Kingdom Protista. Unicellular eukaryotic cells, possessing nuclear membranes. There are two major groups; unicellular algae (e. g., dinoflagellates, coccolithophores, diatoms, silicoflagellates) and protozoans (e. g., foraminifera, radiolarians, and tintinnids).

Dinoflagellates. Photosynthetic plankton, commonly the dominant form in seawater. Certain types are responsible for "red tide" which can poison fish and render mussels inedible. Common as *symbionts* ("zooxanthelae") in foraminifera, radiolarians, corals, sponges, and some mollusks. Their cysts are found as fossils in organic-rich sediments. Phylum Pyrrophyta (**a, b**)

Coccolithophores. Photosynthetic plankton and benthos (the latter rarely as fossils). Calcitic platelets of certain pelagic species are preserved in deep-sea sediments, hence, coccoliths are probably the most abundant type of fossils on Earth (since the Mesozoic). Rockforming (e. g., Cretaceous chalks). Phylum Chrysophyta (**c**)

Diatoms. Photosynthetic plankton and benthos, extremely common in all watery environments. Siliceous skeleton (pill-box principle). Colonial forms abundant in ocean. Characteristic component of sediments in upwelling regions. Rock-forming in Tertiary deposits (diatomite). Phylum Chrysophyta (**d, e**)

Silicoflagellates. Similar to diatoms but with an inner skeleton of silica. Photosynthetic plankton of the ocean. Abundant in fertile regions and their sediments (**f, g**).

Foraminifera. Heterotrophic plankton and benthos, marine only. Calcareous shells. Also agglutinated tests, on the sea floor. Rockforming (e. g., Eocene nummulitic limestones: building material of Egyptian pyramids). Foraminiferal ooze ("Globigerina ooze") is the single most widespread deposit on the face of Earth at present (covers ca. half of the deep-sea floor). Many plankton and reef species have dinoflagellate symbionts (**h, i**)

Radiolarians. Somewhat similar to foraminifera, but with internal skeletons made of silica. Marine plankton only, all depths. Common in pelagic sediments below fertile regions. Rock-forming (radiolarites of late Jurassic). Shallow planktonic forms can have dinoflagellate symbionts. Main groups: nasselarians (opal), spumellarians (opal), phaeodarians (organic-rich opal skeleton). Also there is a distinct group, the acantharians, which have skeletons made of $SrSO_4$. Acantharians are common in tropical surface waters, but do not make fossils (**j, k**)

Tintinnids. Ciliate protozoans with an organic test ("lorica"). Abundant in marine plankton. Loricae are fossilized in organic-rich sediments (upwelling regions) (**l**)

Kingdom Plantae. Multicellular, with walled eukaryotic cells, photosynthetic. There are two major groups: the multicellular algae (e. g., Rhodophyta, Phaeophyta, Chlorophyta) and the higher plants (Metaphyta).

Rhodophyta (Red Algae). Common in the littoral zone, especially in warm regions, and on hard substrates down to very low light levels. Corallinaceae (e. g., *Lithothamnion*) have calcified cell walls, they are important sediment producers in places (**m, n**).

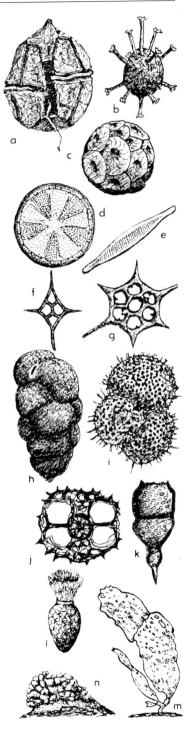

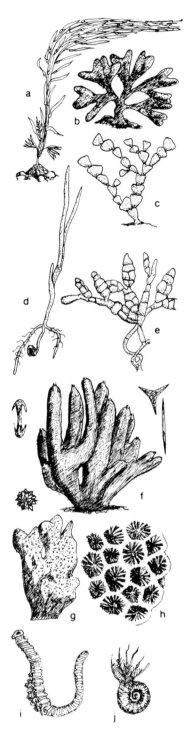

Phaeophyta (Brown Algae). Best known as kelp forests of temperate and cold water coasts, the most highly productive marine biotope. Typical representatives: *Laminaria, Macrocystis, Fucus. Sargassum* is abundant in Sargasso Sea, where it floats on the surface. Delivery of organic sediments to coastal regions (possibly also below Sargasso Sea) (**a, b**)

Chlorophyta (Green Algae). Common at the sea shore (e. g., sea lettuce, *Ulva*). In warm waters abundant lime-secreting forms, e. g., *Halimeda* (Codiaceae) and representatives of the Dasycladaceae which were rock-forming in times past (e. g., alpine Triassic) (**c**)

Metaphyta. The higher plants (mosses, vascular plants) are primarily terrestrial. Exceptions are: the sea grasses (e. g., *Zostera*) in the littoral zone, which are flowering plants. Certain halophytes (*Salicornia* in salt marshes; mangroves in tropical coastal swamps) are important in the intertidal zone (**d**)

Kingdom Fungi. Unicellular and multicellular eukaryotic organisms lacking photosynthetic pigments and depending on nutrition by absorption. There are few marine species (**e**)

Kingdom Animalia. Multicellular eukaryotic organisms without photosynthetic pigments and depending on nutrition by ingestion. Higher forms have sensory neuro-motor systems

Porifera (Sponges). Most species are marine. Common on firm sea floor in fertile area, in all latitudes. Boring sponges *(Cliona)* produce sediment particles, weaken mollusk shells, and erode calcareous substrates. Calcareous and siliceous form can be rock-forming. Siliceous sponge spicules accumulate in great masses on the Antarctic shelf, in places (**f**)

Cnidaria (Coelenterata). Mostly marine. Planktonic forms (hydrozoans, siphonophores, scyphozoans) and benthic forms (hydrozoans, anthozoans) abundant. Stone corals contribute greatly to reef building in tropical regions (*Acropora, Porites, Pocillopora,* etc.). Deep water and water stone corals also exist (e. g., *Lophelia*) but they do not build reefs (**g, h**)

Worms. This common-name category for elongate soft-bodied animals (Vermes) no longer has scientific standing. There are several phyla of worms, such as: *Platyhelminthes,* flatworms (some free-living, but mostly parasites), *Nemertea,* ribbon worms common on the sea floor *Aschelminthes,* round worms (containing the important class of nematodes, which is very abundant in the marine benthos; parasites common), *Phoronida,* phoronid worms (few species, but common), *Sipunculoidea,* peanut worms (few species), *Echiuridea,* spoon worms (few species, deposit feeders), *Annelida,* segment worms (common in plankton and benthos; well-known representative: *Arenicola*), *Pogonophora,* beard worms (lack a digestive tract; tube dwellers, mainly soft bottom deep sea, few species; some live near hot vents, see cover photo),

Hemichordata (pterobranch and acorn worms; ptero-
branchia are common on the sea floor over a large
depth range, tube-building species colony-forming in
places; acorn worms are common in soft sediment,
mainly on shelf).

Worms are important in reworking and processing
surface sediment, and in subsurface irrigation of the
sea floor (enhancing chemical exchange between sediment and water). Filter-feeding worms convert suspended matter to sediment. Tube-building worms can
provide solid substrates for epibenthos to settle on (**i, j,
a, b, c**)

Bryozoans (Moss Animals). Benthic organisms superfi-
cially resembling small colonial hydroids. Many
species secrete a protective and supporting structure
made of carbonate. These structures may be encrusting
rocks or shells, or may stand as small bushes or trees
(**d**)

Brachiopods (Lamp Shells). Common as fossils, but not
so important in the present ocean (30 000 extinct
species, 300 living ones). Benthic organisms, superfi-
cially resembling bivalves. Shells are made of chitino-
phosphatic or of calcareous material (**e**)

Mollusks. Highly diverse group (more than 60 000
species) and of major importance in sediment formation ($CaCO_3$). Includes gastropods (snails including
pteropods), pelecypods (mussels, clams, oysters, etc.),
cephalopods (squids, octopuses, etc.) and the less well-
known classes amphineurans (chitons), scaphopods
(tusk shells), and monoplacophorans, a rare group
thought to be extinct since the Paleozoic before representatives were discovered in the deep sea in the
1950's (**f, g, h, i**).

Arthropods. Most diverse of the animal phyla: arthropods
account for more than 75 % of all animal species. The
great diversity of the insects is responsible; there are
only a few marine insects, which belong to the genus
Halobates, the water strider. The arthropods include
the crustaceans, merostomatans, and pycnogonids (sea
spiders). The crustaceans (shrimps, crabs, lobsters, barnacles, isopods, amphipods, and others) are ubiquitous
in and on the sea floor, reworking sediment and also
contributing to sediment formation, by pelletization,
and by precipitation of phosphatic and calcareous hard
parts. The best-known representative of the merostomatans is the horseshoe crab (**j, k, l**)

Echinoderms. Ubiquitous on the sea floor: echinoids (sea
urchins), asteroids (sea stars), crinoids (sea lilies), holothurians (sea cucumbers), and ophiuroids (brittle
stars). Holothurians ingest and process considerable
amounts of surface sediment in certain regions of the
continental slope and beyond. Virtually all echinoderms precipitate calcium carbonate, which contributes
substantially to shallow water carbonates in places (**m**)

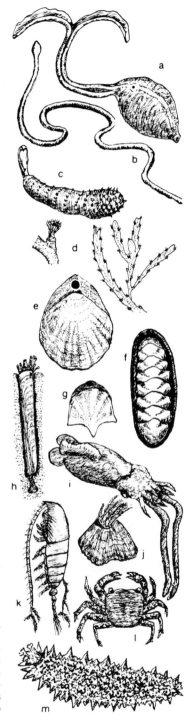

Chordates. Highly diverse phylum including the familiar backboned animals (fishes, amphibians, reptiles, birds, mammals) and also some rather primitive forms: the (benthic) sea squirts and the (planktonic) salps. Salps are important in pelletizing extremely fine-grained suspended matter, which removes it from surface waters and expedites its way to the sea floor (**a, b**)

Organisms Illustrated

Page 328
a spherical, rod-shaped, and flagellated bacteria
b pelagic blue-green alga (*Trichodesmium*, which is red-colored and gives the Red Sea its name)
c colony of a benthic blue-green alga *(Phormidium)*

Page 329
a, b dinoflagellate *(Gonyaulax)* and its resting cyst ("hystrichosphere")
c coccosphere *(Cyclococcolithus)* covered with coccoliths
d, e centric diatom *(Actinoptychus)* and pennate diatom *(Nitzschia)*
f, g common silicoflagellates *(Distephanus, Dictyocha)*
h, i benthic foram *(Verneulinia)* and planktonic foram *(Globigerinoides)*
j, k spumellarian and nasselarian radiolarians *(Octopyle stenozona, Theocorythium trachelium)*
l tintinnid *(Stenosemella)*
m, n red algae *(Rhodymenia, Lithothamnium)*

Page 330
a, b brown algae: giant kelp *(Macrocystis)*, rockweed *(Fucus)*
c calcareous green alga *(Halimeda)*
d sea grass *(Zostera)*
e marine fungus *(Alternaria)*
f calcareous sponge *(Sycon);* also shown are various types of sponge spicules
g benthic calcareous hydrozoan (Millepora)
h stone coral *(Favites)*
i, j annelid worms (the lugworm *Arenicola*, the serpulid *Spirorbis*)

Page 331
a, b, c worms: echiurid *(Boniella)*, nemertine *(Lineus)*, echiurid *(Echiurus)*
d bryozoan colony *(Crisia)*
e brachiopod (terebratulid)
f, g, h, i shell-forming mollusks: chiton or sea cradle *(Nuttallina)*, pelagic snail *(Cavolinia)*, razor clam *(Ensis)*, shell-forming squid *(Spirula)*
j, k, l arthropods: cirriped (barnacle, *Balanus*), copepod *(Calanus)*, shore crab *(Crassipes)*
m holothurian or sea cucumber *(Aspidochirota)*

Page 332
a sea squirt *(Halocynthia)*
b sting ray *(Taemiura)*

Sources for A9: E. Y. Dawson, 1966. Marine botany, Holt Rinehart Winston, New York; B. U. Haq, A. Boersma, 1978, Introduction to marine micropaleontology, Elsevier, New York; A. Remane, V. Storch, U. Welsch, 1976. Systematische Zoologie, Gustav Fischer, Stuttgart; G. G. Simpson, W. S. Beck, 1965, Life – an introduction to biology, 2nd ed., Harcourt Brace Jovanovich, New York; H. U. Sverdrup, M. W. Johnson, R. H. Fleming, 1942, The oceans – their physics, chemistry, and general biology. Prentice-Hall, Englewood Cliffs.

Index of Names

(also see List of Books and Symposia)

Subject Index

abyssal, definition 43, 64
  storms 123, 124
abyssal hills 25
abyssal plains 20, 43
  nature of 66
  illustration 17, 66
acantharians 329
accretionary prism 53, 54
accumulation rates 94
acoustic reflectors 234ff
acoustic swath-mapping 305
Acropora 136, 164, 246, 331
active margins 45, 51
Adriatic Sea 94, 152
Aegean Sea
  volcanism 74
African Rift Valley 46
age of sea floor 36
age of bottom water 234
age-depth curve, equation 23ff
  and sea level 150
agglutinating forams 174
aggregates 157, 187, 220
air gun profiling 57, 59
Albatross see Swedish Deep Sea Expedition
albedo 243, 244
  feedback 253
  - and regression 260
Aleutian Trench 25
algal carbonate 164
  production of 163, 164
algal mats 88, 90, 134
alkali metals 258
alkalinity 73, 89
alkenones 165, 212
allochthons 70
Alvin 182, 183
amphibolites 53
Amphiura (brittle star) 181
  community 182
anaerobic conditions 89, 125, 159, 161, 166
anaerobic sediments 77, 118, 206

Adriatic Sea 94
  upwelling 118, 119
  Baltic Sea 204, 205
Andes, ore deposits in 51, 52
andesites 28, 320
Angola Basin 121
Angola cont. margin 48, 49
anhydrite 90, 319
  hydrothermal 295, 296
annual layers (varves) 93, 94, 209
anoxic events
  Cretaceous ocean 271
  and volcanism 273
Antarctic
  Bottom Water (AABW) 120 ff
  Circumpolar Current 123, 259
  Intermed. Water (AAIW) 121
  icebergs 186
  ice-rafted debris 264
anti-estuarine circulation 124, 125, 202, 203, 207
apatite, formula 319
Aptian (= epoch within Cretaceous)
  salt 49
  volcanism 274
Arabian Sea 125
  oxygen minimum 271
aragonite 86, 194, 319
  in organic hard parts 83
  needles 86
  calcite ratio
  and temperature 194
  formula 319
archaeobacteria 327
arctic environments 200
Arenicola (burrowing worm) 131, 172, 173, 177, 330, 331
Argentine Basin 121
arid basins 201 ff
arid climate 90
arkose, definition 321
arthropods 331